Springer-Verlag Berlin Heidelberg GmbH

Reviews of

146 Physiology Biochemistry and Pharmacology

With 21 Figures and 14 Tables

Springer

ISSN 0303-4240
ISBN 978-3-662-31050-2 ISBN 978-3-540-36207-4 (eBook)
DOI 10.1007/978-3-540-36207-4

Library of Congress-Catalog-Card Number 74-3674

http://www.springer.de

Originally published by Springer-Verlag Berlin Heidelberg New York in 2003.
Softcover reprint of the hardcover 1st edition 2003

Printed on acid-free paper – 27/3150 or 5 4 3 2 1 0

Rev Physiol Biochem Pharmacol (2003) 146:1–53
DOI 10.1007/s10254-002-0001-x

M. A. Jakupec · M. Galanski · B. K. Keppler

Tumour-inhibiting platinum complexes–state of the art and future perspectives

Published online: 4 February 2003

Abstract Thirty years after the onset of the first clinical studies with cisplatin, the development of antineoplastic platinum drugs continues to be a productive field of research. This article reviews the current preclinical and clinical status, including a discussion of the molecular basis for the activity of the parent drug cisplatin and platinum drugs of the second and third generation, in particular their interaction with DNA. Further emphasis is laid on the development of third generation platinum drugs with activity in cisplatin-resistant tumours, particularly on chelates containing 1,2-diaminocyclohexane (DACH) and on the promising and more recently evolving field of non-classic (*trans*- and multinuclear) platinum complexes. The development of oral platinum drugs and drug targeting strategies using liposomes, polymers or low-molecular-weight carriers in order to improve the therapeutic index of platinum chemotherapy are also covered.

Introduction

Broad interest in the pharmacological properties of metal compounds first arose with the pioneering work of Barnett Rosenberg, who in the late 1960s by chance discovered the cytostatic effects of cisplatin and related complexes. By now, cisplatin has become one of the most successful antineoplastic drugs, on the one hand because it made metastatic testicular germ-cell cancer a curable disease in about 90% of cases, and on the other hand for its notable therapeutic efficacy in a broad range of other solid tumours. Driven by the impressive impact of cisplatin on cancer chemotherapy, great efforts have been made to develop new derivatives with improved pharmacological properties, and so cisplatin has become the prototype of a unique class of antineoplastic agents now comprising innumerable derivatives, many of which have been abandoned in preclinical or early clinical stages of their development, while a few others have succeeded in becoming established in routine clinical practice. Among the 33 platinum agents which have entered clinical trials after the onset of clinical studies with cisplatin in the early 1970s (Table 1), only one (carboplatin)

M. A. Jakupec · M. Galanski · B. K. Keppler (✉)
Institute of Inorganic Chemistry, University of Vienna, Währinger Straße 42, 1090 Vienna, Austria
e-mail: keppler@ap.univie.ac.at · Tel.: +43-1-427752602 · Fax: +43-1-427752680

Table 1 Platinum compounds in clinical studies since 1971 in roughly chronological order (based mainly on Lebwohl and Canetta 1998; updated)

Compound	Current status
Entered clinical studies in 1970s	
Cisplatin (CDDP)	Approved worldwide for a broad range of solid neoplasms
PAD	Abandoned during phase I (insufficient solubility)
Platinum uracil blue (PUB)	Abandoned during phase I (cardiac toxicity)
MBA	Abandoned during phase I (one case of severe hypersensitivity)
JM-20 (SHP)	Abandoned during phase I (severe allergic reactions)
JM-74 (PHM)	Abandoned during phase II (nephrotoxicity, inferior activity)
Neo-SHP	Abandoned during phase I (severe allergic reactions)
Neo-PHM	Abandoned during phase II (nephrotoxicity, inferior activity)
BOP	Abandoned during phase I (insufficient solubility)
Entered clinical studies in 1980s	
Carboplatin (CBDCA; JM-8)	Approved worldwide for a broad range of solid neoplasms
Iproplatin (CHIP; JM-9)	Abandoned during phase III (lower activity than carboplatin)
JM-82 (DACCP)	Abandoned during phase II (chemical instability, low activity)
JM-11	Abandoned after pharmacokinetic study
Spiroplatin (TNO-6)	Abandoned during phase II (nephrotoxicity)
PYP	Abandoned during phase I (nephrotoxicity + myelosuppression)
JM-40	Abandoned during phase I (nephrotoxicity)
PHIC	Abandoned during phase I (difficulties in synthesis)
Oxaliplatin (*l*-OHP)	Approved in over 60 countries for metastatic colorectal cancer
Nedaplatin (254-S)	Approved in Japan for several solid neoplasms
CI-973 (NK-121)	Abandoned during phase II (lack of activity)
DWA2114R (miboplatin)	Abandoned during phase III (no advantage over cisplatin)
Enloplatin	Abandoned during phase II (nephrotoxicity)
Zeniplatin	Abandoned during phase II (nephrotoxicity)
L-NDDP (aroplatin)	Phase II
Entered clinical studies in 1990s	
Lobaplatin (D-19466)	Approved in China
Ormaplatin (tetraplatin)	Abandoned during phase I (neurotoxicity)
Cycloplatam	Phase II
JM-216 (satraplatin)	Phase III
SKI 2053R	Approved in South Korea for advanced gastric cancer
ZD0473 (AMD473)	Phase III
SPI-77 (liposomal CDDP)	Phase II
TRK-710	Phase I
BBR3464	Phase II
Entered clinical studies in 2000s	
AP5280	Phase I

has received worldwide approval so far. Four drugs (oxaliplatin, nedaplatin, lobaplatin, SKI2053R) have gained regionally limited approval and another eight continue to be evaluated in clinical studies. Two of these (JM-216, ZD0473) have recently entered phase III studies.

The vast majority of these compounds and all of the clinically established platinum drugs belong to the classic type of uncharged *cis*-configured square-planar platinum(II) complexes described by the general formula *cis*-$[PtA_2X_2]$, A_2 being either two monodentate or one bidentate stable amine ligand(s) and X_2 being either two monodentate or one bidentate anionic leaving ligand(s) (for an overview of the ligand spheres of the clinically studied platinum complexes see Table 2). This type of complex has been favoured since

Table 2 Ligand sphere of clinically studied platinum compounds

Compound	Equatorial ligands		Axial ligands
	Non-leaving (number of ligands)	Leaving (number of ligands)	
Classic platinum(II) compounds			
Cisplatin	Ammine (2)	Chloride (2)	–
SPI-77	Ammine (2)	Chloride (2)	–
Carboplatin	Ammine (2)	1,1-Cyclobutanedicarboxylate	–
Nedaplatin	Ammine (2)	Glycolate	–
AP5280	Ammine (2)	(HPMA-linked) aminomalonate	–
Cycloplatam	Ammine/cyclopentanamine	Malate	–
ZD0473	Ammine/2-methylpyridine	Chloride (2)	–
PAD	Cyclopentanamine (2)	Chloride (2)	–
JM-11	Cyclopropanamine (2)	Chloride (2)	–
JM-40	Ethanediamine	Malonate	–
CI-973	2-Methyl-1,4-butanediamine	1,1-Cyclobutanedicarboxylate	–
DWA2114R	2-Pyrrolidinemethanamine	1,1-Cyclobutanedicarboxylate	–
Enloplatin	Tetrahydro-4H-pyran-4,4-dimethanamine	1,1-Cyclobutanedicarboxylate	–
Zeniplatin	2,2-Bis(aminomethyl)-1,3-propanediol	1,1-Cyclobutanedicarboxylate	–
Lobaplatin	1,2-Cyclobutanedimethanamine	2-Hydroxypropanoate	–
SKI 2053R	2-(1-Methylethyl)-1,3-dioxolane-4,5-dimethanamine	Malonate	–
MBA	1,2-Cyclohexanediamine	Bromoacetate (2)	–
PYP	1,2-Cyclohexanediamine	Pyruvate (2)	–
JM-82	1,2-Cyclohexanediamine	4-Carboxyphthalate	–
PHIC	1,2-Cyclohexanediamine	Isocitrate	–
JM-74 (PHM)	1,2-Cyclohexanediamine	Malonate	–
JM-20 (SHP)	1,2-Cyclohexanediamine	Sulphate	–
L-NDDP	*trans-R,R*-1,2-Cyclohexanediamine	Neodecanoate (2)	–
Neo-SHP	*trans-R,R*-1,2-Cyclohexanediamine	Sulphate	–
TRK-710	*trans-R,R*-1,2-cyclohexanediamine	3-Acetyl-5-methyl-2,4(3H,5H)-furandionate	–
Neo-PHM	*trans-R,R*-1,2-Cyclohexanediamine	Malonate	–
Oxaliplatin	*trans-R,R*-1,2-Cyclohexanediamine	Oxalate	–
Spiroplatin	1,1-Cyclohexanedimethanamine	Sulphate	–
BOP	1,2-Cyclooctanediamine	Bromoacetate (2)	–
Classic platinum(IV) compounds			
Ormaplatin	*trans*-1,2-Cyclohexanediamine	Chloride (2)	Chloride (2)
JM-216	Ammine/cyclohexaneamine	Chloride (2)	Acetate (2)
Iproplatin	Isopropanamine (2)	Chloride (2)	Hydroxyl (2)
Miscellaneous			
BBR3464	Non-classic trinuclear platinum(II) complex		
Platinum uracil blue	Structure undefined		

the early landmark studies by Cleare and Hoeschele on structure–activity relationships within platinum(II) compounds performed soon after the recognition of the antineoplastic properties of cisplatin. These studies (Cleare and Hoeschele 1973a, 1973b) have provided guidelines for the development of new platinum drugs and have already demonstrated the favourable properties of carboplatin, which later emerged as the most successful second-generation platinum drug due to its less severe toxicity.

Another three clinically studied agents belong to a type of octahedral platinum(IV) complexes described by the general formula *cis*-[$PtA_2X_2Y_2$], Y_2 being two other monodentate anionic leaving ligands. Thus, this latter type differs from the former only by the higher oxidation state of the central platinum cation and by the presence of two axial leaving ligands. For the last decade, compounds of this type, which tend to be less reactive than their platinum(II) counterparts, have received attention as a potential means of providing a platinum drug suited for oral administration, and their exploration for this purpose continues. After administration, these agents are eventually transformed into complexes of the former type by reduction of platinum to the +II oxidation state, accompanied by loss of the axial ligands. Therefore, they have to be considered as prodrugs of classic-type platinum(II) agents rather than substantial exceptions from the rules established by Cleare and Hoeschele.

In recent years, the example of oxaliplatin has demonstrated that developmental efforts guided by these rules may even yield drugs with activity in inherently cisplatin-resistant tumours, and despite the discrepancy between the immense number of classic platinum complexes, which have been screened for their antineoplastic activity in preclinical studies, and the scarcity of well-established third-generation platinum drugs, there is no need to assume that the possibilities of succeeding with a classic approach are exhausted. Nevertheless, recent promising developments of non-classic platinum agents, including active *trans*-complexes and especially charged multinuclear complexes, the first of which has already proceeded to the stage of clinical phase II trials, are forcing the research community concerned with the topic to realize that the structural guidelines, which have long been taken for a kind of dogma, have unnecessarily narrowed the perspective.

Despite the tremendous therapeutic progress achieved by cisplatin and its derivatives, it cannot be ignored that even in tumours with high response rates (except testicular germ-cell cancer), therapeutic success is still limited in the vast majority of cases, which either respond incompletely or eventually relapse after complete remission of all apparent signs of disease, accompanied by development of resistance to the drug. However, experience with cell cultures shows that neither inherent nor therapy-induced resistance is absolute. Rather, the concentrations required for therapeutic efficacy are raised to levels which cannot be achieved in the patient due to dose-limiting toxicities. Therefore, strategies to achieve selective drug accumulation (or activation) in tumour tissues in order to overcome the problems of incomplete response and drug resistance have become a major field of research, with some of them currently explored in the clinical setting. Attempts to deliver well-known drugs by appropriate drug delivery devices as well as entirely novel platinum complexes endowed with affinities for specific tumours are within the scope of these strategies summarized under the term “drug targeting”. This important topic will be dealt with in a separate section at the end of this review.

The first generation – cisplatin

Cisplatin (CDDP), *cis*-diamminedichloroplatinum(II), (Fig. 1) has been known as Peyrone’s chloride (denominated after Michele Peyrone, who first synthesized this complex) since the mid nineteenth century. Interest in its potential pharmacological properties arose only in the late 1960s, when Barnett Rosenberg identified this compound and its platinum(IV) analogue, *cis*-diamminetetrachloroplatinum(IV), as strong inhibitors of bacterial proliferation in a set of experiments, which he had performed in order to investigate the effects of electric fields on cell division. Subsequently, their cytostatic activity was examined in tumour-bearing animals, with the encouraging result of complete inhibition of mu-

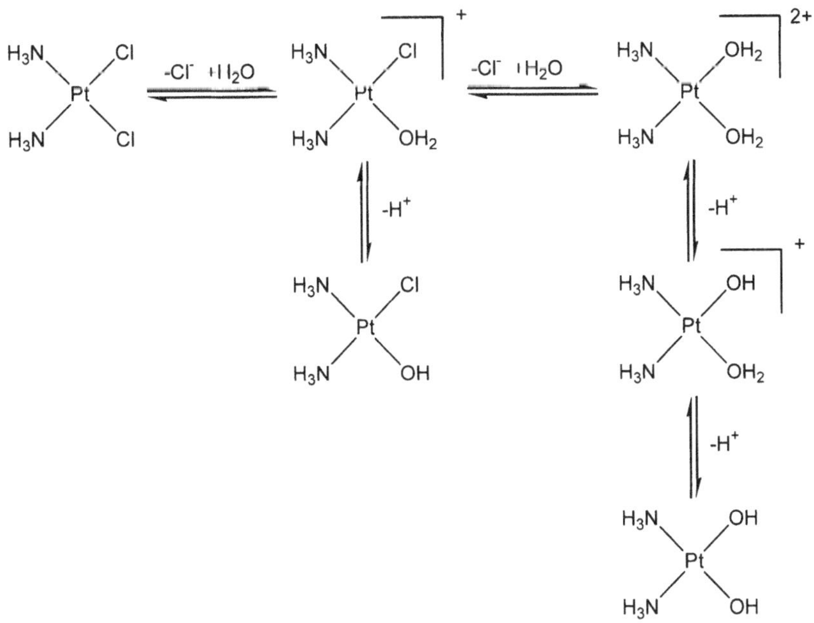

Fig. 1 Structures of cisplatin (*left*) and its hydrolysis products

rine solid sarcoma-180 by cisplatin (Rosenberg et al. 1969; for a review see Rosenberg 1978). Clinical trials of cisplatin were initiated in 1971, but anticancer activity turned out to be associated with severe renal toxicity. This obstacle could be circumvented by pre-hydration and slow infusion, and by the end of the 1970s cisplatin became the mainstay in combination chemotherapy of advanced and metastatic testicular germ-cell cancer (both seminoma and non-seminomatous tumours), which has become a highly curable disease since then. Combinations of cisplatin and etoposide (with or without bleomycin) are the current regimens of choice in this indication. Ovarian cancer is also highly responsive to cisplatin therapy; average progression-free intervals and life spans have been substantially improved by introduction of this drug, but in contrast to testicular germ-cell cancer, advanced disease is rarely cured, because these tumours commonly become resistant in the course of therapy. Cisplatin has also become an essential component of chemotherapy in lung (both small cell and non-small cell), cervical, endometrial, bladder and oesophageal cancer, in squamous cell carcinoma of the head and neck as well as in osteosarcoma. It has also become a therapeutic option in several other solid neoplasms, including liver and gastric cancer, brain tumours, melanoma and soft-tissue sarcomas. The radiosensitizing effect of cisplatin is exploited in combined radiotherapy–chemotherapy, mainly in patients with advanced squamous cell carcinoma of the head and neck or lung cancer (for reviews see O'Dwyer et al. 1999; Highley and Calvert 2000).

Cisplatin is usually administered in chloride-containing solution as short intravenous infusion following either a single-dose or a split-dose schedule. Intraperitoneal administration to ovarian cancer patients and intra-arterial administration to patients with hepatic tumours, melanoma or glioblastoma have been tried in the course of clinical studies in order to increase local drug concentrations, but are not commonly used. In aqueous solution, the chloride ligands of cisplatin are easily replaced by aqua ligands, resulting in much more reactive species. Part of these aqua ligands may be deprotonated to hydroxo ligands, which lowers the reactivity of the complex. These hydrolysis reactions result in a mixture comprising mainly mixed chloro/aqua and chloro/hydroxo, and, in much smaller portions, aqua/hydroxo, diaqua and dihydroxo species (Fig. 1). Hydrolytic activation of cisplatin is to some extent suppressed by high chloride concentrations in extracellular fluids, but is assumed to be promoted by the low chloride concentrations present inside the cell. However, it should be kept in mind that these assumptions are based on equilibria constants, while chemical equilibrium is not attained in biological systems (for a review see Berners-Price and Appleton 2000). Cisplatin reacts with a variety of nitrogen- and sulphur-containing biomolecules by ligand exchange reactions. Whereas binding to nitrogen donors is assumed to require hydrolytic activation of the drug, chloride may be directly exchanged for sulphur donors such as the amino acids methionine or cysteine without the need for prior aquation. In blood a high fraction of cisplatin is bound to plasma proteins, including albumin, transferrin and γ-globulins, and a smaller fraction to erythrocytes (Gullo et al. 1980; Vermorken et al. 1984). Binding to plasma proteins and red blood cells has long been assumed to be completely irreversible and to result in pharmacologically inactive platinum fractions, but findings obtained with refined techniques suggest a close and persistent equilibrium between ultrafiltrable (not protein-bound) and total platinum, with the former amounting to 3–7% of the latter (Gamelin et al. 1995). Nevertheless, the activity of transferrin-bound platinum is only moderate, while the cytotoxicity of cisplatin bound to albumin is quite marginal (Hoshino et al. 1995). Excretion of cisplatin is mainly renal. Residual levels of ultrafiltrable and total platinum steadily increase from course to course (Bonetti et al. 1995; Gamelin et al. 1995).

Although a variety of physiological conditions as well as agonists or antagonists of some known carrier-mediated transport systems influence cellular accumulation of cisplatin, there is no sufficient evidence for an active carrier-mediated uptake mechanism. The fact that cisplatin accumulation is neither saturable nor competitively inhibited by structural analogues suggests that the drug enters the cell through passive diffusion only (for a review see Andrews 2000). Involvement of facilitated diffusion via gated channels has been proposed (Gately and Howell 1993), but the steric demand of cisplatin prohibits entrance through most of the well-characterized ion channels. Moreover, the relatively slow uptake of cisplatin has led to the conclusion that diffusion through the lipid bilayer probably prevails over entrance through water-filled channels (Andrews 2000). However, several findings suchas the modulating effects of ATP depletion or experimental manipulation of carrier-mediated transport systems on uptake of cisplatin remain largely unexplained. It is usually assumed that cisplatin enters the cell mainly in the parent form, but experimentally raising the proportion of aquated species does not affect platinum accumulation in pharmacologically relevant concentrations in vitro despite the electric charge of the aquated species (Jennerwein and Andrews 1994).

A vast number of studies have been performed to investigate the interactions of cisplatin with DNA. Although cisplatin reacts with a wide variety of biomolecules, suggesting the possibility of cellular targets other than DNA such as cell membrane and cytoskeleton constituents (for a review see Wang et al. 1996), it is widely accepted that DNA damage is the decisive effect by which cisplatin and all other platinum drugs exert their tumour-inhibiting effects. This is supported by the following evidence: Impairment of the nucleotide excision repair (NER) system by specific inhibitors such as aphidicolin sensitizes tumour cells to cisplatin in vitro (Masuda et al. 1988, 1990; Beketic-Oreskovic and

Osmak 1995; Sargent et al. 1996) and in vivo (O'Dwyer et al. 1994). This effect is usually more pronounced in cisplatin-resistant than in cisplatin-sensitive cells. Likewise, cells with genetically determined deficiencies in nucleotide excision repair such as those from patients with xeroderma pigmentosum (Fraval et al. 1978; Roberts and Fraval 1978; Plooy et al. 1985; Dijt et al. 1988) or mutant rodent cells (Meyn et al. 1982; Sorenson and Eastman 1988b) are hypersensitive to cisplatin. Moreover, cells from patients with Fanconi's anaemia, which are NER-proficient, but lack the ability to remove a certain kind of cisplatin-induced DNA adduct – the interstrand cross-link – exhibit hypersensitivity to the drug (Plooy et al. 1985; Pera et al. 1987; Dijt et al. 1988). On the other hand, some cell lines with acquired resistance to cisplatin show an enhanced capacity of adduct removal either from the whole genome (Eastman and Schulte 1988; Masuda et al. 1988, 1990; Lai et al. 1988; Parker et al. 1991; Kelland et al. 1992; Oldenburg et al. 1994; Johnson et al. 1994a, 1994b) or from specific, actively transcribed genes (Zhen et al. 1992; Petersen et al. 1996), and this mechanism of resistance may be reversed by inhibitors of repair enzymes.

Cisplatin reacts with DNA preferentially by coordination to the N7 atoms of purine nucleobases. In simple cell-free systems, the rates of both the initial monofunctional binding step and the subsequent closure to a bifunctional adduct are limited by the kinetics of the hydrolysis reaction (Bancroft et al. 1990; Barnham et al. 1995a). However, from the differences in sequence-dependent binding rates of the hydrolysis products of cisplatin to double-stranded oligonucleotides, it has been concluded that the diaqua and/or the aqua/hydroxo complex resulting from hydrolysis of both chloro ligands rather than the chloro/aqua complex seems to be the major species interacting with DNA in vivo (Kozelka et al. 1999; Legendre et al. 2000). Entirely different, hydrolysis-independent routes to DNA adduct formation involving intermediates with reversibly bound thioether ligands have been proposed based on the observation that methionine (Barnham et al. 1995b) and other thioether ligands (van Boom et al. 1999) bound monodentally to the platinum centre may be exchanged for GMP (for a review see Reedijk 1999). However, the following recent findings argue against the assumption that such pathways of DNA platination contribute significantly to the tumour-inhibiting activity of cisplatin: thioether-containing intermediates bind to oligonucleotides with rather slow kinetics (Teuben and Reedijk 2000), while replacement of thioether ligands by thiol-containing ligands in the presence of the latter proceeds much faster and results in species which appear unlikely to serve as intermediates due to their high stability (Teuben et al. 2000).

Bifunctional binding of cisplatin to two bases of either the same or complementary DNA strands results in intra- and interstrand cross-links, respectively. Up to 90% of total platinum-DNA adducts comprise of 1,2-intrastrand cross-links involving adjacent bases. The most frequent kind of adduct, the 1,2-d(GpG) cross-link involving two adjacent guanines, is two to three times more frequent than the 1,2-d(ApG) cross-link involving adenine adjacent to guanine. Low percentages of total platinum-DNA adducts comprise 1,3-intrastrand cross-links involving non-adjacent bases, interstrand cross-links, monofunctional DNA adducts and protein-DNA cross-links (Fichtinger-Schepman et al. 1985, 1987, 1990). All kinds of DNA cross-links exert distorting effects on DNA, which have been extensively studied in double-stranded oligonucleotides (for reviews, see Gelasco and Lippard 1999; Jamieson and Lippard 1999). The different degrees of local bending and unwinding induced by different kinds of cross-links are assumed to influence their further processing.

Intrastrand, but not interstrand cross-links are substrates for the nucleotide excision repair system (Zamble et al. 1996). Since in Chinese hamster ovary cells intrastrand cross-links are preferentially repaired in transcribed compared to non-transcribed genes or non-coding regions (Jones et al. 1991) and within transcribed genes preferentially on the transcribed compared to the non-transcribed DNA strand (May et al. 1993), involvement of a transcription-coupled repair mechanism has been postulated, but its relevance in human cells has been questioned (Cullinane et al. 1999). The mechanism involved in repair of interstrand cross-links remains to be elucidated. There has been some debate on which of the different kinds of DNA adducts is the most detrimental to the cell. While it is widely accepted that monofunctional adducts and DNA-protein cross-links are less critical, either the interstrand or the more abundant intrastrand cross-links have been suspected to mainly account for the cytotoxicity of cisplatin by different authors. However, studies in cells with different kinds of repair deficiencies seem to rule out the possibility that one of these two types of adducts is solely responsible for the effects of cisplatin. In fact, cells which are either more compromised in removal of intrastrand cross-links (such as xeroderma pigmentosum fibroblasts) or of interstrand cross-links (such as Fanconi's anaemia fibroblasts) exhibit similarly enhanced sensitivity to cisplatin (Dijt et al. 1988).

Induction of morphological alterations and dysfunction of mitochondria following exposure to cisplatin has been noted in connection with its nephrotoxicity for some time (Gordon and Gattone 1986; Brady et al. 1990). Depletion of glutathione, resulting in oxidative stress, lipid peroxidation and injury due to free oxygen radicals, has been found involved in the mitochondrial damage of renal proximal tubules (Zhang and Lindup 1993, 1994). Despite some evidence for induction of morphological abnormalities in mitochondria of tumour cells, the potential role of mitochondrial damage in the tumour-inhibiting effects of cisplatin has received comparatively little attention. Apart from metabolic effects like those recognized in renal tissue, mitochondrial DNA (mtDNA) would seem an obvious target for cisplatin. A first example for strongly preferential binding of cisplatin to mtDNA has been found in human malignant melanoma cells, in which an approximately 50-fold higher platination of mtDNA compared with nuclear DNA has been measured, accompanied by a suppression of NADH-ubiquinone reductase activity and a disturbance of ATP generation (Murata et al. 1990). Preferences for mtDNA ranging from 2- to 50-fold have later been found in rodent cells in vitro (Olivero et al. 1995) as well as in normal rodent (Giurgiovich et al. 1997a) and primate tissues in vivo (Giurgiovich et al. 1997b), including fetal tissues after transplacental exposure to cisplatin. The observed differences have been attributed to both higher initial binding to and lack of removal of adducts from mtDNA (Olivero et al. 1997). Damage of mtDNA is particularly enduring, since mitochondria, though not repair-incompetent altogether, exhibit a markedly reduced capacity for removing cisplatin-induced intrastrand cross-links, probably due to lack of the nucleotide excision repair (NER) system, while interstrand cross-links are efficiently removed (LeDoux et al. 1992). However, contrary findings in murine leukaemia cells (Kalinowski et al. 1992) and human ovarian carcinoma cells (Daoud et al. 1995; Woynarowski et al. 1998) indicate that preferential binding of cisplatin to mtDNA is not a feature with universal validity. Moreover, the consequences of alterations in NER capacity for sensitivity or resistance to cisplatin already discussed argue against a predominant role of mitochondrial injury, whether induced by mtDNA damage or by disturbance of metabolism, in the tumour-inhibiting effects of cisplatin.

In cell-free systems, all kinds of DNA cross-links induced by cisplatin inhibit replication (Comess et al. 1992) and transcription (Corda et al. 1993) to a higher extent than monofunctional adducts. Purified DNA polymerases are able to bypass these cross-links on single-stranded oligonucleotides or DNA with low frequencies depending on the kind of cross-link (Comess et al. 1992; Belguise-Valladier et al. 1994). The results obtained from experiments with such simplified model systems certainly do not reflect the situation in intact cells, in which several DNA-binding factors may either inhibit or facilitate proceeding with the replication complex past cisplatin-induced lesions, but cell culture experiments also suggest that most cells are capable of some degree of replicative bypass (Mamenta et al. 1994). As for the bypassing ability of RNA polymerases, there are conflicting findings: the earlier observation that only monofunctional adducts are bypassed by purified wheat germ RNA polymerase II (Corda et al. 1993) is opposed by a more recent report that RNA polymerase II in human cell extracts efficiently bypasses 1,2-d(GpG) cross-links in contrast to 1,3-d(GpTpG) cross-links (Cullinane et al. 1999). However, in intact cells transcriptional bypass has been estimated to occur only with a much lower frequency (Mello et al. 1995) than would be expected if the major 1,2-d(GpG) was bypassed to a high degree. Transcription may also be inhibited indirectly, as has been observed for undamaged genes in the presence of cisplatin-damaged DNA in human cell extracts. This may be explained by the dual involvement of factors such as the TATA binding protein (TBP), TFIIH or other transcription factors in both transcription and repair machineries. These factors may become limiting to transcription initiation if they are required for repair activity in the presence of DNA damage (Vichi et al. 1997; Cullinane et al. 1999).

Cisplatin induces slowing of the S phase, consistent with its inhibitory effect on DNA replication, and a subsequent G_2 arrest, which is transient at low, but persistent at higher drug concentrations. It has been suggested that this G_2 arrest is due to inhibition of transcription of genes essential for passage into mitosis (Sorenson and Eastman 1988a, 1988b). Along with these disturbances of cell cycle progression, cisplatin induces apoptotic cell death (Sorenson et al. 1990; Barry et al. 1990). High concentrations of cisplatin may also induce an S phase arrest followed by apoptosis instead of just slowing progression to the G_2 phase (Ishibashi and Lippard 1998). However, it is not known which of the cellular effects mentioned above trigger the apoptotic process and knowledge about the signal transduction pathways involved in cisplatin-induced apoptosis is far from complete (for a review see Eastman 1999).

The guanosine-rich telomeric DNA sequences, which are constituted of $(TTAGGG)_n$ tandem repeats, as well as the template RNA component of telomerase and the gene coding for this component, have recently received particular attention as potential targets of cisplatin. Cisplatin-induced damage of these sequences may inhibit the activity of telomerase, which is activated in the majority of tumours in contrast to most normal tissues except germ cells and stem cells. This enzyme compensates for the progressive telomere loss in the course of repeated replication, which eventually results in cell senescence and cell death after a critical stage of telomere degradation is reached. Telomerase is therefore supposed to play a role in immortalization of malignant cells and has been proposed as a potential target for tumour therapy. Oligonucleotides consisting of telomeric sequences are known to adopt highly stable quadruplex structures which inhibit telomerase activity. Cisplatin, but even more so its inactive *trans*-configured isomer transplatin, are able to cross-link these structures (Redon et al. 2001). However, the tempting hypothesis that cisplatin may inhibit telomerase by inducing or stabilizing such quadruplex structures in telomeres in the cellular environment remains to be confirmed. Inhibition of telomerase by cisplatin has indeed been observed in human testicular cancer cells in vitro (Burger AM et al. 1997), and a comparatively low concentration of cisplatin has proved to induce substantial telomere loss followed by delayed apoptosis in cervical carcinoma cells. The lack of telomere loss in the fraction of cells surviving continuous exposure to cisplatin has been put forward as further evidence for a causal connection between the observed telomere shortening and apoptosis, but this connection is difficult to establish unequivocally in the presence of other potentially lethal DNA lesions (Ishibashi and Lippard 1998). A correlation between sensitivity to cisplatin and telomerase inhibition could be demonstrated in squamous cell carcinoma cells (Mese et al. 2001). However, in leukaemic cells inhibition of telomerase has been observed only as a comparatively late event during exposure to cytocidal, but not cytostatic concentrations (Akiyama et al. 1999), and within a set of platinum compounds including cisplatin no direct correlation between telomerase-inhibiting activity and cytotoxicity could be found (Metzger and Keppler 1999). Thus, the contribution of telomerase inhibition to the tumour-inhibiting activity of cisplatin remains controversial.

Inherently reduced DNA repair capacity has been recognized for some time as a possible explanation for sensitivity of certain tumours to cisplatin (Bedford et al. 1988; Kelland et al. 1992; Hill et al. 1994; Köberle et al. 1996, 1997) and evidence for reduced activity of specific proteins involved in nucleotide excision repair has been found in some highly sensitive tumour cell lines. Studies in two human testicular tumour cell lines suggest that their sensitivity may be attributed to low levels of the damage recognition protein XPA (Köberle et al. 1999). However, it must be noted that the remarkable sensitivity of testicular germ cell tumours is probably the result of a complex interplay of biochemical characteristics relevant for processing cisplatin and its DNA adducts, which certainly have not been fully elucidated yet. Based on experimental evidence, possible contributions of the following factors are discussed: low levels of detoxification due to low concentrations of glutathione (Masters et al. 1996); expression of testis-specific HMG-domain proteins, which are more effective in inhibiting excision of 1,2-intrastrand adducts than common HMG-domain proteins (Zamble et al. 1996; Ohndorf et al. 1997; Trimmer et al. 1998); overexpression of the mismatch repair protein hMSH2, which is supposed to either shield cisplatin-DNA adducts from repair or trigger unsuccessful and eventually lethal attempts at repair (Mello et al. 1996); a higher susceptibility to the cytotoxic effects of various forms of stress due to low constitutive levels of heat shock protein 27 (Richards et al. 1996); and a higher propensity to undergo apoptosis following damage of DNA (Chresta et al. 1996). As for the contribution of the tumour suppressor p53, evidence is conflicting (for reviews see Zamble and Lippard 1999; Chaganti and Houldsworth 2000): given the pivotal role in activating signal transduction pathways leading to apoptosis, the strikingly high frequency of functional p53 in testicular germ-cell tumours and the identification of p53 mutations in a subset of clinically resistant germ-cell tumours suggest a causal connection (Houldsworth et al. 1998). However, studies in small panels of cell lines failed to establish a correlation between p53 status and susceptibility to cisplatin-induced apoptosis (Burger H et al. 1997, 1998). It has been demonstrated that inactivation of p53 does not necessarily affect cisplatin sensitivity, and evidence for p53-independent apoptotic pathways is provided (Burger H et al. 1999). Likewise, high levels of the p53-regulated apoptosis-promoting Bax protein, accompanied by low or even totally lacking expression of its antagonist Bcl-2, which had been found to be associated with sensitivity in a small panel of testicular germ cell tumour cell lines (Chresta et al. 1996), have not been confirmed as determinants of apoptotic response in another panel (Burger H et al. 1997, 1998).

The second generation – improving the toxicological profile

The nephrotoxicity of cisplatin severely hampered its early clinical development, until pre-hydration and forced diuresis were successfully introduced in order to prevent acute renal failure (Hayes et al. 1977). Further frequent adverse effects include cumulative peripheral sensory neuropathy, which usually is slowly reversible; ototoxicity, affecting high frequency perception due to irreversible damage of the hair cells in the Corti organ; as well as nausea and vomiting, the latter of which may be ameliorated to some extent by continuous administration of serotonin (5-hydroxytryptamine, 5-HT3) receptor antagonists. Thus, cisplatin therapy requires additional medication and is accompanied by troublesome and sometimes irreversible side effects. Much effort has been devoted to the development of chemoprotective agents, which alleviate the side effects on normal tissues without compromising antitumour efficacy. Mainly sulphur-containing agents such as glutathione, sodium thiosulphate, sodium diethyldithiocarbamate (Naddtc), 2-mercaptoethanesulphonate (mesna) and amifostine (WR-2721), but also neutrotropic factors, phosphonate-type antibiotics and scavengers of free oxygen radicals have been explored for their applicability (for a review see Cvitkovic 1998). With the approval of amifostine, only quite recently has a chemoprotectant become widely available for co-administration with cisplatin, which reduces the incidence of nephro- and neurotoxicity to some extent.

Approaches to avoid the toxicities of cisplatin have mainly focussed on development of new derivatives with altered pharmacokinetic and pharmacodynamic properties. Considerable improvement has been achieved by exchange of the labile chloro ligands for comparatively more stable bidentate leaving groups, thereby slowing down hydrolytic activation of the drug. This possibility had been anticipated already by the first extensive studies on structure-activity relationships of platinum complexes summarized by Cleare and Hoeschele, who had demonstrated that the diaqua product of cisplatin is highly toxic and that toxicity decreases with decreasing lability of the leaving ligand, while more stable bidentate leaving ligands can still give rise to active complexes and activity is only lost when very tightly bound ligands are used. These studies have thus revealed a so-called window of activity, which is largely defined by the nature of the bond between platinum and the leaving ligand (Cleare and Hoeschele 1973a, 1973b). Two of the second-generation platinum drugs fulfilling the requirements mentioned above are currently routinely used in cancer chemotherapy: carboplatin has gained worldwide approval and steadily increasing acceptance as a less toxic alternative to cisplatin, while nedaplatin has received only regionally limited approval (Fig. 2). More recently, attempts have been made to provide cisplatin in galenic formulations with more favourable pharmacokinetic properties. The latter will be dealt with in the section "Platinum drug targeting".

Carboplatin

Among the platinum complexes, which have been developed with the aim of overcoming the heavy side effects of cisplatin, carboplatin (CBDCA), *cis*-diammine(1,1-cyclobutanedicarboxylato)platinum(II), (Fig. 2) was the first to prove a more tolerable toxicologi-

Fig. 2a, b Structures of-second generation platinum drugs currently used in cancer chemotherapy: (**a**) carboplatin and (**b**) nedaplatin

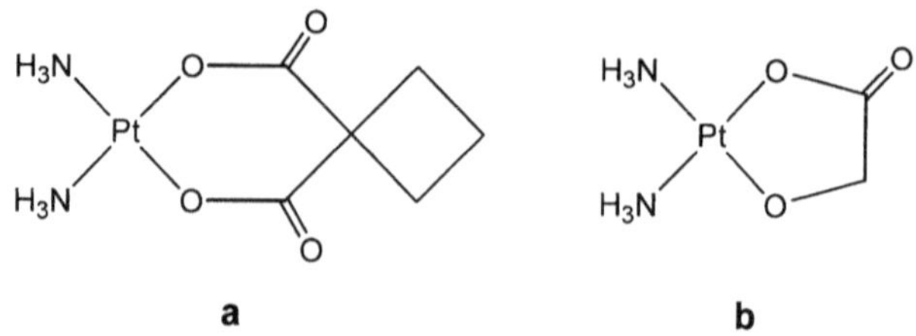

cal profile in clinical trials. It essentially lacks nephrotoxic properties in usual doses and can therefore be given without hydration and diuretics. Furthermore, neurotoxic and ototoxic properties are much less pronounced, gastrointestinal side effects are on average less severe and more easily controlled, and dose is limited by myelosuppression, with thrombocytopaenia usually being more severe than neutropaenia or anaemia. These properties make carboplatin, in contrast to cisplatin, a suitable drug for aggressive high-dose chemotherapy with autologous stem cell support, and the potential clinical benefit of this treatment modality in ovarian and small cell lung cancer as well as in refractory testicular germ-cell cancer has been explored for some time now. Carboplatin has, due to its equivalent efficacy but milder toxicity, largely replaced cisplatin in combination regimens with paclitaxel, which are currently used as standard therapy in ovarian cancer and are also often applied in patients with non-small cell lung cancer. Moreover, carboplatin is often preferred in combination with etoposide in therapy of small cell lung cancer. On the other hand, cisplatin is still the drug of choice for first-line treatment of testicular germ-cell cancer as well as in squamous cell carcinoma of the head and neck and in bladder cancer, which seem to respond to carboplatin to a lesser degree (for reviews see Go and Adjei 1999; O'Dwyer et al. 1999; Highley and Calvert 2000).

The better tolerability of carboplatin has generally been attributed to its higher stability and lower reactivity. Kinetics of hydrolytic activation are substantially slowed down in comparison with cisplatin. The aquation rate of carboplatin in neutral chloride-free phosphate buffer is approximately 100 times lower than that of cisplatin (Knox et al. 1986). Due to substantially reduced binding to plasma proteins, the elimination half-life of the pharmacologically active ultrafiltrable platinum fraction is about three times longer compared to administration of cisplatin, despite higher rates of renal elimination of total platinum. The similarity of the concentration-time curves of carboplatin and ultrafiltrable platinum suggests that the latter fraction essentially consists of the untransformed drug for up to 10 h after administration, and the bulk of excreted platinum is still in the form of carboplatin. Terminal elimination half-life of total platinum, reflecting the long persistence of the protein-bound fraction, is similar to cisplatin (for reviews on the clinical pharmacokinetics of carboplatin see van der Vijgh 1991; Dufull and Robinson 1997).

In cell-free phosphate buffer, the kinetics of DNA damage by carboplatin are much slower compared with cisplatin, in correspondence with their different aquation rates. This difference applies mainly to the initial binding step leading to monofunctional DNA adducts, whereas differences in kinetics of the subsequent ring-closing reaction resulting in bifunctional adducts are much less pronounced (Knox et al. 1986). However, binding experiments with GMP strongly suggest the possibility of direct attack of nucleobases on carboplatin without prior aquation of the drug (Frey et al. 1993; Küng et al. 2001a), and the formation and persistence of a ring-opened species, in which methionine and the cyclobutanedicarboxylate ligand are each bound monodentally, have led to the proposal that such complexes containing monodentate thioether ligands may be important hydrolysis-independent intermediates in the formation of DNA adducts (Barnham et al. 1996). Once bound to DNA in equal amounts, both drugs are equally cytotoxic (Knox et al. 1986) and induce the same kinds of DNA cross-links (Hongo et al. 1994; Blommaert et al. 1995). However, a peculiar discrepancy between relative proportions of the major adducts has been reported, indicating prevalence of 1,3-d(GpNpG) over 1,2-d(GpG) cross-links in DNA of treated cells in contrast to the pattern observed in cell-free DNA binding assays (Blommaert et al. 1995). The reasons for this discrepancy have not been elucidated. Nevertheless these features suggest that taking carboplatin for a mere slow-reacting prodrug of cisplatin may be an incorrect simplification.

The clinical efficacy of carboplatin is much higher than would be expected from its low reactivity towards DNA, as observed in simple cell-free systems. While at least 100-fold higher concentrations of carboplatin than of cisplatin are required to achieve the same rate of platination in isolated DNA (Knox et al. 1986; Blommaert et al. 1995), a 4- to 20-fold higher dose of carboplatin is sufficient to induce equal levels of DNA platination in treated patients (Gill et al. 1991; Blommaert et al. 1993). It is reasonable to assume that the substantially different pharmacokinetics of carboplatin, in particular its lower binding to proteins and longer persistence in plasma, make a large contribution, but it remains unclear whether they are sufficient to solely explain this effect. The possibility of hydrolysis-independent pathways to DNA platination such as those involving thioether-bound intermediates has to be kept in mind.

Nedaplatin

Another second-generation platinum drug, 254-S (nedaplatin), *cis*-diammineglycolatoplatinum(II) (Fig. 2), which has been developed and approved for several indications in Japan, also shows improved toxicological properties compared to cisplatin. With either bolus or 5-day continuous infusion, myelosuppression (thrombocytopaenia and leukopaenia) is dose-limiting and non-haematological toxicities are generally mild (Sasaki et al. 1989; Sasaki et al. 1991). Continuous infusion does not result in reduction, but only in a delay of haematotoxicity. Since the area under the time-concentration curve (AUC) cannot be raised without increasing the myelosuppressive effects, the maximum tolerated dose per course is even lower for the continuous than for the bolus infusion schedule (Sasaki et al. 1991). Bolus infusion has therefore been regarded as favourable and applied in all further clinical studies. The pharmacokinetic properties of nedaplatin strongly resemble those of carboplatin: binding to serum proteins is reduced in comparison to cisplatin with almost 80% of total platinum remaining in free form 2.5 h after drug infusion, and renal excretion is faster compared to cisplatin, with about 50% elimination within the first 8 h (Sasaki et al. 1989).

Clinical activity has been observed in phase II studies in non-small cell lung cancer (Fukuda et al. 1990), testicular cancer, transitional-cell carcinoma of the urinary tract and hormone-refractory prostate cancer (Akaza et al. 1992), head and neck, ovarian, cervical and small cell lung cancer (Ota et al. 1991). Overall pattern of activity and response rates resemble those of cisplatin, but only sporadic response to nedaplatin has been observed in cisplatin-pretreated patients. However, no study directly comparing nedaplatin with carboplatin has been performed yet, so there is no clear evidence for any distinct advantage over carboplatin.

A remarkably high response rate has been observed in oesophageal cancer patients, which was further improved by combination with 5-fluorouracil in a small pilot study (Yoshioka et al. 1999). Activity seems to be comparable to the standard cisplatin + 5-FU regimen. Since in vivo experiments indicate a marked sequence dependency for the effects of the nedaplatin/5-FU combination (Uchida et al. 1998), the schedule applied in the clinical study might have been suboptimal. A combination with vindesine is comparable with cisplatin/vindesine in previously untreated patients with non-small cell lung cancer in terms of efficacy and superior in terms of toxicity, but had only marginal activity in pretreated patients with relapsed or refractory disease (Takigawa et al. 2000). Preclinical synergy has been found for further combinations, including addition of etoposide, cyclophosphamide, gemcitabine or paclitaxel (Yamada et al. 2001 and references cited therein), which remain to be confirmed in the clinical setting.

Abandoned compounds

Iproplatin (CHIP, JM-9), *cis*-dichloro-*trans*-dihydroxybis(isopropylamine)platinum(IV), was extensively studied at the same time as carboplatin and was the first platinum(IV) complex to enter clinical trials, but clearly failed to prove advantageous over carboplatin in terms of response and survival of patients with advanced ovarian cancer in a randomized phase III trial (Trask et al. 1991), which eventually led to cessation of its development. Clinical trials of several other compounds failed to confirm the improved toxicological profile expected from previous animal experiments: platinum uracil blue exhibited unpredicted cardiac toxicity; spiroplatin (TNO-6), (1,1-diaminomethylcyclohexane)sulphatoplatinum(II), and JM-40, (ethane-1,2-diamine)malonatoplatinum(II), failed due to renal toxicity.

After the approval of carboplatin, four analogues containing the cyclobutanedicarbolxylato ligand were selected for clinical examination due to similar or even more favourable toxicological profiles and improved therapeutic index in animal experiments. These complexes contained the following bidentate amine ligands: 2,2-bis(aminomethyl)-1,3-propanediol (zeniplatin), tetrahydro-4H-pyran-4,4-dimethanamine (enloplatin), 2-methyl-1,4-butanediamine (CI-973, NK121) and 1R,2R-pyrrolidinemethanamine (DWA2114R, miboplatin). However, clinical trials have failed to confirm these findings in the case of zeniplatin and enloplatin, which both were found to combine the nephrotoxicity of cisplatin and the myelosuppressive properties of carboplatin. The other two carboplatin analogues, CI-973 (NK121) and DWA2114R (miboplatin), have been abandoned due to inferior activity and the lack of an advantage over cisplatin, respectively (for a review see Lebwohl and Canetta 1998).

The third generation – overcoming cisplatin resistance

Intrinsic resistance of many malignancies limits the range of indications for cisplatin, and acquired resistance diminishes response after repeated courses in many other tumour diseases, eventually leading to failure of cisplatin therapy. Cross-resistance between cisplatin, carboplatin and nedaplatin makes the latter two drugs ineffective in treatment of patients who have failed to respond to cisplatin. Four main mechanisms underlying resistance have been identified:

1. Reduction of drug uptake
2. Increased detoxification by glutathione or metallothionein
3. Increased repair of platinum-DNA adducts
4. Increased tolerance to platinum-DNA adducts

Reduced net uptake of cisplatin seems to be mainly due to reduced drug influx. Cisplatin is not a substrate for multidrug resistance proteins, and the contribution of enhanced drug efflux to reduced net accumulation in cisplatin-resistant cells has been questioned (Andrews 2000). Little is known about the mechanisms by which cells reduce cisplatin influx, but changes in structure, lipid composition and cytoskeleton interactions of cell membranes resulting in decreased permeability are discussed.

Glutathione (γ-glutamylcysteinylglycine) is the most abundant non-protein thiol in cells and is suited to function as a cisplatin scavenger. It is able to compete with formation of monofunctional DNA adducts (Dedon and Borch 1987) and with conversion of monofunctional to bifunctional adducts (Bancroft et al. 1990). In some cases acquired resistance to cisplatin has been found to be associated with elevated levels of glutathione (Wolf et al. 1987; Behrens et al. 1987; Godwin et al. 1992). Like the tripeptide glutathione, the small cysteine-rich protein metallothionein is assumed to act as a detoxifying molecule and is overexpressed in some cisplatin-resistant cells (Kelley et al. 1988).

Resistance to cisplatin may also be conferred by alterations in processing of DNA adducts. Enhanced adduct removal, which is mainly accomplished by the nucleotide excision repair system, was soon recognized as a major cause of resistance in cell lines (see "The first generation – cisplatin"), while enhanced tolerance to DNA adducts has received increasing attention in recent years. At least two mechanisms are involved in increased tolerance to platinum-DNA adducts: deficiencies in the mismatch repair system and enhanced replicative bypass. The mismatch repair (MMR) system recognizes DNA adducts generated by cisplatin, but is not capable of correctly repairing these lesions. As a consequence of futile attempts at repair, lesions may accumulate and single-strand breaks may be induced, thereby increasing the detrimental effects of the primary DNA damage (Duckett et al. 1996). Loss of MMR competence results in low levels of resistance to cisplatin (Aebi et al. 1996; Fink et al. 1996). Replicative bypass, i.e. the ability of the replication complex to synthesize DNA past damaged sites, allows cells to progress through the S phase of the cell cycle despite platinum-induced damage, which may then be repaired afterwards (postreplication repair). This ability of replicative bypass may, due to an unknown mechanism, be enhanced in cells with resistance to cisplatin (Gibbons et al. 1991; Mamenta et al. 1994). These two properties leading to increased adduct tolerance seem to be partially connected, since a higher incidence of replicative bypass has been detected in MMR-deficient than in MMR-proficient cell lines (Vaisman et al. 1998). However, enhanced replicative bypass has also been found in MMR-proficient cells (Mamenta et al. 1994), suggesting the existence of a mechanism independent of MMR status.

These mechanisms have been identified and extensively studied in cell lines with acquired resistance upon chronic exposure to cisplatin, and it soon became obvious that two or more of the mechanisms described may operate simultaneously in the same cell line (Richon et al. 1987; Oldenburg et al. 1994) and that none of them is found in all cisplatin-resistant cells. In vivo experiments suggest that even a reduction in sensitivity as little as twofold may be sufficient to cause failure of treatment (Andrews et al. 1990), and the degree of resistance found in patients is usually low to moderate (Simmonds and McDonald 1994). However, only a few studies have been conducted to examine the crucial question of the clinical relevance of these resistance mechanisms: elevated glutathione levels, but no significant changes in expression of glutathione *S*-transferases have been detected in biopsies of ovarian cancer patients after development of resistance compared to biopsies taken prior to the onset of chemotherapy (Britten et al. 1992). On the other hand, a high activity of glutathione *S*-transferase has been reported to be predictive for resistance to platinum-based chemotherapy in head and neck cancer patients (Nishimura et al. 1996). Expression levels of several components of the nucleotide excision repair system (ERCC1, XPA, XPB) have also been found to correlate with response in ovarian cancer patients, suggesting a role of enhanced DNA repair in clinical resistance to cisplatin (Dabholkar et al. 1994, 2000). Mismatch repair defects are detectable only in a small percentage in untreated ovarian tumours, but cisplatin therapy has been found to induce marked enrichment of DNA mismatch repair-deficient tumour cells in a high percentage of patients. Although these deficiencies confer only low levels of resistance, they may nevertheless contribute to some extent to the frequent decline in sensitivity during cisplatin therapy (Fink et al. 1998).

A great deal of effort has been expended to identify platinum compounds which lack cross-resistance with cisplatin/carboplatin, and several compounds have been chosen for clinical development due to preclinical findings of activity in tumours or tumour cells with intrinsic or acquired resistance to cisplatin. So far, only oxaliplatin (*l*-OHP), which has proved activity in colorectal cancer and has therefore attained approval for this indication in many countries, has shown clinical activity in an intrinsically cisplatin-resistant tumour disease and thus broadened the range of indications for platinum-based therapy.

1,2-Diaminocyclohexane(DACH)-containing platinum complexes

The parent compound of all platinum complexes containing 1,2-diaminocyclohexane (DACH), *cis*-dichloro(1,2-diaminocyclohexane)platinum(II), was recognized as a potent agent in experimental tumour models soon after the onset of clinical studies with cisplatin, but its low solubility in water prohibited intravenous administration and its therapeutic index proved to be lower than that of cisplatin (Connors et al. 1972; Ridgway et al. 1976). Attempts to overcome these problems have mainly focussed on the development of water-soluble derivatives with more stable leaving groups. Derivatives containing sulphate (JM-20, SHP) or malonate (JM-74, PHM) instead of chloride were then among the first investigational platinum drugs, after cisplatin, to enter clinical trials and to prove lack of cross-resistance to cisplatin in vivo (Burchenal et al. 1977). This latter property was later recognized as a general feature of platinum complexes containing DACH as non-leaving ligand and is observed in many, albeit not all experimental tumour models, and a spectrum of activity distinctly different from those of cisplatin and carboplatin has been demonstrated for this class of compounds in vitro (Rixe et al. 1996). Preparation of liposomal formulations of lipophilic derivatives has been pursued as an alternative strategy and has led to the development of L-NDDP, which has been investigated in clinical trials for more than a decade.

1,2-Diaminocyclohexane contains two asymmetric carbon centres and therefore exists in three isomeric forms: *cis-R,S*-DACH and the enantiomers *trans-R,R*- and *trans-S,S*-

DACH. Although differences in activity tend to be rather slight or even undetectable in some cases (Gullotti et al. 1984; Khokhar et al. 1997), the majority of studies on the structure activity relationships of DACH containing platinum compounds in vitro and in vivo have identified the respective *trans-R,R*-DACH isomers as the most active and the *cis*-DACH isomers as the least active (Kidani et al. 1978a; Kidani et al. 1980; Noji et al. 1981; Jennerwein et al. 1991; Kido et al. 1993a, 1993b; Siddik et al. 1994a; Pendyala et al. 1995), while only a few exceptions, partly dependent on tumour model and schedule of drug administration, have been reported (Kidani et al. 1978b; Vollano et al. 1987; Siddik et al. 1993). The superiority of the *trans-R,R*-DACH isomer tends to be even more pronounced in cells with acquired resistance to cisplatin (Jennerwein et al. 1991; Kido et al. 1993a, 1993b; Pendyala et al. 1995). In both *trans*-isomers, the coordination square and cyclohexane are almost coplanar, whereas in the *cis*-isomer they are nearly perpendicular. It has therefore been suggested that the *trans*-isomers may be favoured in approaching DNA, whereas the cyclohexane ring of *cis*-DACH complexes may introduce steric hindrance (Noji et al. 1981). Subtle structural differences between the adducts of the two *trans*-isomers have also been characterized (Inagaki and Kidani 1986; Inagaki et al. 1990). Complexes containing different isomers of DACH produce spectra of DNA adducts very similar to each other and to that induced by cisplatin, but whereas no differences in kinetics of binding and amounts of total platination of isolated DNA had been found initially (Jennerwein et al. 1989), lower rates of monofunctional binding and subsequent rearrangement to bifunctional adducts were later detected for the *cis*-DACH-containing isomer. Evidence for more extensive conformational changes induced by the latter isomer in oligonucleotides further supports the hypothesis that its lower activity is due to steric hindrance of interaction with DNA (Boudný et al. 1992). Slight differences in cellular uptake may also contribute to the observed structure–activity relationship, but due to partial inconsistencies with cytotoxicity, they are not sufficient to fully explain them (Kido et al. 1993a; Pendyala et al. 1995). Differential repair is unlikely to play a significant role, since according to results of a plasmid reactivation assay, only slight, if any, differences exist in repair of DNA adducts generated by platinum coordinated to different isomers of DACH (Jennerwein et al. 1991).

In contrast to some of the earlier representatives of this class of compounds examined in clinical studies (see “Abandoned DACH-containing compounds”), all of the DACH-containing platinum agents currently available in clinical practice (oxaliplatin) or under clinical evaluation (L-NDDP, TRK-710) are applied in isomerically pure form with respect to DACH, containing only the *trans-R,R* isomer (Fig. 3). However, in consideration of the higher neurotoxic potency, as has been demonstrated for oxaliplatin and the *trans-R,R*-DACH-containing isomer of ormaplatin compared to their *trans-S,S*-DACH-containing counterparts by sensory nerve conduction velocity measurements during repeated treatment of rats (Screnci et al. 1997), it is conceivable that the apparent advantage of this choice may be abolished by this toxicity, which is dose-limiting in some DACH-containing platinum complexes such as oxaliplatin and ormaplatin (tetraplatin).

Given the variety and complexity of mechanisms conferring resistance to cisplatin, no simple explanation for the lack of cross-resistance between DACH-containing and diammine platinum drugs observed in many tumour models is to be expected. Differential cellular uptake has been observed in some tumour cells (Richon et al. 1987; Kraker and Moore 1988; Gibbons et al. 1990), but cannot in general account for this phenomenon, since some cell lines with acquired resistance to cisplatin due to reduced uptake also accu-

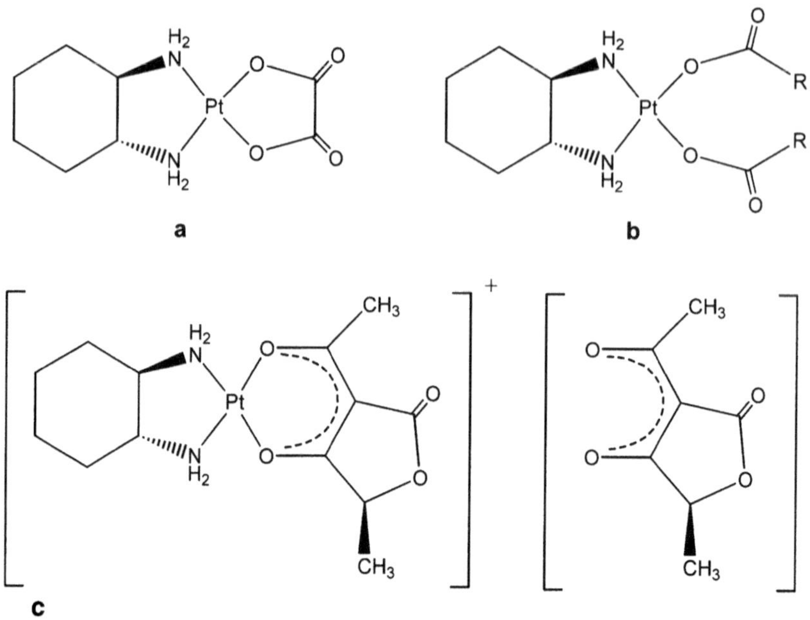

Fig. 3 Structures of *trans-R,R*-1,2-diaminocyclohexane(DACH)-containing platinum drugs: oxaliplatin (**a**), (L-)NDDP (**b**) and TRK-710 (**c**)

mulate DACH-containing compounds such as oxaliplatin and ormaplatin to a lesser extent and are therefore slightly less sensitive than their parent lines, though with a much lower resistance index than to cisplatin (Rixe et al. 1996). However, there is evidence for more efficient incorporation of intracellular platinum into DNA (Schmidt and Chaney 1993). Presence of the bulky DACH ligand has little influence on interaction of the platinum centre with DNA, but has substantial consequences on some aspects of processing of the generated DNA lesions. Neither the rate nor the types or sites of adduct formation are altered, only the rate of conversion of monofunctional to bifunctional adducts is affected in comparison with complexes containing sterically less demanding ethylenediamine instead (Page et al. 1990), and the rate of removal of total DNA adducts and interstrand cross-links is not significantly different from that of cisplatin-induced adducts, even in cells with acquired resistance to cisplatin due to enhanced repair (Gibbons et al. 1990; Schmidt and Chaney 1993; Petersen et al. 1996; Reardon et al. 1999). Since there is no evidence for discrimination by nucleotide excision repair, this mechanism is unlikely to account for the lack of cross-resistance. Nevertheless, presence of the DACH ligand renders platinum-DNA adducts on average more cytotoxic (Schmidt and Chaney 1993), which correlates with a stronger hindrance of replication (Mamenta et al. 1994; Woynarowski et al. 2000). Furthermore, tolerance of platinum-DNA adducts in cell lines, which exhibit resistance to cisplatin due to enhanced replicative bypass or deficiencies in the mismatch repair system, is affected or even abolished (Gibbons et al. 1991; Mamenta et al. 1994; Aebi et al. 1996; Fink et al. 1996; Vaisman et al. 1998). In contrast to cisplatin-induced DNA adducts, those induced by the DACH-containing drug oxaliplatin are not recognized by mismatch repair proteins (Aebi et al. 1996; Fink et al. 1996; Nehmé et al. 1999).

Oxaliplatin

Among the DACH-containing platinum agents, oxaliplatin, (*trans-R,R*-1,2-diaminocyclohexane)oxalatoplatinum(II), (Fig. 3a) has the advantage of being comparatively soluble in water. It has proved to be active in advanced colorectal cancer and to enhance the efficacy of standard 5-fluorouracil (5-FU)/leucovorin therapy in terms of response rate, time to progression and proportion of patients whose metastases become resectable after chemotherapy. Although no significant advantages in terms of overall survival have been found, it should be kept in mind that the greater response rates may translate into improved quality of life and that those patients whose metastases shrink to a resectable size may have much greater benefit than is reflected by median survival. Moderate activity with response rates similar to irinotecan has been observed even in patients with 5-FU-refractory or recurrent disease, and oxaliplatin may, therefore, alternatively to combination with 5-FU/leucovorin in first-line therapy, be suited for second- and third-line therapy of advanced colorectal cancer after failure of prior treatment (for reviews see Cvitkovic and Bekradda 1999; Misset et al. 2000; Armand et al. 2000). Therefore, oxaliplatin was the first platinum drug to show clinical activity in a tumour with primary resistance to cisplatin, and it was the first DACH-containing platinum drug to be approved for clinical use. It is currently registered in many countries in Europe, Asia and Latin America for first-line treatment of metastatic colorectal cancer in combination with 5-FU (plus leucovorin). Furthermore, oxaliplatin has recently been approved in the United States for second-line therapy in combination with 5-FU.

Besides extensive studies in patients with colorectal cancer, oxaliplatin is explored for its potential as a treatment option after failure of cisplatin or carboplatin therapy, and clinical activity has been reported both in relapsed or refractory ovarian cancer (Chollet et al. 1996; Dieras et al. 1998; Piccart et al. 2000) and in refractory germ-cell tumours (Kollmannsberger et al. 2001), suggesting a potential role of oxaliplatin in the therapy of these patient groups. Moreover, clinical activity has been observed in pretreated refractory or relapsed non-Hodgkin's lymphoma (Germann et al. 1999), in anthracycline-resistant metastatic breast cancer (Garufi et al. 2000) and in previously untreated non-small cell lung cancer (Monnet et al. 1998). Results from human tumour cloning assays suggest an even broader spectrum of tumour types responsive to oxaliplatin (Raymond et al. 1998).

Oxaliplatin lacks the nephrotoxicity of cisplatin and has much milder haematological toxicity than carboplatin at therapeutic dosage. It produces gastrointestinal side effects similar to cisplatin, and these were first considered dose-limiting (Mathé et al. 1986), before prophylactic administration of antiemetics allowed dose escalation up to more than threefold levels, at which neurotoxicity prohibited further dose escalation (Extra et al. 1990). This toxicity manifests primarily as cumulative peripheral sensory neuropathy with a symptomatology somewhat different from that induced by cisplatin, and though symptoms regress in most patients, recovery is often slow and incomplete. Neurotoxicity has hampered clinical development of oxaliplatin to some extent, but could be alleviated by protracted or chronomodulated administration (Extra et al. 1998; Levi et al. 2000). The reasons for these peculiar neurotoxic properties have not been elucidated. In comparison with other platinum drugs in rats, neurotoxic potency has been found to roughly correlate with reactivity (measured by the half-life of binding to plasma proteins in vitro), but not with accumulation in peripheral nerves (Screnci et al. 2000). An attempt to explain the neurotoxic effects of oxaliplatin by reduction of the amplitude of sodium currents resulting from immobilization of calcium due to chelation by free oxalate released from the complex (Grolleau et al. 2001) is based on experiments with drug concentrations which are much higher than the peak plasma concentrations achieved with standard dosing and must therefore be considered clinically irrelevant. Moreover, this approach ignores the fact that some biotransformation products which are devoid of the oxalate ligand exhibit similar or even higher neurotoxic potency than oxaliplatin itself (see below).

In contrast to carboplatin, extensive biotransformation, tissue distribution and binding to plasma proteins and red blood cells lead to a rapid decline of parent drug concentrations in plasma ultrafiltrate below detection limits within a few hours (Graham et al. 2000; Allain et al. 2000). Oxaliplatin is converted into several different species by exchange of oxalate for other nucleophiles. Dichloro(*trans-R,R*-1,2-diaminocyclohexane)platinum(II), which displays a higher cytotoxic (Luo et al. 1998) and neurotoxic (Luo et al. 1999d) potency than the parent drug, appears only transiently in plasma due to its high reactivity (Luo et al. 1999a, 1999b). As in the case of cisplatin, the most active monoaqua and diaqua species are assumed to prevail over the dichloro species inside the cells, due to the lower chloride concentrations. The major stable biotransformation products have been identified as methionine- and cysteine-containing species, which essentially lack cytotoxic properties however (Luo et al. 1998), and free 1,2-diaminocyclohexane, which is probably released as a result of trans-labilization by sulphur-containing ligands. Glutathione, which reacts with oxaliplatin to minor biotransformation products, probably contributes to release of DACH (Luo et al. 1999a, 1999b). The exchange of oxalate for methionine results in an inert species, which does not react with GMP (Küng et al. 2001b) and exhibits no biological activity, neither in tumour cells in vitro nor in rat peripheral nerves in vivo (Screnci et al. 1997). Binding to plasma proteins is substantially higher than that of carboplatin and only slightly lower than that of cisplatin (Graham et al. 2000), and the main binding proteins are albumin and γ-globulins (Allain et al. 2000). Oxaliplatin accumulates to a higher extent in red blood cells than cisplatin, and platinum content of erythrocytes declines slowly, with a half-life close to that of erythrocytes themselves (Gamelin et al. 1997). Given the lack of release (Pendyala and Creaven 1993), a high percentage of binding to haemoglobin (Allain et al. 2000) and the inactivity of the major ultrafiltrable biotransformation products which eventually accumulate in the cytoplasm of erythrocytes (Luo et al. 1999b), these are unlikely to serve as a reservoir of cytotoxic species. The remarkably high volume of distribution of ultrafiltrable platinum, a feature oxaliplatin shares with another DACH-containing complex, ormaplatin (tetraplatin), and probably attributable to the lipophilicity of the DACH ligand, suggests enhanced tissue penetration in comparison with cisplatin and carboplatin. Elimination of ultrafiltrable platinum from plasma is equally due to tissue binding and renal excretion, and the rate of urinary elimination is intermediate between those of cisplatin and carboplatin (Graham et al. 2000), but in contrast to the latter, the bulk of urinary platinum after administration of oxaliplatin is composed of biotransformation products (Allain et al. 2000). Unlike cisplatin, oxaliplatin does not accumulate in plasma, but to a higher extent in erythrocytes after repeated cycles (Gamelin et al. 1997).

Oxaliplatin produces adducts on isolated DNA with a similar ratio of GG over AG intrastrand cross-links (Saris et al. 1996), similar sequence specificity and low levels of interstrand cross-link formation (Woynarowski et al. 1998) compared to cisplatin, but with substantially slower kinetics. This difference diminishes to some extent, when DNA binding is compared in intact cells. Though reactivation of drug-exposed plasmids in transfected cells indicates similar repair kinetics, oxaliplatin is more efficient in inhibiting DNA replication than cisplatin (Woynarowski et al. 2000). The similar or even higher cytotoxicity of oxaliplatin despite lower levels of DNA adduct formation compared to cisplatin indicates a higher cytotoxicity of the DNA lesions produced by oxaliplatin, at least to the cells used in the cited studies.

Partly based on preclinical evidence for synergistic interactions with antimetabolites such as 5-fluorouracil (Mathé et al. 1989; Raymond et al. 1997; Fischel et al. 1998) and gemcitabine (Faivre et al. 1999), with topoisomerase I inhibitors such as topotecan (Goldwasser et al. 1999) and irinotecan (Zeghari-Squalli et al. 1999), with the antimitotic drug paclitaxel (Skou et al. 1996) as well as with other platinum drugs such as cisplatin and carboplatin (Mathé et al. 1989; Rixe et al. 1996), various clinical studies exploring the safety and efficacy of these combinations have been or are currently being conducted (for a review see Mani et al. 2000). As a logical consequence of single-agent activities and synergistic effects of each double combination, attempts are being made to integrate oxaliplatin in the combination regimen 5-FU (plus leucovorin) plus irinotecan, which has become available as first-line therapy for metastatic colorectal cancer. Preclinical studies (Fischel et al. 2001) and reports on clinical phase II studies (Calvo et al. 2001; Cure et al. 2001) indicate a high activity, but comparative phase III trials are required to evaluate the potential benefit of the ternary combination.

L-NDDP

L-NDDP, a liposomal formulation of *cis*-(*trans*-*R,R*-1,2-diaminocyclohexane)bis(neodecanoato)platinum(II) (Fig. 3b), was the first liposomal platinum drug to enter clinical trials. Large multilamellar liposomes composed of dimyristoylphosphatidylcholine (DMPC) and dimyristoylphosphatidylglycerol (DMPG) were used to encapsulate the drug, which is an imprecisely defined mixture of at least 18 isomers because of various different structures and configurations of the lipophilic neodecanoato ligands. Toxicological properties and tumour-inhibiting activities of single isomers in mice were found to be on the whole comparable to the isomeric mixture, though some isomers were somewhat more active in cisplatin-resistant murine leukaemia (Khokhar et al. 1991). The liposomal preparation has been used primarily to provide a suitable formulation of this highly lipophilic agent, which is insoluble in water, but its suitability for tumour-targeting purposes , in particular for tumours involving the reticuloendothelial system, has been explored as well. Indeed, L-NDDP accumulates to a higher extent than cisplatin in liver and spleen of mice and rats and is more effective in inhibiting growth of established and in preventing emergence of new liver metastases of reticulosarcoma in mice (Perez-Soler et al. 1987; Vadiei et al. 1992), whereas free NDDP administered by means of solubilizers is generally less active or even inactive in several tumour models (Perez-Soler et al. 1988). The increased activity of the liposomal preparation has been attributed to intraliposomal activation by reaction with DMPG (Perez-Soler and Khokhar 1992).

The dose-limiting toxicity of intravenous L-NDDP is myelosuppression affecting all blood cell lineages and the toxicity profile resembles that of carboplatin (Perez-Soler et al. 1990). Clinical studies have not been restricted to intravenous administration. Its suitability for targeting liver tumours has stimulated attempts at using the hepatic intra-arterial route of administration, but chemical hepatitis has been dose-limiting (Perez-Soler et al. 1993). Furthermore, favourable pharmacokinetic properties of L-NDDP administered into the peritoneal cavity of rats, in particular a much longer peritoneal retention and slower systemic distribution compared to cisplatin (Vadiei et al. 1992), have stimulated studies of intracavitary infusion, both of intrapleural administration in patients with malignant pleural effusions caused by adenocarcinoma of the lung or pleural mesothelioma (Perez-Soler et al. 1997) and intraperitoneal administration in patients with malignant ascites caused by ovarian cancer or peritoneal mesothelioma (Verschraegen et al. 1999). In the former case, the reported dose-limiting toxicity has been chemical pleuritis, and in the latter case post-therapeutic adhesions prevented re-treatment at the highest dose level studied. However, two lethal cases of pneumonia and sepsis, most probably related to the method of infusion, have necessitated alteration of the mode of intrapleural administration in a subsequent phase II study in patients with malignant pleural mesothelioma (Perez-Soler et al. 1999). Nevertheless, a high rate of complete responses has encouraged continuation of studies in patients with pleural mesothelioma.

As far as the mechanisms responsible for the lack of cross-resistance to cisplatin are governed by the properties of the 1,2-diaminocyclohexane ligand, they are expected to be the same as with other DACH-containing platinum drugs (see "1,2-Diaminocyclohexane(DACH)-containing platinum complexes"), but an increased ability of penetrating into cells makes a relevant contribution in the case of L-NDDP: in pairs of cell lines, each consisting of a sensitive parent line and a derivative line with acquired resistance to cisplatin due to reduced net accumulation, uptake of L-NDDP is less affected by the resistance mechanism. Cellular uptake is much less temperature-dependent than that of cisplatin, which has been attributed to the highly lipophilic character of L-NDDP. A potential role of the liposomal carrier in effective delivery of the complex has also been discussed, but remains uncertain. Furthermore, L-NDDP has been found to be generally more cytotoxic, to induce higher levels of DNA platination and to produce more protein-DNA cross-links, but less DNA interstrand cross-links in these cell lines than cisplatin at equimolar concentrations, while at equitoxic concentrations the overall extent of DNA platination induced by both drugs is similar (Han et al. 1993; Han et al. 1994). An alternative explanation for the lack of cross-resistance to cisplatin has been proposed based on the finding that DNA adducts induced by L-NDDP are repaired with significantly lower efficiency than those induced by cisplatin (Yang et al. 1998), which obviously conflicts with findings obtained with other DACH-containing compounds (see "1,2-Diaminocyclohexane(DACH)-containing platinum complexes").

Besides its incompletely defined chemical composition, the major limitation experienced with L-NDDP is the instability of the drug within the liposomes, which has prevented its widespread clinical evaluation. Almost half of the complex is degraded within the first hour after reconstitution of the liposomes in the infusion liquid. Neodecanoic acid is removed by hydrolysis, resulting in activation of the drug (MacLean et al. 2000). These circumstances have actuated a search for more stable derivatives and formulations, but a dilemma has arisen: stability is inversely related to the DMPG content of the liposomal carrier, but on the other hand, DMPG is an essential prerequisite for antitumour activity (Perez-Soler and Khokhar 1992; Perez-Soler et al. 1994). Avoidance of branched in favour of linear leaving groups has been recognized as a possible solution to overcome this problem (Perez-Soler et al. 1994).

TRK-710

In contrast to all other DACH-containing platinum compounds mentioned, TRK 710, (3-acetyl-5-methyl-2,4(3H,5H)-furanedionato-O3,O4)(*trans-R,R*-1,2-diaminocyclohexane) platinum(II) (*S*)3-acetyl-5-methyl-2,4(3H,5H)-furanedionate, (Fig. 3c) is the salt of a cationic complex. Its preclinical evaluation has revealed higher activity compared to cisplatin and a lack of cross-resistance in several tumour models. Lower renal and haematological toxicity, but ototoxicity similar to cisplatin have been reported (Hoshi and Castaner 1997). The compound has been introduced to clinical phase I trials in Japan.

Abandoned DACH-containing compounds

Considering the obvious suitability of platinum complexes containing the 1,2-diaminocyclohexane (DACH) ligand for overcoming at least some forms of cisplatin resistance, it is remarkable that several DACH-containing platinum(II) complexes had entered clinical trials without success before oxaliplatin. These complexes contained either monodentate leaving ligands such as bromoacetate (MBA) or pyruvate (PYP) or bidentate ligands such as sulphate (JM-20, SHP; neo-SHP), malonate (JM-74, PHM; neo-PHM), isocitrate (PHIC) or 4-carboxyphthalate (JM-82, DACCP). Two of these have been studied as racemic mixture containing both *trans-R,R*- and *trans-S,S*-1,2-diaminocyclohexane (JM-20, SHP; JM-74, PHM) and in isomerically pure form containing only *trans-R,R*-1,2-diaminocyclohexane (neo-SHP; neo-PHM). However, all of these compounds were abandoned in early stages of clinical development for various reasons, including unfavourable chemical properties such as insufficient solubility or stability as well as severe toxicities and hypersensitivity reactions (see Table 1; for a review see Lebwohl and Canetta 1998).

Ormaplatin (tetraplatin), tetrachloro(*trans*-1,2-diaminocyclohexane)platinum(IV) containing *trans-R,R*- and *trans-S,S*-DACH in racemic mixture, has been the only platinum(IV) complex among the DACH-containing platinum compounds in clinical trials. Its development was halted because of severe, cumulative and irreversible peripheral neurotoxicity, manifested as sensory polyneuropathy, observed in several phase I studies which had used different administration regimens (Schilder et al. 1994; O'Rourke et al. 1994 and references cited therein). The more severe neurotoxicity of ormaplatin in comparison to oxaliplatin has been attributed to differences in their biotransformation kinetics. Though biotransformation of both drugs yields essentially the same products, dichloro(1,2-diaminocyclohexane)platinum(II) is formed more rapidly and is present in higher concentrations in plasma following administration of ormaplatin to rats (Luo et al. 1999c). This species and its hydrolysis products exhibit a higher neurotoxic potency in rat dorsal root ganglia explant cultures in vitro than either parent compound (Luo et al. 1999d). Modification of the axial ligands in order to slow down reduction and exchange of the equatorial chlorides for other ligands which are very slowly displaced by chloride under physiological conditions have therefore been proposed as strategies to overcome neurotoxicity and possibly improve the toxicological profile, even in comparison to oxaliplatin. Complexes fulfilling these structural criteria were synthesized previously, and some of them exhibit good activity in vivo (Siddik et al. 1994b; Khokhar et al. 1997), but it remains to be seen whether a new derivative with favourable toxicological properties will arise from this class of compounds and proceed to the clinical stage of development.

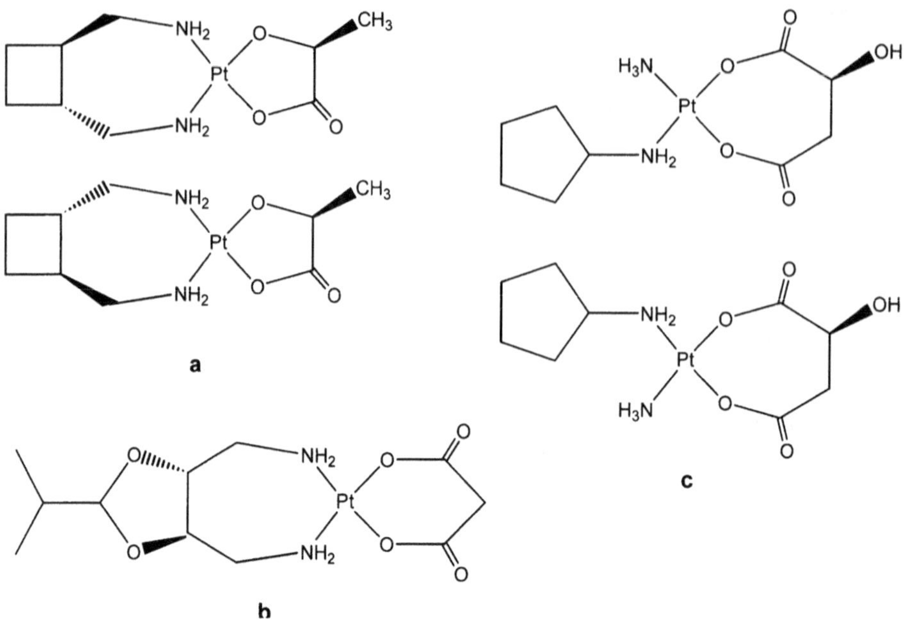

Fig. 4 Structures of lobaplatin isomers (**a**), SKI2053R (**b**) and cycloplatam isomers (**c**)

Lobaplatin

Lobaplatin (D-19466), (1,2-cyclobutanedimethanamine)[(2*S*)-2-hydroxypropanoato(2-)-O1,O2]platinum(II) (Fig. 4a) consists of a 1:1 mixture of two diastereoisomers with *R*,*R*- and *S*,*S*-configuration at the stereocentres of cyclobutanedimethanamine, respectively (Welink et al. 1996). Like carboplatin and oxaliplatin, lobaplatin exhibits a lower reactivity towards isolated DNA compared to cisplatin, but forms GG preferentially over AG intrastrand cross-links with a similar ratio (Saris et al. 1996). Preclinical data have suggested both a favourable toxicological profile and a lack of complete cross-resistance to cisplatin, though the latter applies only to some tumour models (Voegeli et al. 1990; Meijer et al. 1992; Harstrick et al. 1993). Clinical studies have revealed thrombocytopaenia as dose-limiting toxicity, irrespective of the schedule applied, while gastrointestinal toxicity was generally mild and no nephro-, neuro- and ototoxicity has been observed (Fiebig et al. 1991; Gietema et al. 1993a, 1993b). The pharmacokinetic behaviour of lobaplatin is characterized by low binding to serum proteins and a more rapid urinary excretion than carboplatin (Gietema et al. 1993a, 1995). However, there is no clear evidence of clinical activity in cisplatin-resistant tumours. Responses have been reported mainly in patients with ovarian cancer who had relapsed long after initial response to primary platinum-based therapy (Gietema et al. 1995), but no activity has been found in a patient group strictly confined to cisplatin/carboplatin-resistant ovarian cancer; consequently the adequacy of the preclinical tumour models, from which lack of cross-resistance had been predicted, has been questioned (Kavanagh et al. 1995). A response rate of 36% in previously untreated patients with advanced or metastatic oesophageal cancer (Schmoll et al. 1995) has stimulated initiation of a phase III trial comparing the combinations lobaplatin/5-FU versus cisplatin/

5-FU for the treatment of oesophageal carcinoma. Whereas lobaplatin has gained approval in China, its development in the Western world has ceased.

SKI 2053R

SKI 2053R, *cis*-malonato[(4*R*,5*R*)-4,5-bis(aminomethyl)-2-isopropyl-1,3-dioxolane]platinum(II), is structurally related to lobaplatin, but contains an even bulkier amine ligand than the latter (Fig. 4b). The efficacy of SKI 2053R in tumour models exhibits strong similarities to cisplatin, except for retained activity in murine leukaemia with acquired resistance to cisplatin (Kim et al. 1995). Preclinical toxicological studies have suggested that this compound is well tolerated in therapeutically active doses; no impairment of neurological functions and no gastrointestinal side effects have been observed (Kim et al. 1994). Both activity and safety have been confirmed in clinical phase II studies in previously untreated patients with advanced gastric cancer (Kim et al. 1999) and small cell lung cancer (Zang et al. 1999). In gastric cancer, efficacy could apparently be further improved by combination with 5-fluorouracil (Kim et al. 2000). Hepatotoxicity, myelosuppression and nephrotoxicity had been considered dose-limiting (Kim et al. 2001), but since the toxic side effects observed in phase II studies were generally mild, infrequent and transient, the possibility of dose escalation has been emphasized (Kim et al. 1999; Zang et al. 1999). Based on the studies in gastric cancer patients and prior to the completion of ongoing phase III trials, SKI 2053R has received marketing approval in South Korea. So far, there are no reports on clinical activity in cisplatin-resistant tumours.

Cycloplatam

Cycloplatam, amminecyclopentylamine-*S*-(-)-malatoplatinum(II), is a mixture of two isomeric forms (Fig. 4c), but the distinct contributions of these forms to the antineoplastic activity of the mixture are unknown (Drees et al. 1995). In preclinical studies cycloplatam has proved non-cross resistance with cisplatin and lack of nephrotoxicity in a murine leukaemia model (Presnov and Konovalova 1988). Experiments in human tumour cell lines with acquired resistance to cisplatin and their sensitive parental cell lines have confirmed lower resistance factors for cycloplatam compared to cisplatin, and a pattern of activity somewhat different from cisplatin in human tumour xenografts in nude mice has been observed (Drees et al. 1995). Myelosuppression is dose-limiting in the clinical setting. Emesis has been observed frequently, but with no signs of nephrotoxicity. Response rates of 59%, 70% and 22% have been reported from phase II trials in patients with advanced ovarian, breast and prostate cancer, respectively, which had included high portions of previously untreated patients, however (Gorbunova et al. 1995; Ayrapetian 1996; Matveev et al. 1996). The few available reports on clinical activity of cycloplatam are insufficient to definitely judge whether cycloplatam shows any advantage over cisplatin and carboplatin.

Sterically hindered platinum complexes: ZD0473

The search for a platinum compound with reduced susceptibility to inactivation by elevated concentrations of intracellular thiols such as glutathione, a mechanism found active in cells with intrinsic or acquired cisplatin resistance, led to the selection of ZD0473

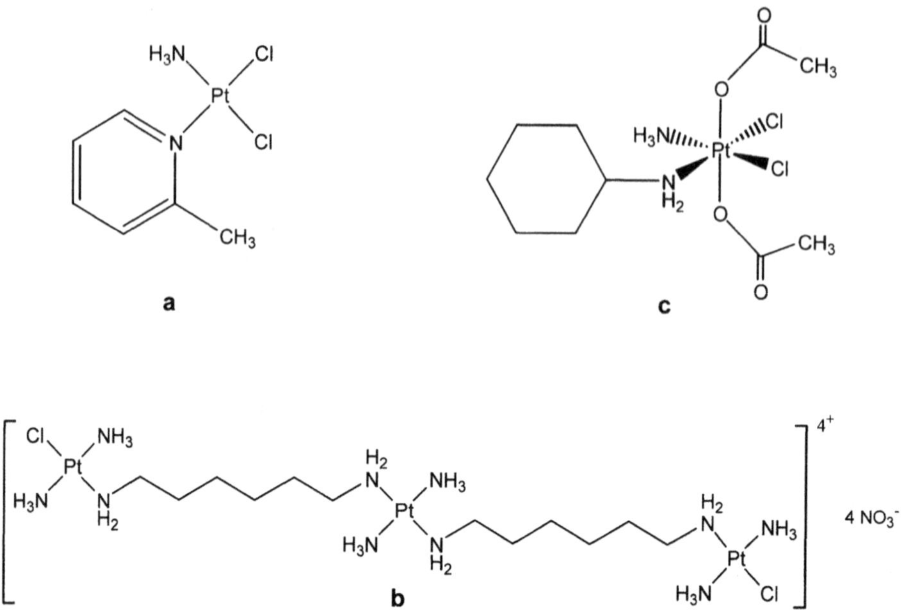

Fig. 5 Structures of the sterically hindered complex ZD0473 (**a**), the trinuclear complex BBR3464 (**b**) and the orally available JM-216 (**c**)

(AMD473), *cis*-amminedichloro(2-methylpyridine)platinum(II), (Fig. 5a) for clinical development. From the crystal structure of this compound it has been concluded that the methyl group situated in the 2-position of the pyridine ring, which is nearly perpendicular to the platinum square plane, sterically hinders an axial approach to the platinum centre and therefore partially protects from associative ligand exchange (Chen et al. 1998). Steric hindrance results in reduced reactivity compared to cisplatin towards sulphur donors such as methionine and thiourea, and binding to DNA is less affected in the presence of glutathione. Reactivity is more generally lowered, as aquation, binding to serum proteins and formation of DNA interstrand cross-links proceed with a lower rate than in the case of cisplatin (Raynaud et al. 1997; Holford et al. 1998a). Comparison of ZD0473 with analogues containing either unsubstituted pyridine or dimethylpyridine revealed that the rate of cross-link formation consistently becomes slower with increasing steric hindrance by bulkier ligands. There is also evidence for an altered sequence specificity of DNA adduct formation compared with that of cisplatin (Holford et al. 1998a), and model experiments indicate that steric hindrance by the methylpyridine group influences binding to DNA, as formation of a single stereoisomer is favoured, when ZD0473 binds to a double-stranded oligonucleotide containing two adjacent guanosines, whereas little stereoselectivity is found for reactions with a corresponding single-stranded oligonucleotide (Chen et al. 1999). Induction of p53 tumour suppressor protein is slower, but lasts longer than after exposure to cisplatin. Furthermore, ZD0473 showed consistently lower resistance factors than cisplatin in pairs of cisplatin-sensitive and -resistant cell lines, representing different mechanisms of resistance and a distinct pattern of response from all other platinum agents in the cell line panel of the NCI (Holford et al. 1998b).

When administered as bolus intravenous infusion, a carboplatin-like toxicity profile is observed with dose-limiting myelosuppression (predominantly thrombocytopaenia), generally mild non-haematological toxicities and no signs of nephro- and neurotoxicity (Trigo et al. 1999; Stevenson et al. 2001). Phase II studies are or have been conducted in various neoplasms, including ovarian, lung (small cell and non-small cell), prostate, bladder, cervical and breast cancer and mesothelioma. Moderate response rates have been found in platinum-pretreated patients with refractory or relapsed ovarian (Gore et al. 2001) and small cell lung cancer (Bonomi et al. 2001), but only cases of disease stabilization in non-small cell lung cancer (Treat et al. 2001). Initiation of a phase III study in patients with ovarian cancer has been announced recently. Based on preclinical findings, which indicate synergistic interaction with paclitaxel and either synergistic or antagonistic interaction with gemcitabine, topotecan and vinorelbine, depending on the cell line (Rogers et al. 2000), several phase I studies of combination regimens are or have been conducted (O'Dwyer et al. 2001; Cosaert et al. 2001; Gatzemeier et al. 2001). An interesting feature of ZD0473 has been revealed by in vivo experiments using oral administration. Tumour-inhibiting effects are preserved and the therapeutic index is improved to values comparable to JM-216 (see "Oral platinum drugs") due to reduced toxicity (Raynaud et al. 1997). It is thus the first platinum(II) drug to prove good oral availability and activity, and an oral formulation for clinical use is awaited.

Non-classic platinum complexes

Development of antineoplastic platinum compounds has long been guided by the structure–activity relationships established from comparison of cisplatin, transplatin and a series of related pairs of isomers, which had suggested that the *cis*-configuration is an essential prerequisite for tumour-inhibiting activity. Furthermore, it has been concluded that the complex should be neutral to be active (Cleare and Hoeschele 1973a, 1973b). However, reports on notable exceptions to these rules have accumulated over the past decade, and one of these non-classic platinum complexes, the charged trinuclear compound BBR3464, is currently being evaluated in clinical studies.

Trans-complexes

Transplatin, *trans*-diamminedichloroplatinum(II), has soon been recognized to be inactive in vivo and to be less cytotoxic than cisplatin in vitro. Divergent explanations, each based on experimental findings, have been presented: stronger inhibition of DNA replication (Johnson et al. 1980) and transcription (Mello et al. 1995) by cisplatin than by transplatin adducts, without differences in rates of adduct removal (Roberts and Friedlos 1987) or, in contrast, more rapid repair of transplatin adducts (Cicarelli et al. 1985; Heiger-Bernays et al. 1990), consistent with the inability of high-mobility group protein HMG1 to recognize transplatin adducts (Pil and Lippard 1992). Slower closure of monofunctional to bifunctional transplatin adducts and therefore higher susceptibility to destabilization by glutathione has also been proposed as a contributing factor (Eastman and Barry 1987; Bancroft et al. 1990). In each case, structural differences of the DNA adducts have been found responsible for these observations. The most obvious difference in this respect is the inability of transplatin to produce 1,2-intrastrand cross-links, which are the most frequent adducts formed by cisplatin. From another point of view, it has been argued that transplatin is generally more reactive than cisplatin and therefore more susceptible to rapid inactivation (Farrell 1996; Wong and Giandomenico 1999; Natile and Coluccia 2001).

However, *trans*-configured complexes with equivalent or even higher activity compared to their *cis*-configured counterparts both in vitro and in vivo have been known for some time now (Farrell et al. 1989). One class of active *trans*-complexes contains planar aromatic amines such as pyridine, *N*-methylimidazole, thiazole or quinoline. Remarkably, several of these complexes retain their activity in cells with acquired resistance to cisplatin and show an activity pattern in the cell line panel of the NCI distinctly different from cisplatin (Farrell et al. 1992; Van Beusichem and Farrell 1992). Reduced reactivity with biomolecules such as glutathione and thus reduced susceptibility to inactivation as a consequence of steric hindrance of associative ligand exchange by the bulky planar groups has been put forward as an explanation for the remarkable activity of these compounds in comparison to transplatin (Farrell 1996). Furthermore, the adducts produced by these complexes may be more effective than those produced by transplatin. A higher portion of interstrand cross-links relative to total DNA adducts (Zou et al. 1993; Zákovská et al. 1998; Brabec et al. 2000) and formation of bifunctional DNA adducts and DNA-Pt-protein adducts dissimilar to those produced by *cis*-configured complexes (Farrell 1996) may account for the higher activity compared to the *cis*-isomers and for the lack of cross-resistance with cisplatin. In the case of *trans*-amminedichloroquinolineplatinum(II), model calculations and experimental evidence suggest the formation of pseudobifunctional adducts, i.e. basically monofunctional adducts which induce bending of DNA due to intercalation of the quinoline ligand (Zákovská et al. 1998).

A second series of active *trans*-complexes contains iminoethers (Coluccia et al. 1993, 1995). Like the aromatic amines contained in the complexes mentioned above, iminoethers have a planar geometry and introduce steric hindrance to ligand exchange reactions. Binding to DNA is slower compared to cisplatin, but proceeds for a longer time, thus eventually resulting in similar levels of DNA platination (Coluccia et al. 1995). These *trans*-complexes are more active than the corresponding *cis*-isomers both in vitro and in vivo (Coluccia et al. 1995) and lack cross-resistance with cisplatin in vitro (Coluccia et al. 1999a, 1999b; Boccarelli et al. 1999). In contrast to *trans*-complexes containing aromatic amines, the complexes containing two *trans*-standing iminoether ligands show a much lower interstrand cross-linking efficiency than cisplatin and transplatin, both in cellular and in isolated DNA (Coluccia et al. 1995). On the other hand, they produce high numbers of monofunctional adducts which do not evolve into DNA cross-links (Zaludova et al. 1997; Boccarelli et al. 1999), but are much more resistant to destabilization by sulphur donors than adducts formed by transplatin (Brabec et al. 1996). Since transcription of DNA is nevertheless terminated by *trans*-bis(iminoether) complexes with a similar efficiency as by cisplatin, it has been concluded that stable monofunctional adducts are the lesions most likely responsible for the cytotoxic activity of this class of compounds (Natile and Coluccia 1999).

A series of platinum(IV) complexes containing axial hydroxide ligands and one bulky aliphatic amine *trans* to ammine has also proved activity in vivo. Some of these lack cross-resistance to cisplatin in vitro and even retain activity in cisplatin-resistant murine tumours, whereas the platinum(II) analogues tested are inactive in vivo (Kelland et al. 1995; Goddard et al. 1996). The latter finding is particularly remarkable, because it has generally been assumed that platinum(IV) complexes require reduction to platinum(II) species in order to exert their antitumour effects (see also "Oral platinum drugs"). JM335, *trans*-ammine(dichlorocyclohexylamine)dihydroxoplatinum(IV), the most extensively studied complex within this series, is capable of forming DNA interstrand cross-links, and induction of single-strand breaks has been observed, but only in one of two cell lines (Mellish et al. 1995). The specific activity of JM335 as compared to its *cis*-isomer and cisplatin in an intrinsically cisplatin-resistant ovarian carcinoma cell line correlates with the inability of gene-specific repair of JM335-induced DNA adducts (O'Neill et al. 1999).

The most recently discovered class of active *trans*-complexes comprises platinum(II) compounds with asymmetric ramified aliphatic amines described by the general formula *trans*-[$PtCl_2$(amine)(isopropylamine)] (Montero et al. 1999; Perez et al. 1999). Higher cytotoxic activity in *ras*-transformed than in normal murine keratinocytes in vitro provides evidence for a higher therapeutic index than cisplatin, but activity in vivo has not been reported yet.

Dinuclear and trinuclear complexes

The first attempts to produce an antineoplastic platinum compound containing more than one platinum centre have been based on the original cisplatin structure. Flexible alkyldiamine chains have been used to link two platinum moieties, each containing two *cis*-configured reactive sites bound to chloride (Farrell et al. 1988). These complexes are capable of forming DNA interstrand cross-links by binding one platinum unit to each DNA strand (Roberts et al. 1989). Since each platinum moiety is able to bind to DNA bifunctionally, simultaneous inter- and intrastrand cross-linking, resulting in complex adducts are conceivable, but remarkably tumour-inhibiting activity is retained when one chloride of each platinum unit is replaced by an ammine ligand, resulting in a charged complex with each platinum moiety containing only one reactive site (Farrell et al. 1990; Farrell 2000). The classic structure–activity relationships for platinum drugs are violated in another respect, as the *trans,trans*-configured complexes, but not their *cis,cis*-configured counterparts, overcome cisplatin resistance in murine leukaemia cells in vitro. Higher levels of DNA binding have been found for the former isomers (Farrell et al. 1995). The first step in binding of the positively charged complexes to DNA is electrostatic interaction with the negatively charged DNA backbone. In accordance with classic platinum drugs, aquations of the first and the second chlorides are the rate-limiting steps in formation of monofunctional and bifunctional adducts, respectively, but these proceed with a higher rate than in the case of cisplatin (Cox et al. 2001). In contrast to classic platinum drugs and depending on the length of the linker, dinuclear complexes are capable of forming interstrand cross-links that span distances of several nucleotides (Zou et al. 1994; Cox et al. 2001).

Whereas simple extension of the alkyldiamine chain proved to be unfavourable because this reduces water solubility, potency could be enhanced by introduction of a further positively charged platinum unit into the flexible aliphatic diamine linker (for a review of structure-activity relationships see Farrell et al. 1999). This modification leads to trinuclear complexes with improved water solubility and higher affinity to DNA due to their higher capacity of hydrogen bonding and electrostatic interaction. From this entirely new class of platinum complexes BBR3464, *trans*-{bis[*trans*-diamminechloroplatinum(II)(μ-1,6-hexanediamine)]}diammineplatinum(II) tetranitrate (Fig. 5b) has been selected for clinical development. This compound shows a generally higher potency than cisplatin and a substantially different pattern of response in the cell line panel of the NCI (Manzotti et al. 2000). Activity is retained or even enhanced in cell lines and human tumour xenografts with acquired or intrinsic resistance to cisplatin (Roberts et al. 1999; Perego et al. 1999a, 1999b; Manzotti et al. 2000) including p53-mutant xenografts (Pratesi et al. 1999), confirming the ability to overcome different mechanisms of cisplatin resistance. Surprisingly, transfection of cells lacking p53 function with the wild-type *p53* gene reduces their sensitivity to BBR3464, the contrary of the effect on cisplatin sensitivity. The observation that BBR3464 is capable of inducing apoptosis by a p53-independent pathway in a lung carcinoma cell line (Pratesi et al. 1999) has not been paralleled in astrocytoma cells (Servidei et al. 2001), however.

BBR3464 is taken up by cells to a higher extent than cisplatin (Perego et al. 1999b; Riccardi et al. 2001) and binds more rapidly to DNA, but as to the relative number of interstrand cross-links compared to total DNA adducts, there are conflicting reports. Whereas BBR3464 has been found to produce a higher portion of interstrand cross-links than cisplatin in plasmid DNA (Brabec et al. 1999), the reverse has been observed in DNA isolated from pre-exposed tumour cells (Roberts et al. 1999; Perego et al. 1999b). However, the interstrand cross-links formed by BBR3464 persist longer than those induced by cisplatin and are therefore considered less susceptible to repair. This is consistent with the finding of a more prolonged cytostatic effect, as reflected by more persistent perturbations of cell cycle progression at equitoxic dose levels in vitro and higher specific growth delay values in vivo (Manzotti et al. 2000). The lethal DNA lesions induced by BBR3464 are unlikely to be repaired by the nucleotide excision and the mismatch repair system, since deficiencies of these repair systems do not alter sensitivity to this compound compared to repair-proficient parental cells (Perego et al. 1998; Perego et al. 1999a). On the other hand, intrastrand cross-links formed by BBR3464 are not recognized by high mobility group 1 proteins, but are efficiently removed by nucleotide excision repair and are therefore less likely to be relevant for tumour-inhibiting effects (Zehnulova et al. 2001). Two phase I studies using different schedules of administration have revealed myelosuppression (neutropaenia) and gastrointestinal side effects (diarrhoea) to be dose-limiting, while no significant nephro- or neurotoxicity has been observed (Calvert et al. 2000; Sessa et al. 2000). Some activity has been observed in these studies already, and phase II trials using a single intermittent schedule in patients with lung (both non-small cell and small cell), gastric and ovarian cancer are ongoing.

Dinuclear platinum complexes containing a charged polyamine linker such as spermine or spermidine instead of the central platinum unit are another class of exceptionally potent compounds capable of interacting with DNA in a similar manner as BBR3464 (Farrell 2000). Blocking the polyamine linker by cleavable carbamate groups is currently pursued as a strategy to prepare a pro-drug with an improved therapeutic index (Hegmans et al. 2001). Initiation of clinical studies of a dinuclear polyamine-bridged complex has been announced for the near future.

Oral platinum drugs

Despite the high interest in development of an orally available platinum drug suited for outpatient treatment, only one platinum compound designed for oral administration has been applied in the clinical setting on a trial basis so far: JM-216 (satraplatin), bis(acetato)amminedichloro(cyclohexylamine)platinum(IV), (Fig. 5c) has been chosen from a series of related dicarboxylate platinum(IV) complexes because of an antitumour activity at least comparable to intravenous carboplatin, relatively low emetogenic potency and favourable physicochemical properties, including solubility in water and stability in acidic solution. A more than threefold higher intestinal absorption compared to carboplatin in mice and a therapeutic index more than tenfold higher compared to intraperitoneal administration and threefold higher than that of intraperitoneal cisplatin in subcutaneous murine plasmocytoma has supported its potential as an oral drug (Kelland 2000). In vitro activity of JM-216 is hardly affected by transport-determined acquired resistance to cisplatin in pairs of ovarian and cervical carcinoma cell lines (Kelland et al. 1993a; Mellish et al. 1993), which has been attributed to the higher lipophilicity of JM-216 (Sharp et al. 1995). Furthermore, loss of mismatch repair does not confer resistance to JM-216 (Fink et al. 1996). However, the general pattern of response across cell lines and human ovarian carcinoma xenografts is similar to cisplatin and carboplatin (Kelland et al. 1993a, 1993b).

In phase I trials, a split-dose schedule has proven favourable to administration as a single bolus because of saturable intestinal absorption (McKeage et al. 1997). A toxicological profile similar to carboplatin, with dose-limiting thrombocytopaenia and leukopaenia, mild gastrointestinal toxicity and a lack of nephro- and neurotoxicity observed in preclinical studies, has been confirmed by clinical experience. After administration, JM-216 is completely converted into at least six different compounds. Part of the parent drug undergoes reduction accompanied with a loss of the axial acetate ligands to *cis*-amminedichloro(cyclohexylamine)platinum(II), which is the major species detected in plasma. Furthermore, platinum(IV) species with one or both chlorides replaced by hydroxo ligands and two unidentified metabolites, but no parent drug has been detected (Raynaud et al. 1996). Phase II studies have revealed activity in previously untreated patients with small cell lung cancer (Fokkema et al. 1999) and in patients with prostate cancer who presented with progressive disease after antiandrogen withdrawal (Peereboom et al. 1997), but no significant activity in non-small cell lung cancer (Judson et al. 1997; Fokkema et al. 1998). A phase III trial with and without prednisone in hormone-refractory prostate cancer has been initiated. Synergistic interaction with orally administered etoposide observed in leukaemic mice (Rose 1997) and radiosensitizing effects both in vitro and in vivo, probably due to inhibition of repair of radiation-induced DNA damage (van der Vaart et al. 1997; Amorino et al. 1999; Amorino et al. 2000) have stimulated clinical phase I studies of combinations with oral etoposide and radiotherapy, respectively. In patients with locoregionally advanced lung or oesophageal cancer treated with JM-216 and concomitant radiotherapy of the chest, severe haematological toxicity occurred at half the maximum tolerated dose of single JM-216 (Hoffman et al. 1998). Further regimens examined in phase I studies include combinations with intravenous paclitaxel (Jones et al. 1999) and oral uracil/ftorafur plus leucovorin (DeMario et al. 1999). The latter combination has been considered unfavourable, because dose-limiting nausea and vomiting occurred at concentrations substantially below the maximum tolerated doses of the single drugs despite antiemetic medication.

Development of orally available platinum drugs has focused on octahedral coordinated platinum(IV) complexes, because they are generally more inert to hydrolysis and other ligand exchange reactions than platinum(II) compounds and are thus less likely to be degraded in the gastrointestinal tract. The degree of inertness depends both on the axial and equatorial ligands. For example, among a homologous series of ethylenediamineplatinum(IV) complexes, those containing carboxylato or hydroxo ligands in axial position proved to be more inert to reduction than that containing axial chlorine, consistent with the different electron-withdrawing power of these ligands. In addition, the more readily reduced complexes have been found to bind more readily to DNA (Ellis et al. 1995) and to exert a stronger antiproliferative effect in vitro (Choi et al. 1998). Similar observations have been made within a series of cycloalkylamineplatinum(IV) complexes (Yoshida et al. 1994). Furthermore, comparison of pairs of platinum(IV) complexes with different amine ligands, but identical axial ligands, suggests a direct correlation between reduction rate and cytotoxicity (Choi et al. 1998). Extracellular or intracellular reduction to the divalent oxidation state has generally been considered a prerequisite for reactivity towards DNA. This assumption has been somewhat challenged, since direct reaction of platinum(IV) species with 5'-GMP (Roat and Reedijk 1993; Talman et al. 1997; Galanski and Keppler 2000) and DNA (Brabec et al. 1986; Nováková et al. 1995) has been observed in cell-free model experiments. However, reactions of platinum(IV) species with 5'-GMP or DNA have been noted to proceed generally much slower than those of platinum(II) analogues. Thus, the contribution of platinum(IV) species to the biological activity of platinum(IV) drugs remains uncertain.

Platinum-drug targeting

The majority of patients treated with platinum drugs (except those with testicular germ-cell cancer) experience only incomplete, if any, response, with systemic toxicity preventing administration of higher doses. Strategies to modify drugs in such a way that they are selectively accumulated and/or activated in tumour tissue would make it possible to obtain higher drug concentrations in the tumour without enhancing their toxic side effects and would therefore be of great benefit.

In the context of platinum drugs, strategies based on the so-called enhanced permeability and retention (EPR) effect have recently proceeded to the early clinical stage of development. The discovery of the EPR effect has stimulated development of various concepts and numerous investigational products which aim at exploiting this effect for tumour-targeted drug delivery. The EPR effect leads to enhanced extravasation of macromolecules and particles of similar size and their accumulation in the interstitium of solid tumours as a consequence of several factors: defective architecture of the vascular endothelium (large gaps in endothelial cell–cell junctions, lack of smooth muscle layer, etc.), impaired lymphatic drainage and increased production of permeability mediators (bradykinin, nitric oxide, VEGF, prostaglandins, matrix metalloproteinases, etc.) in neovascularized tumour tissue in contrast to healthy tissue. Consequently, conjugation of drugs to a large diversity of biological and synthetic polymer carriers and encapsulation of drugs in microparticles such as liposomes are extensively explored as a means of achieving accumulation in tumour tissue. The EPR effect, originally described for serum proteins such as albumin, transferrin and immunoglobulin G, has been confirmed for several polymer-conjugated antineoplastic agents (for reviews see Maeda et al. 2000; Maeda 2001). Attempts to deliver platinum drugs involve liposomal or synthetic polymer carriers (discussed below), but also extend to appropriate biological carrier molecules such as albumin and other proteins (Zimmermann et al. 1999).

Further approaches to drug targeting involve the use of low-molecular-weight carriers with specific affinity for structures which are associated with certain tumour tissues. Targeting of mammary, endometrium and prostate carcinoma by preparing platinum complexes with affinity for steroid hormone receptors and targeting of bone tumours by linking platinum moieties to ligands with affinity for calcified tissue are the most extensively studied strategies.

Liposomal platinum drugs

An advanced liposome technology, which overcomes the problem of instability of other liposomal preparations of cisplatin and which prevents rapid removal of liposomes by the reticuloendothelial system by means of incorporation of methoxypolyethyleneglycol, has been applied to encapsulate cisplatin. This formulation (SPI-77) has been designed to increase delivery to tumour cells by retarding renal excretion and prolonging retention in the circulatory system, to reduce toxicity in comparison to free cisplatin and to render hydration and antiemetics unnecessary. However, cases of severe hypersensitivity reactions to liposomal constituents have necessitated extended prophylaxis using corticosteroids and antihistamines (DeMario et al. 1998). Apart from a certain risk of hypersensitivity, clinical experience so far has confirmed the expected improvement of toxicological properties, as it has the expected pharmacokinetic behaviour, including reduced volume of distribution, lower total and renal clearance, increased elimination half-life and AUC values. Either no or only low-level nephro-, oto-, neuro-, gastrointestinal and haematological toxicity have been reported in adult (Schellens et al. 1999) and in paediatric patients pre-treated with hydrocortisone and chlorpheniramine and given the same dose range of SPI-77, which has not been escalated to the maximally tolerated dose (Veal et al. 2001). However, no responses to treatment have been observed in any of the patients studied, even though the given doses contained up to three times more cisplatin than is usually administered with conventional solutions. Moreover, results from preclinical studies are somewhat conflicting. Whereas SPI-77 exhibited superior activity compared to free cisplatin in some murine tumour models (Newman et al. 1999), another study suggests that cisplatin is insufficiently released from the liposomes, since much fewer DNA adducts are formed in spite of higher accumulation in tumour tissue when SPI-77 is compared to administration of the equal dose of free cisplatin to tumour-bearing mice (Zamboni et al. 2000). While clinical studies have proceeded to phase II, the need for a modified formulation ensuring a proper balance between encapsulation and release has been emphasized (Veal et al. 2001).

Polymer-conjugated platinum drugs

AP5280 (Fig. 6a) is the product of another approach aiming at accumulation of a platinum complex in tumour tissue by means of the EPR effect. It contains platinum moieties linked by peptide spacers to *N*-(2-hydroxypropyl)methacrylamide (HPMA). Promising results obtained with a similar HPMA-doxorubicin conjugate in phase I studies have stimulated extension of the concept to other chemotherapeutic agents, including paclitaxel, camptothecin and platinum compounds (Duncan et al. 2001). The polymer has been optimized to a size large enough to take advantage of the EPR effect and small enough to ensure renal elimination of the drug fraction remaining in the circulatory system; the peptide spacer (Gly-Phe-Leu-Gly) is designed to be largely stable throughout the whole process of delivery to tumour cells, but to be cleaved by lysosomal proteases inside the cells. Studies on a series of HPMA copolymer platinates preceding the selection of AP5280 for clinical development have explored different modes of linking the platinum-containing moiety to the peptide spacer. This spacer may terminate either (1) in an amino species such as ethylene-

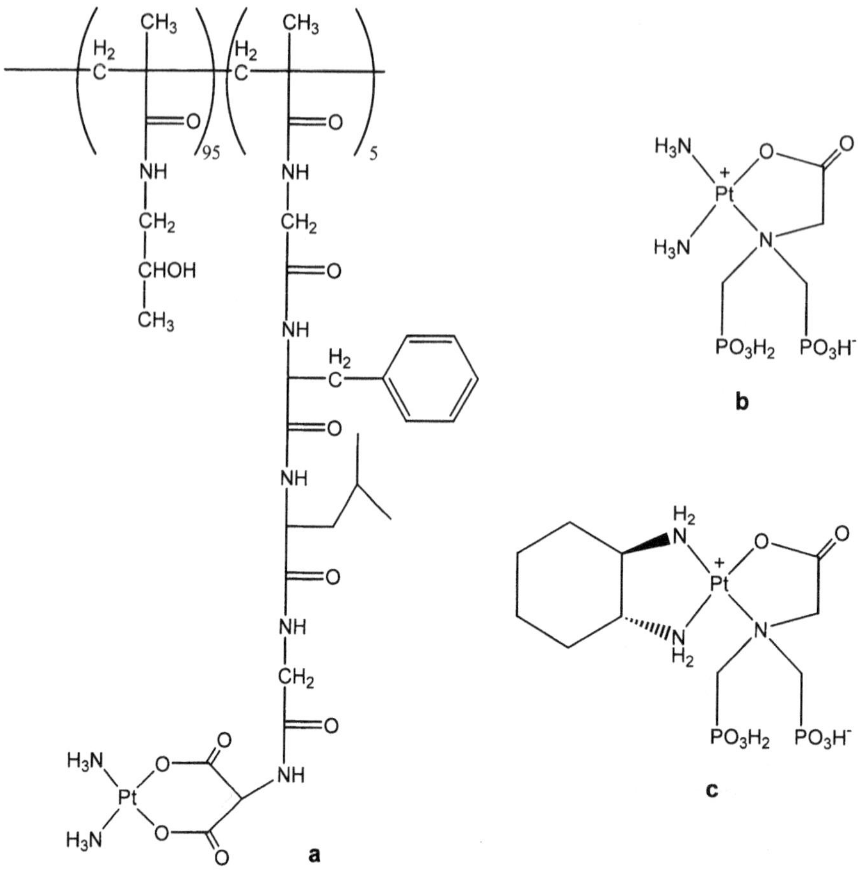

Fig. 6 Structures of the polymer-conjugate AP5280 (**a**) and the osteotropic phosphonate complexes KP735 (**b**) and KP1363 (**c**)

diamine, which is supposed to form a stable bond to a *cis*-dichloroplatinum(II) moiety, or (2) in a carboxylate, which is supposed to form a hydrolytically labile bond to a *cis*-diammineplatinum(II) moiety. In vivo experiments suggest that only the former type of conjugate, which is largely stable to hydrolysis and is designed to release the platinum-containing moiety only upon endocytotic uptake and intracellular enzymatic degradation, shows distinct advantages over free cisplatin in terms of therapeutic index (Gianasi et al. 1999). Nevertheless, AP5280 is more closely related to the latter type of conjugate, as the peptide spacer terminates in a malonate group, but this may serve as a bidentate ligand and thus be less susceptible to hydrolysis than a single carboxylate ligand (cf. Evagorou et al. 1998; Gianasi et al. 1999). However, it has been noted that binding of platinum to other potential coordination sites offered by the peptide spacer such as amide nitrogen cannot be ruled out (Stewart et al. 1999). AP5280 has entered phase I studies recently, and the development of HPMA conjugates containing a diaminocyclohexaneplatinum(II) moiety corresponding to oxaliplatin has been announced.

Platinum drugs containing low-molecular-weight carriers

Steroid hormone receptors are expressed by the majority of mammary, endometrium and prostate carcinomas. Even though a high percentage of these tumours progress from hormone-dependent to hormone-independent growth, steroid receptor expression is often retained. Therefore, attempts are made to exploit these receptors as a target in order to achieve selectivity of platinum drugs for these tumours (for reviews see Bednarski et al. 1990; Angerer 1993). Platinum complexes with low affinity for the oestrogen receptor have been produced by coordination to diphenylethylenediamine ligands derived from the synthetic non-steroidal oestrogen diethylstilbestrol. Several of these complexes strongly inhibit growth of hormone-dependent rat mammary tumours. (Anti)oestrogenic effects seem to prevail over a DNA damaging mode of action in those compounds which exhibit selectivity for hormone-sensitive tumours (Karl et al. 1988; Schlemmer et al. 1999; Gust et al. 2000), whereas a lack of endocrine effects is mostly associated with either a loss of selectivity or even a total loss of activity (Gust and Schönenberger 1993; Gust et al. 1995). Complexes containing 2-phenylindole ligands show a higher affinity for the oestrogen receptor and a marked selectivity for hormone-dependent tumours. An antioestrogenic mode of action is assumed and interaction with DNA following internalization of an oestrogen receptor-bound complex has been proposed (Knebel and Angerer 1991). A third series of complexes, which contains triphenylethylene ligands derived from tamoxifen, exhibits no selectivity for oestrogen receptor-positive tumour cells in vitro, and there are no reports on in vivo activity available (Bérubé et al. 1993, He et al. 1995; Bérubé et al. 1997). An approach to exploit luteinizing hormone-releasing hormone (LH-RH) receptors as a target by linking platinum moieties to short-chain peptide analogues to LH-RH (Bajusz et al. 1989; Janáky et al. 1992) has not been developed further.

Diverse ligands containing amino phosphonic acid groups have been used in order to produce platinum drugs with selective activity in primary and secondary bone tumours (for two examples see Fig. 6b, c). Due to the high affinity of phosphonic acid for calcium, these complexes are endowed with osteotropic properties comparable to the bisphosphonates currently applied for therapy of various bone disorders such as Paget's disease, osteoporosis and tumour-induced osteolysis and hypercalcaemia. Indeed, accumulation of such complexes in bone has been confirmed by autoradiography, and model experiments on the binding behaviour to mono- and dinucleotides show that calcium is able to promote adduct formation (Galanski et al. 1999; Bloemink et al. 1999). NMR studies have demonstrated that binding to nucleotides is associated with release of the phosphonate ligand, irrespective of its N-donor function (Bloemink et al. 1994). Activity has been proved in an orthotopically transplantable rat osteosarcoma model which disseminates to the lung, producing lethal osteoid-forming metastases (Klenner et al. 1993). A growth-inhibitory activity in the primary tumour comparable to cisplatin, but a higher increase in life span, suggesting delayed growth of lung metastases, has been observed with some of these complexes (Klenner et al. 1990a, 1990b). Current attempts to optimize therapeutic effects involve both modification of the phosphonate-containing leaving group and modification of the non-leaving amine group in order to produce a drug which lacks cross-resistance to cisplatin (Slaby et al. 1999).

Conclusions and perspectives

Since the establishment of carboplatin as a less toxic and at least in some indications equally effective alternative to cisplatin, interest in development of analogues only for the sake of improving tolerability in comparison to cisplatin has waned considerably. The two remaining objectives, finding new agents suited for the treatment of cisplatin-resistant tumours and endowing platinum drugs with higher tumour selectivity, have had different impacts on recent clinical development.

On the one hand, the integration of the diaminocyclohexane-containing drug oxaliplatin within the treatment options for colorectal cancer has proved that broadening the range of indications for platinum-based therapy is feasible with a classic cisplatin derivative. Furthermore, protection against inactivation reactions by introduction of other bulky non-leaving ligands, as in the sterically hindered complex ZD0473, may be a promising strategy to this end. Probably the most exciting recent developments are those of non-classic complexes, such as the charged multinuclear BBR3464, violating the established structure–activity relationships, which had long been taken as a kind of dogma and therefore rather seem to have hampered some innovative and promising developments, yet without having prevented them. Hitherto, no conclusive picture arises from investigations into the mechanisms of action of the recently identified classes of active *trans*-complexes, but in any case, they call for a critical reappraisal of former explanations for the erroneously assumed general inactivity of *trans*-configured platinum complexes. It remains to be seen whether the new leads will give rise to effective drugs with spectra of activity distinctly different from those of cisplatin and carboplatin.

On the other hand, attempts to improve tumour selectivity and thus to increase therapeutic effects in tumours, which usually respond only incompletely to the platinum drugs and currently available formulations, are comparatively scarce, and only two investigational drugs developed with this objective have entered clinical trials in recent years. Considering the substantial improvements of therapeutic efficacy they may offer, greater efforts should be devoted to the development of such approaches.

References

Aebi S, Kurdi-Haidar B, Gordon R, Cenni B, Zheng H, Fink D, Christen RD, Boland CR, Koi M, Fishel R, Howell SB (1996) Loss of DNA mismatch repair in acquired resistance to cisplatin. Cancer Res 56:3087–3090

Akaza H, Togashi M, Nishio Y, Miki T, Kotake T, Matsumura Y, Yoshida O, Aso Y (1992) Phase II study of cis-diammine(glycolato)platinum, 254-S, in patients with advanced germ-cell testicular cancer, prostatic cancer, and transitional-cell carcinoma of the urinary tract. Cancer Chemother Pharmacol 31:187–192

Akiyama M, Horiguchi-Yamada J, Saito S, Hoshi Y, Yamada O, Mizoguchi H, Yamada H (1999) Cytostatic concentrations of anticancer agents do not affect telomerase activity of leukaemic cells in vitro. Eur J Cancer 35:309–315

Allain P, Heudi O, Cailleux A, Le Bouil A, Larra F, Boisdron-Celle M, Gamelin E (2000) Early biotransformations of oxaliplatin after its intravenous administration to cancer patients. Drug Metab Dispos 28:1379–1384

Amorino GP, Freeman ML, Carbone DP, Lebwohl DE, Choy H (1999) Radiopotentiation by the oral platinum agent, JM216: role of repair inhibition. Int J Radiat Oncol Biol Phys 44:399–405

Amorino GP, Mohr PJ, Hercules SK, Pyo H, Choy H (2000) Combined effects of the orally active cisplatin analog, JM216, and radiation in antitumor therapy. Cancer Chemother Pharmacol 46:423–426

Andrews PA (2000) Cisplatin accumulation. In: Kelland LR, Farrell N (eds) Platinum-based drugs in cancer therapy. Humana, Totowa, pp 89–113

Andrews PA, Jones JA, Varki NM, Howell SB (1990) Rapid emergence of acquired cis-dichlorodiammineplatinum(II) resistance in an in vivo model of human ovarian carcinoma. Cancer Commun 2:93–100

Angerer EV (1993) Platinum complexes with specific activity against hormone-dependent tumors. In: Keppler BK (ed) Metal complexes in cancer chemotherapy. VCH, Weinheim, pp 73–83

Armand JP, Boige V, Raymond E, Fizazi K, Faivre S, Ducreux M (2000) Oxaliplatin in colorectal cancer: an overview. Semin Oncol 27, Suppl 10:96–104

Ayrapetian LG (1996) New analog of platinum—cycloplatam—treatment of breast cancer. Proc Int Congress on Anti-Cancer Chemother 6:403

Bajusz S, Janaky T, Csernus VJ, Bokser L, Fekete M, Srkalovic G, Redding TW, Schally AV (1989) Highly potent metallopeptide analogues of luteinizing hormone-releasing hormone. Proc Natl Acad Sci U S A 86:6313–6317

Bancroft DP, Lepre CA, Lippard SJ (1990) 195Pt NMR kinetic and mechanistic studies of cis- and trans-diamminedichloroplatinum(II) binding to DNA. J Am Chem Soc 112:6860–6871

Barnham KJ, Berners-Price SJ, Frenkiel TA, Frey U, Sadler PJ (1995a) Platination pathways for reactions of cisplatin with GG single-stranded and double-stranded decanucleotides. Angew Chem Int Ed Engl 34:1874–1877

Barnham KJ, Djuran MI, Murdoch P del S, Ranford JD, Sadler PJ (1995b) l-Methionine increases the rate of reaction of 5'-guanosine monophosphate with the anticancer drug cisplatin: mixed-ligand adducts and reversible methionine binding. J Chem Soc Dalton Trans 1995:3721–3726

Barnham KJ, Djuran MI, Murdoch P del S, Ranford JD, Sadler PJ (1996) Ring-opened adducts of the anticancer drug carboplatin with sulfur amino acids. Inorg Chem 35:1065–1072

Barry MA, Behnke CA, Eastman A (1990) Activation of programmed cell death (apoptosis) by cisplatin, other anticancer drugs, toxins and hyperthermia. Biochem Pharmacol 40:2353–2362

Bedford P, Fichtinger-Shepman AMJ, Shellard SA, Walker MC, Masters JRW, Hill BT (1988) Differential repair of platinum-DNA adducts in human bladder and testicular tumor continuous cell lines. Cancer Res 48:3019–3024

Bednarski PJ, Gust R, Spruss T, Knebel N, Otto A, Farbel M, Koop R, Holler E, Angerer EV, Schönenberger H (1990) Platinum compounds with estrogen receptor affinity. Cancer Treat Rev 17:221–231

Behrens BC, Hamilton TC, Masuda H, Grotzinger KR, Whang-Peng J, Louie KG, Knutsen T, McKoy WM, Young RC, Ozols RF (1987) Characterization of a cis-diamminedichloroplatinum(II)-resistant human ovarian cancer cell line and its use in evaluation of platinum analogues. Cancer Res 47:414–418

Beketic-Oreskovic L, Osmak M (1995) Modulation of resistance to cisplatin by amphotericin B and aphidicolin in human larynx carcinoma cells. Cancer Chemother Pharmacol 35:327–333

Belguise-Valladier P, Maki H, Sekiguchi M, Fuchs RP (1994) Effect of single DNA lesions on in vitro replication with DNA polymerase III holoenzyme. Comparison with other polymerases. J Mol Biol 236:151–164

Berners-Price SJ, Appleton TG (2000) The chemistry of cisplatin in aqueous solution. In: Kelland LR, Farrell N (eds) Platinum-based drugs in cancer therapy. Humana, Totowa pp 335

Bérubé G, Wheeler P, Ford CHJ, Gallant M, Tsaltsas Z (1993) Synthesis and preliminary in vitro biological activity of non-steroidal cytotoxic estrogens designed for the treatment of breast cancer. Can J Chem 71:1327–1333

Bérubé G, He Y, Groleau S, Séné A, Thérien HM, Caron M (1997) The synthesis and characterization of new triphenylethylene platinum(II) complexes. Inorg Chim Acta 262:139–145

Bloemink MJ, Dorenbos JP, Heetebrij RJ, Keppler BK, Reedijk J, Zahn H (1994) New antitumor platinum compounds linked to amino phosphonic acids which lose the phosphonate and tertiary amine ligand upon binding to nucleic acids. Inorg Chem 33:1127–1132

Bloemink MJ, Diederen JJH, Dorenbos JP, Heetebrij RJ, Keppler BK, Reedijk J (1999) Calcium ions do accelerate the DNA binding of new antitumor-active platinum aminophosphonate complexes. Eur J Inorg Chem 1999:1655–1657

Blommaert FA, Michael C, Terheggen PMAB, Muggia FM, Kortes V, Schornagel JH, Hart AAM, den Engelse L (1993) Drug-induced DNA modification in buccal cells of cancer patients receiving carboplatin and cisplatin combination chemotherapy, as determined by an immunocytochemical method: interindividual variation and correlation with disease response. Cancer Res 53:5669–5675

Blommaert FA, van Dijk-Knijnenburg HCM, Dijt FJ, den Engelse L, Baan RA, Berends F, Fichtinger-Shepman AMJ (1995) Formation of DNA adducts by the drug carboplatin: different nucleotide sequence preferences in vitro and in cells. Biochemistry 34:8474–8480

Boccarelli A, Coluccia M, Intini FP, Natile G, Locker D, Leng M (1999) Cytotoxicity and DNA binding mode of new platinum-iminoether derivatives with different configuration at the iminoether ligands. Anti-Cancer Drug Des 14:253–264

Bonetti A, Franceschi T, Apostoli P, Messori A, Sperotto L, Cetto GL, Molino A, Leone R (1995) Cisplatin pharmacokinetics using a five-day schedule during repeated courses of chemotherapy in germ cell tumors. Ther Drug Monit 17:25–32

Bonomi P, Modiano M, Cornett P, Koehler M, Schiller J (2001) Phase II trial to assess the activity of ZD0473 in patients with small-cell lung cancer who have failed one prior platinum-based chemotherapy regimen. Proc Am Soc Clin Oncol 20:abstract no. 1284

Boudný V, Vrána O, Gaucheron F, Kleinwachter V, Leng M, Brabec V (1992) Biophysical analysis of DNA modified by 1,2-diaminocyclohexane platinum(II) complexes. Nucl Acids Res 20:267–272

Brabec V, Vrana O, Kleinwachter V (1986) Tetravalent platinum complexes can exert their antitumor effect via direct reaction with DNA. Stud Biophys 114:199–207

Brabec V, Vrána O, Nováková O, Kleinwächter V, Intini FP, Coluccia M, Natile G (1996) DNA adducts of antitumor trans-[PtCl2 (E-imino ether)2]. Nucleic Acids Res 24:336–341

Brabec V, Kaspárková J, Vrána O, Nováková O, Cox JW, Qu Y, Farrell N (1999) DNA modifications by a novel bifunctional trinuclear platinum phase I anticancer agent. Biochemistry 38:6781–6790

Brabec V, Neplechova K, Kasparkova J, Farrell N (2000) Steric control of DNA interstrand cross-link sites of trans platinum complexes: specificity can be dictated by planar nonleaving groups. J Biol Inorg Chem 5:364–368

Brady HR, Kone BC, Stromski ME, Zeidel ML, Giebisch G, Gullans SR (1990) Mitochondrial injury: an early event in cisplatin toxicity to renal proximal tubules. Am J Physiol 258:F1181–1187

Britten RA, Green JA, Warenius HM (1992) Cellular glutathione (GSH) and glutathione S-transferase (GST) activity in human ovarian tumor biopsies following exposure to alkylating agents. Int J Radiat Oncol Biol Phys 24:527–531

Burchenal JH, Kalaher K, O'Toole T, Chisholm J (1977) Lack of cross-resistance between certain platinum coordination compounds in mouse leukemia. Cancer Res 37:3455–3457

Burger AM, Double JA, Newell DR (1997) Inhibition of telomerase activity by cisplatin in human testicular cancer cells. Eur J Cancer 33:638–644

Burger H, Nooter K, Boersma AWM, Kortland CJ, Stoter G (1997) Lack of correlation between cisplatin-induced apoptosis, p53 status and expression of Bcl-2 family proteins in testicular germ cell tumour cell lines. Int J Cancer 73:592–599

Burger H, Nooter K, Boersma AWM, Kortland CJ, Stoter G (1998) Expression of p53, Bcl-2 and Bax in cisplatin-induced apoptosis in testicular germ cell tumour cell lines. Br J Cancer 77:1562–1567

Burger H, Nooter K, Boersma AWM, Van Wingerden KE, Looijenga LHJ, Jochemsen AG, Stoter G (1999) Distinct p53-independent apoptotic cell death signalling pathways in testicular germ cell tumour cell lines. Int J Cancer 81:620628

Calvert H, Sessa C, Carmichael J, Calvert P, Zucchetti M, Marsoni S, Van den Bosch S, Robinson A, Verdi E, Camboni G (2000) An intermittent phase I and pharmacokinetic study of BBR3464, a novel cationic triplatinum complex. Ann Oncol 11, Suppl 4:132

Calvo E, Cortes J, Rodriguez J, Rebollo J, Martin-Algarra S, Garcia-Foncillas J, Salgado E, Gonzalez-Cao M, Aramendia JM, Fernandez-Hidalgo O, Brugarolas A (2001) Irinotecan, oxaliplatin plus 5-FU/leucovorin in advanced colorectal cancer. Proc Am Soc Clin Oncol 20: abstract no. 542

Chaganti RSK, Houldsworth J (2000) Genetics and biology of adult human germ cell tumors. Cancer Res 60:1475–1482

Chen Y, Guo Z, Parsons S, Sadler PJ (1998) Stereospecific and kinetic control over the hydrolysis of a sterically hindered platinum picoline anticancer complex. Chem Eur J 4:672–676

Chen Y, Parkinson JA, Guo Z, Brown T, Sadler PJ (1999) A new platinum anticancer drug forms a highly stereoselective adduct with duplex DNA. Angew Chem Int Ed 38:2060–2063

Choi S, Filotto C, Bisanzo M, Delaney S, Lagasee D, Whitworth JL, Jusko A, Li C, Wood NA, Willingham J, Schwenker A, Spaulding K (1998) Reduction and anticancer activity of platinum(IV) complexes. Inorg Chem 37:2500–2504

Chollet P, Bensmaïne MA, Brienza S, Deloche C, Curé H, Caillet H, Cvikovic E (1996) Single agent activity of oxaliplatin in heavily pretreated advanced epithelial ovarian cancer. Ann Oncol 7:1065–1070

Chresta CM, Masters JRW, Hickman JA (1996) Hypersensitivity of human testicular tumors to etoposide-induced apoptosis is associated with functional p53 and a high Bax:Bcl-2 ratio. Cancer Res 56:1834–1841

Ciccarelli RB, Solomon MJ, Varshavsky A, Lippard SJ (1985) In vivo effects of cis- and trans-diamminedichloroplatinum(II) on SV40 chromosomes: differential repair, DNA protein cross-linking, and inhibition of replication. Biochemistry 24:7533–7540

Cleare MJ, Hoeschele JD (1973a) Antitumor platinum compounds. Relation between structure and activity. Plat Met Rev 17:2–13

Cleare MJ, Hoeschele JD (1973b) Studies on the antitumor activity of group VIII transition metal complexes. Part I. Platinum(II) complexes. Bioinorg Chem 2:187–210

Coluccia M, Nassi A, Loseto F, Boccarelli A, Mariggio MA, Giordano D, Intini FP, Caputo P, Natile G (1993) A trans-platinum complex showing higher antitumor activity than the cis congeners. J Med Chem 36:510–512

Coluccia M, Boccarelli A, Mariggio MA, Cardellicchio N, Caputo P, Intini FP, Natile G (1995) Platinum(II) complexes containing iminoethers: a trans platinum antitumour agent. Chem-Biol Interact 98:251–266

Coluccia M, Nassi A, Boccarelli A, Giordano D, Cardellicchio N, Locker D, Leng M, Sivo M, Intini FP, Natile G (1999a) In vitro and in vivo antitumour activity and cellular pharmacological properties of new platinum-iminoether complexes with different configuration at the iminoether ligands. J Inorg Biochem 77:31–35

Coluccia M, Nassi A, Boccarelli A, Giordano D, Cardellicchio N, Intini FP, Natile G, Barletta A, Paradiso A (1999b) In vitro antitumour activity and cellular pharmacological properties of the platinum-iminoether complex trans-[PtCl2{E-HN=C(OMe)Me}2]. Int J Oncol 15:1039–1044

Comess KM, Burstyn JN, Essigmann JM, Lippard SJ (1992) Replication inhibition and translesion synthesis on templates containing site-specifically placed cis-diamminedichloroplatinum(II) DNA adducts. Biochemistry 31:3975–3990

Connors TA, Jones M, Ross WCJ, Braddock PD, Khokhar AR, Tobel ML (1972) New platinum complexes with antitumour activity. Chem-Biol Interact 5:415–424

Corda Y, Job C, Anin MF, Leng M, Job D (1993) Spectrum of DNA-platinum adduct recognition by prokaryotic and eukaryotic DNA-dependent RNA polymerases. Biochemistry 32:8582–8588

Cosaert J, Douillard JY, Barbarot V (2001) ZD0473 phase I trial in combination with vinorelbine in patients with refractory solid malignancies. Proc Am Soc Clin Oncol 20: abstract no. 2114

Cox JW, Berners-Price SJ, Davies MS, Qu Y, Farrell N (2001) Kinetic analysis of the stepwise formation of a long-range DNA interstrand cross-link by a dinuclear platinum antitumor complex: evidence for aquated intermediates and formation of both kinetically and thermodynamically controlled conformers. J Am Chem Soc 123:1316–1326

Cullinane C, Mazur SJ, Essigmann JM, Phillips DR, Bohr VA (1999) Inhibition of RNA polymerase II transcription in human cell extracts by cisplatin DNA damage. Biochemistry 38:6204–6212

Cure H, Forestier D, Chevalier V, Berger V, Bons JM, Mouret-Reynier MA, Chipponi J, Chollet P, Pezet D (2001) Phase II trial of chronomodulated (CM) quadritherapy combining 5-fluorouracil (5-FU), L-folinic acid (L-FA), oxaliplatin (L-OHP) and irinotecan (CPT-11) in previously untreated patients with metastatic colorectal cancer (MCC). Proc Am Assoc Cancer Res 42:620

Cvitkovic E (1998) Cumulative toxicities from cisplatin therapy and current cytoprotective measures. Cancer Treat Rev 24:265–281

Cvitkovic E, Bekradda M (1999) Oxaliplatin: a new therapeutic option in colorectal cancer. Semin Oncol 26:647–662

Dabholkar M, Vionnet J, Bostick-Bruton F, Yu JJ, Reed E (1994) Messenger RNA levels of XPAC and ERCC1 in ovarian cancer tissue correlate with response to platinum-based chemotherapy. J Clin Invest 94:703–708

Dabholkar M, Thornton K, Vionnet J, Bostick-Bruton F, Yu JJ, Reed E (2000) Increased mRNA levels of xeroderma pigmentosum complementation group B (XPB) and Cockayne's syndrome complementation group B (CSB) without increased mRNA levels of multidrug-resistance gene (MDR1) or metallothionein-II (MT-II) in platinum-resistant human ovarian cancer tissues. Biochem Pharmacol 60:1611–1619

Daoud SS, Clements MK, Small CL (1995) Polymerase chain reaction analysis of cisplatin-induced mitochondrial DNA damage in human ovarian carcinoma cells. Anti-Cancer Drugs 6:405–412

Dedon PC, Borch RF (1987) Characterization of the reactions of platinum antitumor agents with biologic and nonbiologic sulfur-containing nucleophiles. Biochem Pharmacol 36:1955–1964

DeMario MD, Vogelzang NJ, Janisch L, Tonda M, Amantea MA, Pendyala L, Ratain MJ (1998) A phase I study of liposome-formulated cisplatin (SPI-77 R) given every 3 weeks in patients with advanced cancer. Proc Am Soc Clin Oncol 17:abstract no. 883

DeMario MD, Ratain MJ, Vogelzang NJ, Mani S, Vokes EE, Fleming GF, Melton K, Johnson S, Benner S, Lebwohl D (1999) A phase I study of oral uracil/ftorafur (UFT) plus leucovorin and bis-acetato-ammine-dichloro-cyclohexylamine-platinum IV (JM-216) each given over 14 days every 28 days. Cancer Chemother Pharmacol 43:385–388

Dieras V, Bougnoux P, Petit T, Chollet P, Borel C, Husseini F, Goupil A, Kerbrat P, Misset JL, Flesch M, Fandi R, Bensmaïne MA, Beuzeboc P (1998) Oxaliplatin (L-OHP) phase II study in platinum (PT) pretreated advanced ovarian cancer (AOC): preliminary results. Proc Am Soc Clin Oncol 17:abstract no. 1405

Dijt FJ, Fichtinger-Shepman AMJ, Berends F, Reedijk J (1988) Formation and repair of cisplatin-induced adducts to DNA in cultured normal and repair-deficient human fibroblasts. Cancer Res 48:6058–6062

Drees M, Dengler WM, Hendriks HR, Kelland LR, Fiebig HH (1995) Cycloplatam: a novel platinum compound exhibiting a different spectrum of anti-tumour activity to cisplatin. Eur J Cancer 31A:356–361

Duckett DR, Drummond JT, Murchie AIH, Reardon JT, Sancar A, Lilley DMJ, Modrich P (1996) Human MutSα recognizes damaged DNA base pairs containing O6-methylguanine, O4-methylthymine, or the cisplatin-d(GpG) adduct. Proc Natl Acad Sci U S A 93:6443–6447

Duffull SB, Robinson BA (1997) Clinical pharmacokinetics and dose optimisation of carboplatin. Clin Pharmacokinet 33:161–183

Duncan R, Gac-Breton S, Keane R, Musila R, Sat YN, Satchi R, Searle F (2001) Polymer-drug conjugates, PDEPT and PELT: basic principles for design and transfer from the laboratory to clinic. J Controlled Release 74:135–146

Eastman A (1999) The mechanism of action of cisplatin: from adducts to apoptosis. In: Lippert B (ed) Cisplatin. Chemistry and biochemistry of a leading anticancer drug. Verlag Helvetica Chimica Acta, Zürich; Wiley-VCH, Weinheim pp 111–134

Eastman A, Barry MA (1987) Interaction of trans-diamminedichloroplatinum(II) with DNA: formation of monofunctional adducts and their reaction with glutathione. Biochemistry 26:3303–3307

Eastman A, Schulte N (1988) Enhanced DNA repair as a mechanism of resistance to cis-diamminedichloroplatinum(II). Biochemistry 27:4730–4734

Ellis LT, Er HM, Hambley TW (1995) The influence of the axial ligands of a series of platinum(IV) anticancer complexes on their reduction to platinum(II) and reaction with DNA. Austr J Chem 48:793–806

Evagorou EG, Duncan R, Buckley RG, Gianasi E (1998) Polymer-platinum compounds. Pat. No. WO9847537

Extra JM, Espie M, Calvo F, Ferme C, Mignot L, Marty M (1990) Phase I study of oxaliplatin in patients with advanced cancer. Cancer Chemother Pharmacol 25:299–303

Extra JM, Marty M, Brienza S, Misset JL (1998) Pharmacokinetics and safety profile of oxaliplatin. Semin Oncol 25, Suppl 5:1322

Faivre S, Raymond E, Woynarowski JM, Cvitkovic E (1999) Supraadditive effect of 2',2'-difluorodeoxycytidine (gemcitabine) in combination with oxaliplatin in human cancer cell lines. Cancer Chemother Pharmacol 44:117–123

Farrell N (1996) Current status of structure-activity relationships of platinum anticancer drugs: activation of the trans geometry. Met Ions Biol Syst 32:603–639

Farrell N (2000) Polynuclear charged platinum compounds as a new class of anticancer agents. Toward a new paradigm. In: Kelland LR, Farrell N (eds) Platinum-based drugs in cancer therapy. Humana, Totowa, pp 321–338

Farrell NP, de Almeida SG, Skov KA (1988) Bis(platinum) complexes containing two cis-diammine units. Synthesis and initial DNA-binding studies. J Am Chem Soc 110:5018–5019

Farrell N, Ha TTB, Souchard JP, Wimmer FL, Cros S, Johnson NP (1989) Cytostatic trans-platinum(II) complexes. J Med Chem 32:2240–2241

Farrell N, Qu Y, Feng L, Van Houten B (1990) Comparison of chemical reactivity, cytotoxicity, interstrand cross-linking and DNA sequence specificity of bis(platinum) complexes containing monodentate or bidentate coordination spheres with their monomeric analogues. Biochemistry 29:9522–9531

Farrell N, Kelland LR, Roberts JD, Van Beusichem M (1992) Activation of the trans geometry in platinum antitumor complexes: a survey of the cytotoxicity of trans complexes containing planar ligands in murine L1210 and human tumor panels and studies on their mechanism of action. Cancer Res 52:5065–5072

Farrell N, Appleton TG, Qu Y, Roberts JD, Soares Fontes AP, Skov KA, Wu P, Zou Y (1995) Effects of geometric isomerism and ligand substitution in bifunctional dinuclear platinum complexes on binding properties and conformational changes in DNA. Biochemistry 34:15480–15486

Farrell N, Qu Y, Bierbach U, Valsecchi M, Menta E (1999) Structure-activity relationships within di- and trinuclear platinum phase-I clinical anticancer agents. In: Lippert B (ed) Cisplatin. Chemistry and biochemistry of a leading anticancer drug. Verlag Helvetica Chimica Acta, Zürich; Wiley-VCH, Weinheim, pp 479–496

Fichtinger-Schepman AMJ, van der Veer JL, den Hartog JHJ, Lohman PHM, Reedijk J (1985) Adducts of the antitumor drug cis-diamminedichloroplatinum(II) with DNA: formation, identification, and quantitation. Biochemistry 24: 707–713

Fichtinger-Schepman AMJ, van Oosterom AT, Lohman PHM, Berends F (1987) cis-Diamminedichloroplatinum(II)-induced DNA adducts in peripheral leukocytes from seven cancer patients: quantitative immunochemical detection of the adduct induction and removal after a single dose of cis-diamminedichloroplatinum(II). Cancer Res 47:3000–3004

Fichtinger-Schepman AMJ, van der Velde-Visser SD, van Dijk-Knijnenburg HCM, van Oosterom AT, Baan RA, Berends F (1990) Kinetics of the formation and removal of cisplatin-DNA adducts in blood cells and tumour tissue of cancer patients receiving chemotherapy: comparison with in vitro adduct formation. Cancer Res 50:7887–7894

Fiebig HH, Mross K, Henß H, Aulenbacher P, Queißer W (1991) Phase I study of the new platinum complex D 19466 on a single intermittent schedule. Eur J Cancer 27, Suppl 2:S197

Fink D, Nebel S, Aebi S, Zheng H, Cenni B, Nehmé A, Christen RD, Howell SB (1996) The role of DNA mismatch repair in platinum drug resistance. Cancer Res 56:4881–4886

Fink D, Nebel S, Norris PS, Baergen RN, Wilczynski SP, Costa MJ, Haas M, Cannistra SA, Howell SB (1998) Enrichment for DNA mismatch repair-deficient cells during treatment with cisplatin. Int J Cancer 77:741–746

Fischel JL, Etienne MC, Formento P, Milano G (1998) Search for the optimal schedule for the oxaliplatin/5-fluoruracil association modulated or not by folinic acid: preclinical data. Clin Cancer Res 4:2529–2535

Fischel JL, Rostagno P, Formento P, Dubreuil A, Etienne MC, Milano G (2001) Ternary combination of irinotecan, fluorouracil-folinic acid and oxaliplatin: results on human colon cancer cell lines. Br J Cancer 84:579–585

Fokkema E, Lunenberg J, van Putten JWG, van Rijswijk REN, Weil C, Groen HJM (1998) Randomized phase II study of oral JM216 versus intravenous (iv) cisplatin in non-small cell lung cancer (NSCLC): preliminary results. Proc Am Soc Clin Oncol 17:abstract no. 1858

Fokkema E, Groen HJM, Bauer J, Uges DRA, Weil C, Smith IE (1999) Phase II study of oral platinum drug JM216 as first-line treatment in patients with small-cell lung cancer. J Clin Oncol 17:3822–3827

Fraval HNA, Rawlings CJ, Roberts JJ (1978) Increased sensitivity of UV-repair-deficient human cells to DNA bound platinum products which unlike thymidine dimers are not recognized by an endonuclease extracted from Micrococcus luteus. Mutat Res 51:121–132

Frey U, Ranford JD, Sadler PJ (1993) Ring-opening reactions of the anticancer drug carboplatin: NMR characterization of cis-[Pt(NH3)2(CBDCA-O)(5'-GMP-N7)] in solution. Inorg Chem 32:1333–1340

Fukuda M, Shinkai T, Eguchi K, Sasaki Y, Tamura T, Ohe Y, Kojima A, Oshita F, Hara K, Saijo N (1990) Phase II study of (glycolate-O,O') diammineplatinum(II), a novel platinum complex, in the treatment of non-small-cell lung cancer. Cancer Chemother Pharmacol 26:393–396

Galanski M, Keppler BK (2000) Is reduction required for antitumour activity of platinum(IV) compounds? Characterisation of a platinum(IV)-nucleotide adduct [enPt(OCOCH3)3(5'-GMP)] by NMR spectroscopy and ESI-MS. Inorg Chim Acta 300–302:783–789

Galanski M, Slaby S, Keppler BK (1999) The mode of action of antitumor platinum compounds linked to amino phosphonic acids with particular activity against bone malignancies and the influence of calcium ions onto the DNA binding behavior. Contrib Oncol 54:435–438

Gamelin E, Allain P, Maillart P, Turcant A, Delva R, Lortholary A, Larra F (1995) Long-term pharmacokinetic behavior of platinum after cisplatin administration. Cancer Chemother Pharmacol 37:97–102

Gamelin E, Le Brouil A, Boisdron-Celle M, Turcant A, Delva R, Cailleux A, Krikorian A, Brienza S, Cvitkovic E, Robert J, Larra F, Allain P (1997) Cumulative pharmacokinetic study of oxaliplatin, administered every three weeks, combined with 5-fluorouracil in colorectal cancer patients. Clin Cancer Res 3:891–899

Garufi C, Nisticö C, Brienza S, Vaccaro A, D'Ottavio A, Zappala AR, Aschelter AM, Terzoli E (2000) Single-agent oxaliplatin in pretreated advanced breast cancer patients: a phase II study. Ann Oncol 12:179–182

Gately DP, Howell SB (1993) Cellular accumulation of the anticancer agent cisplatin: a review. Br J Cancer 67:1171–1176

Gatzemeier U, Anthoney DA, Groth G, Kaye SB, Cosaert J (2001) A phase I dose-escalation study of ZD0473 combined with paclitaxel in refractory solid malignancies. Proc Am Soc Clin Oncol 20:abstract no. 2775

Gelasco A, Lippard SJ (1999) Anticancer activity of cisplatin and related complexes. Top Biol Inorg Chem 1:1–43

Germann N, Brienza S, Rotarski M, Emile FJ, Di Palma M, Musset M, Reynes M, Soulie P, Cvitkovic E, Misset JL (1999) Preliminary results on the activity of oxaliplatin (L-OHP) in refractory/recurrent non-Hodgkin's lymphoma patients. Ann Oncol 10:351–354

Gianasi E, Wasil M, Evagorou EG, Keddle A, Wilson G, Duncan R (1999) HPMA copolymer platinates as novel antitumour agents: in vitro properties, pharmacokinetics and antitumour activity in vivo. Eur J Cancer 35:994–1002

Gibbons GR, Page JD, Mauldin SK, Husain I, Chaney SG (1990) Role of carrier ligand in platinum resistance in L1210 cells. Cancer Res 50:6497–6501

Gibbons GR, Kaufmann WK, Chaney SG (1991) Role of DNA replication in carrier-ligand-specific resistance to platinum compounds in L1210 cells. Carcinogenesis 12:2253–2257

Gietema JA, De Vries EGE, Sleijfer DT, Willemse PHB, Guchelaar HJ, Uges DRA, Aulenbacher P, Voegeli R, Mulder NH (1993a) A phase I study of 1,2-diamminomethyl-cyclobutane-platinum (II)-lactate (D-19466; lobaplatin) administered daily for 5 days. Br J Cancer 67:396–401

Gietema JA, Guchelaar HJ, de Vries EGE, Aulenbacher P, Sleijfer DT, Mulder NH (1993b) A phase I study of lobaplatin (D-19466) administered by 72 h continuous infusion. Anti-Cancer Drugs 4:51–55

Gietema JA, Veldhuis GJ, Guchelaar HJ, Willemse PHB, Uges DRA, Cats A, Boonstra H, van der Graaf WTA, Sleijfer DT, de Vries EGE, Mulder NH (1995) Phase II and pharmacokinetic study of lobaplatin in patients with relapsed ovarian cancer. Br J Cancer 71:1302–1307

Gill I, Muggia FM, Terheggen PM, Michael C, Parker RJ, Kortes V, Grunberg S, Christian MC, Reed E, den Engelse L (1991) Dose-escalation study of carboplatin (day 1) and cisplatin (day 3): tolerance and relation to leukocyte and buccal cell platinum-DNA adducts. Ann Oncol 2:115–121

Giurgiovich AJ, Diwan BA, Olivero OA, Anderson LM, Rice JM, Poirier MC (1997a) Elevated mitochondrial cisplatin-DNA adduct levels in rat tissues after transplacental cisplatin exposure. Carcinogenesis 18:93–96

Giurgiovich AJ, Anderson LM, Jones AB, Dove LF, Moskal DJ, Rice JM, Olivero OA, Poirier MC (1997b) Transplacental cisplatin exposure induces persistent fetal mitochondrial and genomic DNA damage in patas monkeys. Reprod Toxicol 11:95–100

Go RS, Adjei AA (1999) Review of the comparative pharmacology and clinical activity of cisplatin and carboplatin. J Clin Oncol 17:409–422

Goddard PM, Orr RM, Valenti MR, Barnard CFJ, Murrer BA, Kelland LR, Harrap KR (1996) Novel trans platinum complexes: comparative in vitro and in vivo activity against platinum-sensitive and resistant murine tumours. Anticancer Res 16:33–38

Godwin AK, Meister A, O'Dwyer PJ, Huang CS, Hamilton TC, Anderson ME (1992) High resistance to cisplatin in human ovarian cancer cell lines is associated with marked increase of glutathione synthesis. Proc Natl Acad Sci U S A 89:3070–3074

Goldwasser F, Bozec L, Zeghari-Squalli N, Misset JL (1999) Cellular pharmacology of the combination of oxaliplatin with topotecan in the IGROV-1 human ovarian cancer cell line. Anti-Cancer Drugs 10:195–201

Gorbunova VA, Smirnova NB, Orel NE (1995) Phase II study of cycloplatam in advanced ovarian carcinoma. Proc Int Congress on Anti-Cancer Chemother 5:382

Gordon JA, Gattone VH (1986) Mitochondrial alterations in cisplatin-induced acute renal failure. Am J Physiol 250:F991–998

Gore M, Atkinson RJ, Dirix L, Rischin D, Beale P, Harnet P, Hacking D, Cure H, Cosaert J (2001) ZD0473 phase II monotherapy trial in second-line ovarian cancer. Proc Am Soc Clin Oncol 20:abstract no. 2501

Graham MA, Lockwood GF, Greenslade D, Brienza S, Bayssas M, Gamelin E (2000) Clinical pharmacokinetics of oxaliplatin: a critical review. Clin Cancer Res 6:1205–1218

Grolleau F, Gamelin L, Boisdron-Celle M, Lapied B, Pelhate M, Gamelin E (2001) A possible explanation for a neurotoxic effect of the anticancer agent oxaliplatin on neuronal voltage-gated sodium channels. J Neurophysiol 85:2293–2297

Gullo JJ, Litterst CL, Maguire PJ, Sikic BI, Hoth DF, Woolley PV (1980) Pharmacokinetics and protein binding of cis-dichlorodiammine platinum (II) administered as a one hour or as a twenty hour infusion. Cancer Chemother Pharmacol 5:21–26

Gullotti M, Pasini A, Ugo R, Filippeschi S, Marmonti L, Spreafico F (1984) Enantiomeric cisplatin analogues: an investigation on their activity towards tumors in mice. Inorg Chim Acta 91:223–227

Gust R, Schönenberger H (1993) Synthesis and evaluation of the anti-mammary tumor activity and of the estrogenic side effects of [1,2-bis(2,6-dihalo-3-hydroxyphenyl)ethylenediamine]platinum(II) complexes. Eur J Med Chem 28:103–115

Gust R, Karl J, Faderl M, Schönenberger H (1995) Reduction of the estrogenic side effects of the mammary tumor-inhibiting drug [1,2-bis(2,6-dichloro-4-hydroxyphenyl)ethylenediamine]dichloroplatinum(II) by variation of ring substituents. Arch Pharm 328:457–463

Gust R, Faderl M, Schönenberger H (2000) [Aqua-1-(2,6-dichloro-4-hydroxyphenyl)-2-phenylethylenediamine]sulfatoplatinum(II) complexes with variable substituents in the 2-phenyl ring 3. Investigation of breast cancer inhibiting properties. J Cancer Res Clin Oncol 126:647–654

Han I, Ling YH, Al-Baker S, Khokhar AR, Perez-Soler R (1993) Cellular pharmacology of liposomal cis-bis-neodecanoato-trans-R,R-1,2-diaminocyclohexaneplatinum(II) in A2780/S and A2780/PDD cells. Cancer Res 53:4913–4919

Han I, Nguyen T, Yang LY, Khokhar AR, Perez-Soler R (1994) Cellular accumulation and DNA damage induced by liposomal cis-bis-neodecanoato-trans-R,R-1,2-diaminocyclohexaneplatinum(II) in LoVo and LoVo/PDD cells. Anti-Cancer Drugs 5:64–68

Harstrick A, Bokemeyer C, Scharnofske M, Hapke G, Reile D, Schmoll HJ (1993) Preclinical activity of a new platinum analogue, lobaplatin, in cisplatin-sensitive and -resistant human testicular, ovarian, and gastric carcinoma cell lines. Cancer Chemother Pharmacol 33:43–47

Hayes DM, Cvitkovic E, Golbey RB, Scheiner E, Helson L, Krakoff IH (1977) High dose cis-platinum diammine dichloride: amelioration of renal toxicity by mannitol diuresis. Cancer 39:1372–1381

He Y, Groleau S, Gaudreault RC, Caron M, Thérien HM, Bérubé G (1995) Synthesis and in vitro biological evaluation of new triphenylethylene platinum(II) complexes. Bioorg Med Chem Lett 5:2217–2222

Hegmans A, Qu Y, Kelland LR, Roberts JD, Farrell N (2001) Novel approaches to polynuclear platinum pro-drugs. Selective release of cytotoxic platinum-spermidine species through hydrolytic cleavage of carbamates. Inorg Chem 40:6108–6114

Heiger-Bernays WJ, Essigmann JM, Lippard SJ (1990) Effect of the antitumor drug cis-diamminedichloroplatinum(II) and related platinum complexes on eukaryotic DNA replication. Biochemistry 29:8461–8466

Highley MS, Calvert AH (2000) Clinical experience with cisplatin and carboplatin. In: Kelland LR, Farrell N (eds) Platinum-based drugs in cancer therapy. Humana, Totowa, pp 171–194

Hill BT, Scanlon KJ, Hansson J, Harstrick A, Pera M, Fichtinger-Schepman AM, Shellard SA (1994) Deficient repair of cisplatin-DNA adducts identified in human testicular teratoma cell lines established from tumours from untreated patients. Eur J Cancer 30A:832–837

Hoffman P, Mauer A, Haraf D, Masters G, Lebwohl D, Krauss S, Golomb H, Vokes E (1998) Oral JM-216 plus concomitant radiotherapy for patients with advanced malignancies of the chest. Proc Am Soc Clin Oncol 17:abstract no. 1878

Holford J, Raynaud F, Murrer BA, Grimaldi K, Hartley JA, Abrams M, Kelland LR (1998a) Chemical, biochemical and pharmacological activity of the novel sterically hindered platinum co-ordination complex, cis-[amminedichloro(2-methylpyridine)] platinum(II) (AMD473). Anti-Cancer Drug Des 13:118

Holford J, Sharp SY, Murrer BA, Abrams M, Kelland LR (1998b) In vitro circumvention of cisplatin resistance by the novel sterically hindered platinum complex AMD473. Br J Cancer 77:366–373

Hongo A, Seki S, Akiyama K, Kudo T (1994) A comparison of in vitro platinum-DNA adduct formation between carboplatin and cisplatin. Int J Biochem 26:1009–1016

Hoshi A, Castañer J (1997) TRK-710. Drugs Future 22:38–39

Hoshino T, Misaki M, Yamamoto M, Shimizu H, Ogawa Y, Toguchi H (1995) In vitro cytotoxicities and in vivo distribution of transferrin-platinum(II) complex. J Pharm Sci 84:216–221

Houldsworth J, Xiao H, Murty VVVS, Chen W, Ray B, Reuter VE, Bosl GJ, Chaganti RSK (1998) Human male germ cell tumor resistance to cisplatin is linked to TP53 gene mutation. Oncogene 16:2345–2349

Inagaki K, Kidani Y (1986) Differences in binding of (1,2-cyclohexanediamine)platinum(II) isomers with d(GpG). Inorg Chem 25:1–3

Inagaki K, Nakahara H, Alink M, Kidani Y (1990) Characterization of (1,2-cyclohexanediamine)platinum(II) isomers and their d(GpG) adducts by means of proton NMR spectroscopy. A minor structural change induced by the isomers. Inorg Chem 29:4496–4500

Ishibashi T, Lippard SJ (1998) Telomere loss in cells treated with cisplatin. Proc Natl Acad Sci USA 95:4219–4223

Jamieson ER, Lippard SJ (1999) Structure, recognition, and processing of cisplatin-DNA adducts. Chem Rev 99:2467–2498

Janáky T, Juhász A, Rékási Z, Serfözö P, Pinski J, Bokser L, Srkalovic G, Milanovic S, Redding TW, Halmos G, Nagy A, Schally AV (1992) Short-chain analogs of luteinizing hormone-releasing hormone containing cytotoxic moieties. Proc Natl Acad Sci U S A 89:10203–10207

Jennerwein M, Andrews PA (1994) Drug accumulation and DNA platination in cells exposed to aquated cisplatin species. Cancer Lett 81:215–220

Jennerwein MM, Eastman A, Khokhar A (1989) Characterization of adducts produced in DNA by isomeric 1,2-diaminocyclohexaneplatinum(II) complexes. Chem-Biol Interact 70:39–49

Jennerwein MM, Eastman A, Khokhar AR (1991) The role of DNA-repair in resistance of L1210 cells to isomeric 1,2-diaminocyclohexaneplatinum complexes and ultraviolet irradiation. Mutat Res 254:89–96

Johnson NP, Hoeschele JD, Rahn RO, O'Neill JP, Hsie AW (1980) Mutagenicity, cytotoxicity, and DNA binding of platinum(II)-chloroammines in Chinese hamster ovary cells. Cancer Res 40:1463–1468

Johnson SW, Perez RP, Godwin AK, Yeung AT, Handel LM, Ozols RF, Hamilton TC (1994a) Role of platinum-DNA adduct formation and removal in cisplatin resistance in human ovarian cancer cell lines. Biochem Pharmacol 47:689–697

Johnson SW, Swiggard PA, Handel LM, Brennan JM, Godwin AK, Ozols RF, Hamilton TC (1994b) Relationship between platinum-DNA adduct formation and removal and cisplatin cytotoxicity in cisplatin-sensitive and -resistant human ovarian cancer cells. Cancer Res 54:5911–5916

Jones JC, Zhen W, Reed E, Parker RJ, Sancar A, Bohr VA (1991) Gene-specific formation and repair of cisplatin intrastrand adducts and interstrand cross-links in Chinese hamster ovary cells. J Biol Chem 266:7101–7107

Jones SF, Greco FA, Hainsworth JD, Burris HA, Calvert SA, Bulanhagui C, Lebwohl DE (1999) Phase I study of JM-216 in combination with paclitaxel in patients with advanced malignancies. Proc Am Soc Clin Oncol 18:abstract no. 831

Judson IR, Cerny T, Epelbaum R, Dunlop D, Smyth J, Schaefer B, Roelvink M, Kaplan S, Hanauske A (1997) Phase II trial of the oral platinum complex JM216 in non-small-cell lung cancer: an EORTC early clinical studies group investigation. Ann Oncol 8:604–606

Kalinowski DP, Illenye S, Van Houten B (1992) Analysis of DNA damage and repair in murine leukemia L1210 cells using a quantitative polymerase chain reaction. Nucl Acids Res 20:3485–3494

Karl J, Gust R, Spruss T, Schneider MR, Schönenberger H, Engel J, Wrobel KH, Lux F, Trebert Haeberlin S (1988) Ring-substituted [1,2-bis(4-hydroxyphenyl)ethylenediamine]dichloroplatinum(II) complexes: compounds with a selective effect on the hormone-dependent mammary carcinoma. J Med Chem 31:72–83

Kavanagh JJ, Edward CL, Freedman RS, Finnegan MB, Balat O, Tresukosol D, Burk K, Loechner S, Hord M, Franklin JL, Kudelka AP (1995) A trial of lobaplatin (D-19466) in platinum-resistant ovarian cancer. Gynecol Oncol 58:106–109

Kelland LR (2000) New platinum drugs. The pathway to oral therapy. In: Kelland LR, Farrell N (eds) Platinum-based drugs in cancer therapy. Humana, Totowa, pp 299–319

Kelland LR, Mistry P, Abel G, Freidlos F, Loh SY, Roberts JJ, Harrap KR (1992) Establishment and characterization of an in vitro model of acquired resistance to cisplatin in a human testicular nonseminomatous germ cell line. Cancer Res 52:1710–1716

Kelland LR, Abel G, McKeage MJ, Jones M, Goddard PM, Valenti M, Murrer BA, Harrap KR (1993a) Preclinical antitumor evaluation of bis-acetato-ammine-dichloro-cyclohexylamine platinum(IV): an orally active platinum drug. Cancer Res 53:2581–2586

Kelland LR, Jones M, Gwynne JJ, Valenti M, Murrer BA, Barnard CFJ, Vollano JF, Giandomenico CM, Abrams MJ, Harrap KR (1993b) Antitumour activity of orally administered ammine/amine platinum (IV) dicarboxylate complexes against a panel of human ovarian carcinoma xenografts. Int J Oncol 2:1043–1048

Kelland LR, Barnard CFJ, Evans IG, Murrer BA, Theobald BRC, Wyer SB, Goddard PM, Jones M, Valenti M, Bryant A, Rogers PM, Harrap KR (1995) Synthesis and in vitro and in vivo antitumor activity of a series of trans platinum antitumor complexes. J Med Chem 38:3016–3024

Kelley SL, Basu A, Teicher BA, Hacker MP, Hamer DH, Lazo JS (1988) Overexpression of metallothionein confers resistance to anticancer drugs. Science 241:1813–1815

Khokhar AR, Al-Baker S, Brown T, Perez-Soler R (1991) Chemical and biological studies on a series of lipid-soluble (trans-(R,R)- and -(S,S)-1,2-diaminocyclohexane)platinum(II) complexes incorporated in liposomes. J Med Chem 34:325–329

Khokhar AR, Al-Baker S, Shamsuddin S, Siddik ZH (1997) Chemical and biological studies on a series of novel (trans-(1R,2R)-, trans-(1S,2S)-, and cis-1,2-diaminocyclohexane)platinum(IV) carboxylate complexes. J Med Chem 40:112–116

Kidani Y, Noji M, Tsukagoshi S, Tashiro T (1978a) Antitumor activity of water-soluble platinum(II) complexes of 1,2-cyclohexanediamine isomers. Gann 69:263–265

Kidani Y, Inagaki K, Iigo M, Hishi A, Kuretani K (1978b) Antitumor activity of 1,2-diaminocyclohexane-platinum complexes against sarcoma-180 ascites form. J Med Chem 21:1315–1318

Kidani Y, Noji M, Tashiro T (1980) Antitumor activity of platinum(II) complexes of 1,2-diaminocyclohexane isomers. Gann 71:637–643

Kido Y, Khokhar AR, Siddik ZH (1993a) Differential cytotoxicity, uptake and DNA binding of tetraplatin and analogous isomers in sensitive and resistant cancer cell lines. Anti-Cancer Drugs 4:251–258

Kido Y, Khokhar AR, Al-Baker S, Siddik ZH (1993b) Modulation of cytotoxicity and cellular pharmacology of 1,2-diaminocyclohexane platinum(IV) complexes mediated by axial and equatorial ligands. Cancer Res 53:4567–4572

Kim DK, Ahn JS, Ryu G, Kim KH, Park CW, Kim MS, Chung MH, Shin SG, Suh YH, Kim YS (1994) General pharmacology of cis-malonato[(4R,5R)-4,5-bis(aminomethyl)-2-isopropyl-1,3-dioxolane]platinum(II). Arzneimittelforsch/Drug Res 44:1080–1088

Kim DK, Kim HT, Cho YB, Tai JH, Ahn JS, Kim TS, Kim KH, Hong WS (1995) Antitumor activity of cis-malonato[(4R,5R)-4,5-bis(aminomethyl)-2-isopropyl-1,3-dioxolane]platinum(II), a new platinum analogue, as an anticancer agent. Cancer Chemother Pharmacol 35:441–445

Kim NK, Im SA, Kim DW, Lee MH, Jung CW, Cho EK, Lee JT, Ahn JS, Heo DS, Bang YJ (1999) Phase II clinical trial of SKI-2053R, a new platinum analog, in the treatment of patients with advanced gastric adenocarcimona. Cancer 86:1109–1115

Kim NK, Bang Y, Heo DS, Kang Y, Kang WK, Kim H, Kim H, Kim JS, Kim HT, Roh JK (2000) SKI2053R and 5-fluorouracil (5-FU) in advanced gastric adenocarcinoma: phase II clinical trial. Proc Am Soc Clin Oncol 19:abstract no. 1277

Kim NK, Kim TY, Shin SG, Park YI, Lee JA, Cho YB, Kim KH, Kim DK, Heo DS, Bang YJ (2001) A phase I study of cis-malonato[(4R,5R)-4,5-bis(aminomethyl)-1,3-dioxolane] platinum(II) in patients with advanced malignancies. Cancer 91:1549–1556

Klenner T, Wingen F, Keppler BK, Krempien B, Schmähl D (1990a) Anticancer-agent-linked phosphonates with antiosteolytic and antineoplastic properties: a promising perspective in the treatment of bone-related malignancies? J Cancer Res Clin Oncol 116:341–350

Klenner T, Valenzuela-Paz P, Keppler BK, Angres G, Scherf HR, Wingen F, Amelung F, Schmähl D (1990b) Cisplatin-linked phosphonates in the treatment of the transplantable osteosarcoma in vitro and in vivo. Cancer Treat Rev 17:253–259

Klenner T, Valenzuela-Paz P, Amelung F, Münch H, Zahn H, Keppler BK, Blum H (1993) Platinum phosphonato complexes with particular activity against bone malignancies. An evaluation of an experimental tumor model highly predictive for the clinical situation. In: Keppler BK (ed) Metal complexes in cancer chemotherapy. VCH, Weinheim, pp 85–127

Knebel NG, Angerer EV (1991) 2-Phenylindole-linked [2-(aminoalkyl)pyridine]dichloroplatinum(II): complexes with a selective action on estrogen receptor positive mammary tumors. J Med Chem 34:2145–2152

Knox RJ, Friedlos F, Lydall DA, Roberts JJ (1986) Mechanism of cytotoxicity of anticancer platinum drugs: evidence that cis-diamminedichloroplatinum(II) and cis-diammine(1,1-cyclobutanedicarboxylato)platinum(II) differ only in the kinetics of their interaction with DNA. Cancer Res 46:1972–1979

Köberle B, Payne J, Grimaldi KA, Hartley JA, Masters JRW (1996) DNA repair in cisplatin-sensitive and resistant human cell lines measured in specific genes by quantitative polymerase chain reaction. Biochem Pharmacol 52:1729–1734

Köberle B, Grimaldi KA, Sunters A, Hartley JA, Kelland LR, Masters JRW (1997) DNA repair capacity and cisplatin sensitivity of human testis tumour cells. Int J Cancer 70:551–555

Köberle B, Masters JRW, Hartley JA, Wood RD (1999) Defective repair of cisplatin-induced DNA damage caused by reduced XPA protein in testicular germ cell tumours. Curr Biol 9:273–276

Kollmannsberger C, Rick O, Beyer J, Derigs G, Mayer F, Schleucher N, Schoeffski P, Sosada M, Spott C, Kanz L, Bokemeyer C (2001) Oxaliplatin in patients (pts) with cisplatin-refractory germ-cell cancer (GCT). Proc Am Soc Clin Oncol 20:abstract no. 770

Kozelka J, Legendre F, Reeder F, Chottard JC (1999) Kinetic aspects of interactions between DNA and platinum complexes. Coord Chem Rev 190–192:61–82

Kraker AJ, Moore CW (1988) Accumulation of cis-diamminedichloroplatinum(II) and platinum analogues by platinum-resistant murine leukemia cells in vitro. Cancer Res 48:9–13

Küng A, Zenker A, Galanski A, Keppler BK (2001a) Capillary electrophoretic study of carboplatin and analogues with nucleoside monophosphates, di- and trinucleotides. J Inorg Biochem 83:181–186

Küng A, Strickmann DB, Galanski M, Keppler BK (2001b) Comparison of the binding behavior of oxaliplatin, cisplatin and analogues to 5'-GMP in the presence of sulfur containing molecules by means of capillary electrophoresis and electrospray mass spectrometry. J Inorg Biochem 86:691–698

Lai GM, Ozols RF, Smyth JF, Young RC, Hamilton TC (1988) Enhanced DNA repair and resistance to cisplatin in human ovarian cancer. Biochem Pharmacol 37:4597–4600

Lebwohl D, Canetta R (1998) Clinical development of platinum complexes in cancer therapy: an historical perspective and an update. Eur J Cancer 34:1522–1534

LeDoux SP, Wilson GL, Beecham EJ, Stevnsner T, Wassermann K, Bohr VA (1992) Repair of mitochondrial DNA after various types of DNA damage in Chinese hamster ovary cells. Carcinogenesis 13:1967–1973

Legendre F, Bas V, Kozelka J, Chottard JC (2000) A complete kinetic study of GG versus AG platination suggests that the doubly aquated derivatives of cisplatin are the actual DNA binding species. Chem Eur J 6:2002–2010

Levi F, Metzger G, Massari C, Milano G (2000) Oxaliplatin. Pharmacokinetics and chronopharmacological aspects. Clin Pharmacokinet 38:1–21

Luo FR, Wyrick SD, Chaney SG (1998) Cytotoxicity, cellular uptake, and cellular biotransformations of oxaliplatin in human colon carcinoma cells. Oncol Res 10:595–603

Luo FR, Yen TY, Wyrick SD, Chaney SG (1999a) High-performance liquid chromatographic separation of the biotransformation products of oxaliplatin. J Chromatogr B, 724:345–356

Luo FR, Wyrick SD, Chaney SG (1999b) Biotransformations of oxaliplatin in rat blood in vitro. J Biochem Mol Toxicol 13:159–169

Luo FR, Wyrick SD, Chaney SG (1999c) Pharmacokinetics and biotransformations of oxaliplatin in comparison with ormaplatin following a single bolus intravenous injection in rats. Cancer Chemother Pharmacol 44:19–28

Luo FR, Wyrick SD, Chaney SG (1999d) Comparative neurotoxicity of oxaliplatin, ormaplatin, and their biotransformation products utilizing a rat dorsal root ganglia in vitro explant culture model. Cancer Chemother Pharmacol 44:29–38

MacLean DS, Khokhar AR, Tyle P, Perez-Soler R (2000) Intraliposomal chemical activation patterns of liposomal cis-bis-neodecanoato-trans-R,R-1,2-diaminocyclohexane platinum (II) (L-NDDP) a potential antitumour agent. J Microencapsulat 17:307–322

Maeda H (2001) The enhanced permeability and retention (EPR) effect in tumor vasculature: the key role of tumor-selective macromolecular drug targeting. Adv Enzyme Regul 41:189–207

Maeda H, Wu J, Sawa T, Matsumura Y, Hori K (2000) Tumor vascular permeability and the EPR effect in macromolecular therapeutics: a review. J Controlled Release 65:271–284

Mamenta EL, Poma EE, Kaufmann WK, Delmastro DA, Grady HL, Chaney SG (1994) Enhanced replicative bypass of platinum-DNA adducts in cisplatin-resistant human ovarian carcinoma cell lines. Cancer Res 54:3500–3505

Mani S, Manalo J, Bregman D (2000) Novel combinations with oxaliplatin. Oncology 14, Suppl 11:52–58

Manzotti C, Pratesi G, Menta E, Di Domenico R, Cavalletti E, Fiebig HH, Kelland LR, Farrell N, Polizzi D, Supino R, Pezzoni G, Zunino F (2000) BBR 3464: a novel triplatinum complex, exhibiting a preclinical profile of antitumor efficacy different from cisplatin. Clin Cancer Res 6:2626–2634

Masters JRW, Thomas R, Hall AG, Hogarth L, Matheson EC, Cattan AR, Lohrer H (1996) Sensitivity of testis tumour cells to chemotherapeutic drugs: role of detoxifying pathways. Eur J Cancer 32A:1248–1253

Masuda H, Ozols RF, Lai GM, Fojo A, Rothenberg M, Hamilton TC (1988) Increased DNA repair as a mechanism of acquired resistance to cis-diamminedichloroplatinum(II) in human ovarian cancer cell lines. Cancer Res 48:5713–5716

Masuda H, Tanaka T, Matsuda H, Kusaba I (1990) Increased removal of DNA-bound platinum in a human ovarian cancer cell line resistant to cis-diamminedichloroplatinum(II). Cancer Res 50:1863–1866

Mathé G, Kidani Y, Triana K, Brienza S, Ribaud P, Goldschmidt E, Ecstein E, Despax R, Musset M, Misset JL (1986) A phase I trial of trans-diaminocyclohexane oxalato-platinum(l-OHP). Biomed Pharmacother 40:372–376

Mathé G, Kidani Y, Segiguchi M, Eriguchi M, Fredj G, Peytavin G, Misset JL, Brienza S, de Vassals F, Chenu E, Bourut C (1989) Oxalato-platinum or l-OHP, a third-generation platinum complex: an experimental and clinical appraisal and preliminary comparison with cis-platinum and carboplatinum. Biomed Pharmacother 43:237–250

Matveev BP, Gorbunova VA, Smirnova NB, Zhumagazin ZD, Bukharin BV, Orel NF (1996) Cycloplatam in the treatment of patients with disseminated prostatic cancer. Urol Nefrol (Mosk) 1996:34–36

May A, Nairn RS, Okumoto DS, Wassermann K, Stevnsner T, Jones JC, Bohr VA (1993) Repair of individual DNA strands in the hamster dihydrofolate reductase gene after treatment with ultraviolet light, alkylating agents, and cisplatin. J Biol Chem 268:1650–1657

McKeage MJ, Raynaud F, Ward J, Berry C, O'Dell D, Kelland LR, Murrer B, Santabárabara P, Harrap KR, Judson IR (1997) Phase I and pharmacokinetic study of an oral platinum complex given daily for 5 days in patients with cancer. J Clin Oncol 15:2691–2700

Meijer C, Mulder NH, Timmer-Bosscha H, Sluiter WJ, Meersma GJ, De Vries EGE (1992) Relationship of cellular glutathione to the cytotoxicity and resistance of seven platinum compounds. Cancer Res 52:6885–6889

Mellish KJ, Kelland LR, Harrap KR (1993) In vitro platinum drug chemosensitivity of human cervical squamous cell carcinoma cell lines with intrinsic and acquired resistance to cisplatin. Br J Cancer 68:240–250

Mellish KJ, Barnard CFJ, Murrer BA, Kelland LR (1995) DNA-binding properties of novel cis- and trans platinum-based anticancer agents in 2 human ovarian carcinoma cell lines. Int J Cancer 62:717–723

Mello JA, Lippard SJ, Essigmann JM (1995) DNA adducts of cis-diamminedichloroplatinum(II) and its trans isomer inhibit RNA polymerase II differentially in vivo. Biochemistry 34:14783–14791

Mello JA, Acharya S, Fishel R, Essigmann JM (1996) The mismatch-repair protein hMSH2 binds selectively to DNA adducts of the anticancer drug cisplatin. Chem Biol 3:579–589

Mese H, Ueyama Y, Suzuki A, Nakayama S, Sasaki A, Hamakawa H, Matsumura T (2001) Inhibition of telomerase activity as a measure of tumor cell killing by cisplatin in squamous cell carcinoma cell line. Chemotherapy 47:136–142

Metzger SW, Keppler BK (1999) Potential in inhibition of telomerase activity and cytotoxicity of different platinum compounds. 8th Int Symp Platinum and Other Metal Coordination Compounds in Cancer Chemotherapy: 4.P14

Meyn RE, Jenkins SF, Thompson LH (1982) Defective removal of DNA cross-links in a repair-deficient mutant of Chinese hamster cells. Cancer Res 42:3106–3110

Misset JL, Bleiberg H, Sutherland W, Bekradda M, Cvitkovic E (2000) Oxaliplatin clinical activity: a review. Crit Rev Oncol Hematol 35:75–93

Monnet I, Brienza S, Hugret F, Voisin S, Gastiaburu J, Saltiel JC, Soulié P, Armand JP, Cvitkovic E, de Cremoux H (1998) Phase II study of oxaliplatin in poor-prognosis non-small cell lung cancer (NSCLC). Eur J Cancer 34:1124–1127

Montero EI, Diaz S, González-Vadillo AM, Pérez JM, Alonso C, Navarro-Ranninger C (1999) Preparation and characterization of novel trans-[PtCl2(amine)(isopropylamine)] compounds: cytotoxic activity and apoptosis induction in ras-transformed cells. J Med Chem 42:4264–4268

Murata T, Hibasami H, Maekawa S, Tagawa T, Nakashima K (1990) Preferential binding of cisplatin to mitochondrial DNA and suppression of ATP generation in human malignant melanoma cells. Biochem Int 20:949–955

Natile G, Coluccia M (1999) trans-Platinum compounds in cancer therapy: a largely unexplored strategy for identifying novel antitumor platinum drugs. Top Biol Inorg Chem 1:73–98

Natile G, Coluccia M (2001) Current status of trans-platinum compounds in cancer therapy. Coord Chem Rev 216–217:383–410

Nehmé A, Baskaran R, Nebel S, Fink D, Howell SB, Wang JYJ, Christen RD (1999) Induction of JNK and c-Abl signalling by cisplatin and oxaliplatin in mismatch repair-proficient and -deficient cells. Br J Cancer 79:1104–1110

Newman MS, Colbern GT, Working PK, Engbers C, Amantea MA (1999) Comparative pharmacokinetics, tissue distribution, and therapeutic effectiveness of cisplatin encapsulated in long-circulating, pegylated liposomes (SPI-077) in tumor-bearing mice. Cancer Chemother Pharmacol 43:1–7

Nishimura T, Newkirk K, Sessions RB, Andrews PA, Trock BJ, Rasmussen AA, Montgomery EA, Bischoff EK, Cullen KJ (1996) Immunohistochemical staining for glutathione S-transferase predicts response to platinum-based chemotherapy in head and neck cancer. Clin Cancer Res 2:1859–1865

Noji M, Okamoto K, Kidani Y (1981) Relation of conformation to antitumor activity of platinum(II) complexes of 1,2-cyclohexanediamine and 2-(aminomethyl)cyclohexylamine isomers against leukemia P388. J Med Chem 24:508–515

Nováková O, Vrána O, Kiseleva VI, Brabec V (1995) DNA interactions of antitumor platinum(IV) complexes. Eur J Biochem 228:616–624

O'Dwyer PJ, Moyer JD, Suffness M, Harrison SD Jr, Cysyk R, Hamilton TC, Plowman J (1994) Antitumor activity and biochemical effects of aphidicolin glycinate (NSC 303812) alone and in combination with cisplatin in vivo. Cancer Res 54:724–729

O'Dwyer PJ, Stevenson JP, Johnson SW (1999) Clinical status of cisplatin, carboplatin, and other platinum-based antitumor drugs. In: Lippert B (ed) Cisplatin. Chemistry and biochemistry of a leading anticancer drug. Verlag Helvetica Chimica Acta, Zürich; Wiley-VCH, Weinheim, pp 31–69

O'Dwyer PJ, Stevenson JP, Redlinger M, Sun W, Giantonio B, Hahn S, Algazy K, Burnite M, Koehler M (2001) Phase I trial of the novel platinum analogue ZD0473 in combination with gemcitabine for patients with advanced cancers. Proc Am Soc Clin Oncol 20:abstract no. 1354

Ohndorf UM, Whitehead JP, Raju NL, Lippard SJ (1997) Binding of tsHMG, a mouse testis-specific HMG-domain protein, to cisplatin-DNA adducts. Biochemistry 36:14807–14815

Oldenburg J, Begg AC, van Vugt MJH, Ruevekamp M, Schornagel JH, Pinedo HM, Los G (1994) Characterization of resistance mechanisms to cis-diamminedichloroplatinum(II) in three sublines of the CC531 colon adenocarcinoma cell line in vitro. Cancer Res 54:487–493

Olivero OA, Semino C, Kassim A, Lopez-Larraza DM, Poirier MC (1995) Preferential binding of cisplatin to mitochondrial DNA of Chinese hamster ovary cells. Mutat Res 346:221–230

Olivero OA, Chang PK, Lopez-Larraza DM, Semino-Mora MC, Poirier MC (1997) Preferential formation and decreased removal of cisplatin-DNA adducts in Chinese hamster ovary cell mitochondrial DNA as compared to nuclear DNA. Mutat Res 391:79–86

O'Neill CF, Hunakova L, Kelland LR (1999) Cellular pharmacology of cis and trans pairs of platinum complexes in cisplatin-sensitive and -resistant human ovarian carcinoma cells. Chem-Biol Interact 123:11–29

O'Rourke TJ, Weiss GR, New P, Burris HA, Rodriguez G, Eckhart J, Hardy J, Kuhn JG, Fields S, Clark GM, Von Hoff DD (1994) Phase I clinical trial of ormaplatin (tetraplatin, NSC 363812). Anti-Cancer Drugs 5:520–526

Ota K, Miyake H, Inuyama I et al. (1991) Clinical evaluation of cis-diammine (glycolato) plat 254-S, a new anticancer platinum complex. Proc NCI-EORTC Symp New Drugs in Cancer Therapy 7:340

Page JD, Husain I, Sancar A, Chaney SG (1990) Effect of the diaminocyclohexane carrier ligand on platinum adduct formation, repair, and lethality. Biochemistry 29:1016–1024

Parker RJ, Eastman A, Bostick-Bruton F, Reed E (1991) Acquired cisplatin resistance in human ovarian cancer cells is associated with enhanced repair of cisplatin-DNA lesions and reduced drug accumulation. J Clin Invest 87:772–777

Peereboom DM, Wood L, Connell C, Spisak J, Smith D, Liebwohl D, Bukowski RM (1997) Phase II trial of oral BMS-182751 (JM-216) in hormone refractory prostate cancer (HRPC). Proc Am Soc Clin Oncol 16:abstract no. 1210

Pendyala L, Creaven PJ (1993) In vitro cytotoxicity, protein binding, red blood cell partitioning, and biotransformation of oxaliplatin. Cancer Res 53:5970–5976

Pendyala L, Kidani Y, Perez R, Wilkes J, Bernacki RJ, Creaven PJ (1995) Cytotoxicity, cellular accumulation and DNA binding of oxaliplatin isomers. Cancer Lett 97:177–184

Pera MF, Friedlos F, Mills J, Roberts JJ (1987) Inherent sensitivity of cultured human embryonal carcinoma cells to adducts of cis-diamminedichloroplatinum(II) on DNA. Cancer Res 47:6810–6813

Perego P, Zunino F, Carenini N, Giuliani F, Spinelli S, Howell SB (1998) Sensitivity to cisplatin and platinum-containing compounds of Schizosaccharomyces pombe rad mutants. Mol Pharmacol 54:213–219

Perego P, Caserini C, Gatti L, Carenini N, Romanelli S, Supino R, Colangelo D, Viano I, Leone R, Spinelli S, Pezzoni G, Manzotti C, Farrell N, Zunino F (1999a) A novel trinuclear platinum complex overcomes cisplatin resistance in an osteosarcoma cell system. Mol Pharmacol 55:528–534

Perego P, Gatti L, Caserini C, Supino R, Colangelo D, Leone R, Spinelli S, Farrell N, Zunino F (1999b) The cellular basis of the efficacy of the trinuclear platinum complex BBR 3464 against cisplatin-resistant cells. J Bioinorg Chem 77:59–64

Pérez JM, Montero EI, González AM, Alvarez-Valdés A, Alonso C, Navarro-Ranninger C (1999) Apoptosis induction and inhibition of H-ras overexpression by novel trans-[PtCl2(isopropylamine)(amine')] complexes. J Inorg Biochem 77:37–42

Perez-Soler R, Khokhar AR (1992) Lipophilic cisplatin analogues entrapped in liposomes: role of intraliposomal drug activation in biological activity. Cancer Res 52:6341–6347

Perez-Soler R, Khokhar AR, Lopez-Berestein G (1987) Treatment of prophylaxis of experimental liver metastases of M5076 reticulosarcoma with cis-bis-neodecanoato-trans-R,R-1,2-diaminocyclohexaneplatinum(II) encapsulated in multilamellar vesicles. Cancer Res 47:6462–6466

Perez-Soler R, Yang LY, Drewinko B, Lauterzstain J, Khokhar AR (1988) Increased cytotoxicity and reversal of resistance to cis-diamminedichloroplatinum(II) with entrapment of cis-bis-neodecanoato-trans-R,R-1,2-diaminocyclohexaneplatinum(II) in multilamellar lipid vesicles. Cancer Res 48:4509–4512

Perez-Soler R, Lopez-Berestein G, Lautersztain J, Al-Baker S, Francis K, Micias-Kiger D, Raber MN, Khokhar AR (1990) Phase I clinical and pharmacological study of liposome-entrapped cis-bis-neodecanoato-trans-R,R-1,2-diaminocyclohexane platinum(II). Cancer Res 50:4254–4259

Perez-Soler R, Siddik ZH, Chase J, Al-Baker S, Khikhar AR, Pazdur R (1993) Phase I clinical and pharmacology study of liposome-entrapped cis-bis-neodecanoato-trans-R,R-1,2-diaminocyclohexane platinum(II) (l-NDDP). Proc Am Soc Clin Oncol 12:153

Perez-Soler R, Han I, Al-Baker S, Khokhar AR (1994) Lipophilic platinum complexes entrapped in liposomes: improved stability and preserved antitumor activity with complexes containing linear alkyl carboxylato leaving groups. Cancer Chemother Pharmacol 33:378–384

Perez-Soler R, Shin DM, Siddik ZH, Murphy WK, Huber M, Lee JS, Khokhar AR, Hong WK (1997) Phase I clinical and pharmacological study of liposome-entrapped NDDP administered intrapleurally in patients with malignant pleural effusions. Clin Cancer Res 3:373–379

Perez-Soler R, Walsh GL, Swisher SG, Shin HJC, Kemp BL, Terry K, Khuri FR, Fossella FV, Hong WK, Lopez-Berestein G, Khokhar AR, Shin DM (1999) Phase II study of a liposome-entrapped cisplatin analog (L-NDDP) administered in patients with malignant pleural mesothelioma. Proc Am Soc Clin Oncol 18:abstract no. 1626

Petersen LN, Mamenta EL, Stevsner T, Chaney SG, Bohr VA (1996) Increased gene specific repair of cisplatin induced interstrand crosslinks in cisplatin resistant cell lines, and studies on carrier ligand specificity. Carcinogenesis 17:2597–2602

Piccart MJ, Green JA, Lacave AJ, Reed N, Vergote I, Benedetti-Panici P, Bonetti A, Kristeller-Tome V, Fernandez CM, Curran D, Van Glabekke M, Lacombe D, Pinel MC, Pecorelli S (2000) Oxaliplatin or paclitaxel in patients with platinum-pretreated advanced ovarian cancer: a randomized phase II study of the European Organization for Research and Treatment of Cancer Gynecology Group. J Clin Oncol 18:1193–1202

Pil PM, Lippard SJ (1992) Specific binding of chromosomal protein HMG1 to DNA damaged by the anticancer drug cisplatin. Science 256:234–237

Plooy ACM, van Dijk M, Berends F, Lohman PHM (1985) Formation and repair of DNA interstrand cross-links in relation to cytotoxicity and unscheduled DNA synthesis induced in control and mutant human cells treated with cis-diamminedichloroplatinum(II). Cancer Res 45:4178–4184

Pratesi G, Perego P, Polizzi D, Righetti SC, Supino R, Caserini C, Manzotti C, Giuliani FC, Pezzoni G, Tognella S, Spinelli S, Farrell N, Zunino F (1999) A novel charged trinuclear platinum complex effective against cisplatin-resistant tumours: hypersensitivity of p53-mutant human tumour xenografts. Br J Cancer 80:1912–1919

Presnov MA, Konovalova AL (1988) Cycloplatam and oxoplatin the new antitumor platinum compounds of the second generation. Arch Geschwulstforsch 58:43–49

Raymond E, Buquet-Fagot C, Djelloul S, Mester J, Cvitkovic E, Allain P, Louvet C, Gespach C (1997) Antitumor activity of oxaliplatin in combination with 5-fluoruracil and the thymidylate synthase inhibitor AG337 in human colon, breast and ovarian cancers. Anti-Cancer Drugs 8:876–885

Raymond E, Lawrence R, Izbicka E, Faivre S, Von Hoff DD (1998) Activity of oxaliplatin against human tumor colony-forming units. Clin Cancer Res 4:1021–1029

Raynaud FI, Mistry P, Donaghue A, Poon GK, Kelland LR, Barnard CFJ, Murrer BA, Harrap KR (1996) Biotransformation of the platinum drug JM216 following oral administration to cancer patients. Cancer Chemother Pharmacol 38:155–162

Raynaud FI, Boxall FE, Goddard PM, Valenti M, Jones M, Murrer BA, Abrams M, Kelland LR (1997) cis-Amminedichloro(2-methylpyridine) platinum(II) (AMD473), a novel sterically hindered platinum complex: in vivo activity, toxicology, and pharmacokinetics in mice. Clin Cancer Res 3:2063–2074

Reardon JT, Vaisman A, Chaney SG, Sancar A (1999) Efficient nucleotide excision repair of cisplatin, oxaliplatin, and bis-aceto-ammine-dichloro-cyclohexylamine-platinum(IV) (JM216) platinum intrastrand DNA diadducts. Cancer Res 59:3968–3971

Redon S, Bombard S, Elizondo-Riojas MA, Chottard JC (2001) Platination of the (T2G4)4 telomeric sequence: a structural and cross-linking study. Biochemistry 40:8463–8470

Reedijk J (1999) Why does cisplatin reach guanine-N7 with competing S-donor ligands available in the cell? Chem Rev 99:2499–2510

Riccardi A, Meco D, Ferlini C, Servidei T, Carelli G, Segni G, Manzotti C, Riccardi R (2001) In vitro and in vivo antitumor activity of the novel trinuclear platinum complex BBR 3464 in neuroblastoma. Cancer Chemother Pharmacol 47:498–504

Richards EH, Hickey E, Weber L, Masters JRW (1996) Effect of overexpression of the small heat shock protein HSP27 on the heat and drug sensitivities of human testis tumor cells. Cancer Res 56:2446–2451

Richon VM, Schulte N, Eastman A (1987) Multiple mechanisms of resistance to cis-diamminedichloroplatinum(II) in murine leukemia L1210 cells. Cancer Res 47:2056–2061

Ridgway H, Hall LM, Speer RJ, Stewart DP, Edwards GR, Hill JM (1976) Antitumor activity of cis-dichloro-(1,2-diaminocyclohexane)platinum(II). J Clin Hematol Oncol 6:11–16

Rixe O, Ortuzar W, Alvarez M, Parker R, Reed E, Paull K, Fojo T (1996) Oxaliplatin, tetraplatin, cisplatin, and carboplatin: spectrum of activity in drug-resistant cell lines and in the cell lines of the National Cancer Institute's anticancer drug screen panel. Biochem Pharmacol 52:1855–1865

Roat RM, Reedijk J (1993) Reaction of mer-trichloro(diethylenetriamine)platinum(IV) chloride, (mer-[Pt(dien)Cl3]Cl), with purine nucleosides and nucleotides results in formation of platinum(II) as well as platinum(IV) complexes. J Inorg Biochem 52:263–274

Roberts JD, Van Houten B, Qu Y, Farrell NP (1989) Interaction of novel bis(platinum) complexes with DNA. Nucleic Acids Res 17:9719–9733

Roberts JD, Peroutka J, Beggiolin G, Manzotti C, Piazzoni L, Farrell N (1999) Comparison of cytotoxicity and cellular accumulation of polynuclear platinum complexes in L1210 murine leukemia cell lines. J Inorg Biochem 77:47–50

Roberts JJ, Fraval HNA (1978) The interaction of antitumor platinum compounds with cellular DNA in cultured cells and animal tissues: relationship to DNA cellular repair processes. Biochimie 60:869–877

Roberts JJ, Friedlos F (1987) Differential toxicity of cis- and trans-diamminedichloroplatinum(II) toward mammalian cells: lack of influence of any difference in the rates of loss of their DNA-bound adducts. Cancer Res 47:31–36

Rogers PM, Boxall F, Allott CP, Stephens T, Kelland LR (2000) In vitro combination studies with the sterically hindered platinum drug, ZD0473, in cisplatin-sensitive and resistant human ovarian carcinoma cell lines. Proc 11th NCI EORTC AACR Symp, abstract no. 352

Rose WC (1997) Combination chemotherapy involving orally administered etoposide and JM-216 in murine tumor models. Cancer Chemother Pharmacol 40:51–56

Rosenberg B (1978) Platinum complexes for the treatment of cancer. Interdiscip Sci Rev 3:134–147

Rosenberg B, Van Camp L, Trosko JE, Mansour VH (1969) Platinum compounds: a new class of potent antitumour agents. Nature 222:385–386

Sargent JM, Elgie AW, Williamson CJ, Taylor CG (1996) Aphidicolin markedly increases the platinum sensitivity of cells from primary ovarian tumours. Br J Cancer 74:1730–1733

Saris CP, van der Vaart PJ, Rietbroek RC, Blommaert FA (1996) In vitro formation of DNA adducts by cisplatin, lobaplatin and oxaliplatin in calf thymus DNA in solution and in cultured human cells. Carcinogenesis 17:2763–2769

Sasaki Y, Tamura T, Eguchi K, Shinkai T, Fujiwara Y, Fukuda M, Ohe Y, Bungo M, Horichi N, Niimi S, Minato K, Nakagawa K, Saijo N (1989) Pharmacokinetics of (glycolato-O,O')-diammine platinum(II), a new platinum derivative, in comparison with cisplatin and carboplatin. Cancer Chemother Pharmacol 23:243–246

Sasaki Y, Amano T, Morita M, Shinkai T, Eguchi K, Tamura T, Ohe Y, Kojima A, Saijo N (1991) Phase I study and pharmacological analysis of cis-diammine(glycolato)platinum (254-S; NSC 375101D) administered by 5-day continous intravenous infusion. Cancer Res 51:1472–1477

Schellens JHM, Groenewegen G, Meerum Terwogt JM, Maliepaard M, Tibben MM, Pluim D, ten Bokkel Huinink WW, Scot M, Welbank H, Blijham GH, Beijnen JH (1999) Phase I and pharmacologic study of Stealth® liposomal encapsulated cisplatin (SPI-077). Proc Am Soc Clin Oncol 18:abstract no. 774

Schilder RJ, LaCreta FP, Perez RP, Johnson SW, Brennan JM, Rogatko A, Nash S, McAleer C, Hamilton TC, Roby D, Young RC, Ozols RF, O'Dwyer PJ (1994) Phase I and pharmacokinetic study of ormaplatin (tetraplatin, NSC 363812) administered on a day 1 and day 8 schedule. Cancer Res 54:709–717

Schlemmer R, Spruß T, Bernhardt G, Schönenberger H (1999) [meso-1,2-Bis(2,6-dichloro-4-hydroxyphenyl)ethylenediamine]dichloroplatinum(II), a compound with a specific activity on hormone-sensitive breast cancers—evidence for a diethylstilbestrol-like mode of action. Arch Pharm Med Chem 332:59–69

Schmidt W, Chaney SG (1993) Role of carrier ligand in platinum resistance of human carcinoma cell lines. Cancer Res 53:799–805

Schmoll HJ, Kohne CH, Papageorgiou E, Luft S, Harstrick A, Bachmann P, Hilgard P, Schubert U (1995) Single agent lobaplatin is active in patients with esophageal squamous cell carcinoma: a phase II evaluation. Proc Am Soc Clin Oncol 14:abstract no. 483

Screnci D, Er HM, Hambley TW, Galettis P, Brouwer W, McKeage MJ (1997) Stereoselective peripheral sensory neurotoxicity of diaminocyclohexane platinum enantiomers related to ormaplatin and oxaliplatin. Br J Cancer 76:502–510

Screnci D, McKeage MJ, Galettis P, Hambley TW, Palmer BD, Baguley BC (2000) Relationships between hydrophobicity, reactivity, accumulation and peripheral nerve toxicity of a series of platinum drugs. Br J Cancer 82:966–972

Servidei T, Ferlini C, Riccardi A, Meco D, Scambia G, Segni G, Manzotti C, Riccardi R (2001) The novel trinuclear platinum complex BBR3464 induces a cellular response different from cisplatin. Eur J Cancer 37:930–938

Sessa C, Capri G, Gianni L, Peccatori F, Grasselli G, Bauer J, Zucchetti M, Viganò L, Gatti A, Minoia C, Liati P, Van den Bosch S, Bernareggi A, Camboni G, Marsoni S (2000) Clinical and pharmacological phase I study with accelerated titration design of a daily times five schedule of BBR3464, a novel cationic triplatinum complex. Ann Oncol 11:977–983

Sharp SY, Rogers PM, Kelland LR (1995) Transport of cisplatin and bis-acetato-ammine-dichlorocyclohexylamine platinum(IV) (JM216) in human ovarian carcinoma cell lines: identification of a plasma membrane protein associated with cisplatin resistance. Clin Cancer Res 1:981–989

Siddik ZH, Al-Baker S, Burditt TL, Khokhar AR (1993) Differential antitumor activity and toxicity of isomeric 1,2-diaminocyclohexane platinum (II) complexes. J Cancer Res Clin Oncol 120:12–16

Siddik ZH, Al-Baker S, Thai G, Khokhar AR (1994a) Antitumor activity of isomeric 1,2-diaminocyclohexane platinum(IV) complexes. J Cancer Res Clin Oncol 120:409–414

Siddik ZH, Al-Baker S, Thai G, Khokhar AR (1994b) Modulatory effects of axial and equatorial ligands on antitumor activities of trans-1R,2R-diaminocyclohexane platinum(IV) complexes. Anti-Cancer Drug Des 9:139–151

Simmonds AP, McDonald EC (1984) Ovarian carcinoma cells in culture: assessment of drug sensitivity by clonogenic assay. Br J Cancer 50:317–326

Skou G, Mangold G, Dexter D, Von Hoff D (1996) Comparison of an oxaliplatin-taxol versus a carboplatin-taxol regimens in the treatment of the MV-522 human lung tumor xenograft. Proc Am Ass Cancer Res 37:375

Slaby S, Galanski M, Metzger SW, Keppler BK (1999) Design of platinum complexes with high affinity for osteosarcoma and bone metastases—structure-activity relationship. Contrib Oncol 54:201–205

Sorenson CM, Eastman A (1988a) Mechanism of cis-diamminedichloroplatinum(II)-induced cytotoxicity: role of G2 arrest and DNA double-strand breaks. Cancer Res 48:4484–4488

Sorenson CM, Eastman A (1988b) Influence of cis-diamminedichloroplatinum(II) on DNA synthesis and cell cycle progression in excision repair proficient and deficient Chinese hamster ovary cells. Cancer Res 48:6703–6707

Sorenson CM, Barry MA, Eastman A (1990) Analysis of events associated with cell cycle arrest at G2 phase and cell death induced by cisplatin. J Natl Cancer Inst 82:749–755

Stevenson JP, Blair IA, Redlinger M, Sun W, Oe T, Koehler M, Roberts DW, Malone MD, O'Dwyer PJ (2001) Pharmacokinetic/pharmacodynamic trial of the novel platinum compound ZD0473 administered as an iv infusion every 21 days. Proc Am Soc Clin Oncol 20:abstract no. 450

Stewart DR, Shannon KF, Callahan EH, Jacob JE, Juthy CL, Buckley RG, Duncan R (1999) Optimization of the synthesis and purification of a polymer-based carboplatin analog. 8th Int Symp Platinum and Other Metal Coordination Compounds in Cancer Chemotherapy:1.09

Takigawa N, Segawa Y, Ueoka H, Kiura K, Tabata M, Shibayama T, Takata I, Miyamoto H, Eguchi K, Harada M (2000) Combination of nedaplatin and vindesine for treatment of relapsed or refractory non-small-cell lung cancer. Cancer Chemother Pharmacol 46:272–278

Talman EG, Brüning W, Reedijk J, Spek AL, Veldman N (1997) Crystal and molecular structures of asymmetric cis- and trans-platinum(II/IV) compounds and their reactions with DNA fragments. Inorg Chem 36:854–861

Teuben JM, Reedijk J (2000) Reaction of DNA oligonucleotides with [Pt(dien)GSMe]2+ (GSMe = S-methylated glutathione) and cis-[Pt(NH3)2(GSMe)2]2+: evidence of oligonucleotide platination via sulfur-coordinated platinum intermediates. J Biol Inorg Chem 5:463–468

Teuben JM, Rodriguez I, Zubiri M, Reedijk J (2000) Glutathione readily replaces the thioether on platinum in the reaction with [Pt(dien)(GSMe)]2+ (GSMe = S-methylated glutathione); a model study for cisplatin-protein interactions. J Chem Soc Dalton Trans 2000:369–372

Trask C, Silverstone A, Ash CM, Earl H, Irwin C, Bakker A, Tobias JS, Souhami RL (1991) A randomized trial of carboplatin versus iproplatin in untreated advanced ovarian cancer. J Clin Oncol 9:1131–1137

Treat J, Huang C, Damjanou N, Jahanzeb M, Edelman M, Cosaert J, Koehler M (2001) Phase II monotherapy trial of ZD0473 as second-line therapy in non-small cell lung cancer. Proc Am Soc Clin Oncol 20:abstract no. 2808

Trigo JM, Beale P, Judson IR, Raynaud F, Rees C, Milan D, Wolf L, Walker R, Hanwell J, Giandomenico C (1999) Phase I and pharmacokinetic study of cis-amminedichloro(2-methylpyridine)platinum(II) (ZD0473), a novel sterically hindered platinum complex, in patients with advanced solid malignancies. Proc Am Soc Clin Oncol 18:abstract no. 648

Trimmer EE, Zamble DB, Lippard SJ, Essigmann JM (1998) Human testis-determining factor SRY binds to the major DNA adduct of cisplatin and a putative target sequence with comparable affinities. Biochemistry 37:352–362

Uchida N, Takeda Y, Hojo K, Maekawa R, Sugita K, Yoshioka T (1998) Sequence-dependent antitumour efficacy of combination chemotherapy of nedaplatin, a novel platinum complex, with 5-fluorouracil in an in vivo murine tumour model. Eur J Cancer 34:1796–1801

Vadiei K, Siddik ZH, Khokhar AR, Al-Baker S, Sampedro F, Perez-Soler R (1992) Pharmacokinetics of liposome-entrapped cis-bis-neodecanoato-trans-R,R-1,2-diaminocyclohexane platinum(II) and cisplatin given i.v. and i.p. in the rat. Cancer Chemother Pharmacol 30:365–369

Vaisman A, Varchenko M, Umar A, Kunkel TA, Risinger JI, Barrett JC, Hamilton TC, Chaney SG (1998) The role of hMLH1, hMSH3, and hMSH6 defects in cisplatin and oxaliplatin resistance: correlation with replicative bypass of platinum-DNA adducts. Cancer Res 58:3579–3585

Van Beusichem M, Farrell N (1992) Activation of the trans geometry in platinum antitumor complexes. Synthesis, characterization and biological activity of complexes with planar ligands pyridine, N-methylimidazole, thiazole and quinoline. The crystal and molecular structure of trans-dichlorobis(thiazole)platinum(II). Inorg Chem 31:634–639

Van Boom SSGE, Chen BW, Teuben JM, Reedijk J (1999) Platinumthioether bonds can be reverted by guanineN7 bonds in Pt(dien)2+ model adducts. Inorg Chem 38:1450–1455

Van der Vaart PJM, Klaren HM, Hofland I, Begg AC (1997) Oral platinum analog JM216, a radiosensitizer in oxic murine cells. Int J Radiat Biol 72:675–683

Van der Vijgh WJF (1991) Clinical pharmacokinetics of carboplatin. Clin Pharmacokinet 21:242–261

Veal GJ, Griffin MJ, Price E, Parry A, Dick GS, Little MA, Yule SM, Morland B, Estlin EJ, Hale JP, Pearson ADJ, Welbank H, Boddy AV (2001) A phase I study in paediatric patients to evaluate the safety and pharmacokinetics of SPI-77, a liposome encapsulated formulation of cisplatin. Br J Cancer 84:1029–1035

Vermorken JB, van der Vijgh WJF, Klein I, Hart AAM, Gall HE, Pinedo HM (1984) Pharmacokinetics of free and total platinum species after short-term infusion of cisplatin. Cancer Treat Rep 68:505–513

Verschraegen CF, Mansfield PF, Feig BW, Steger M, Wu QP, MacLean D, Siddik Z, Perez-Soler R, Hu W, Kudelka AP, Kavanagh JJ, Khokhar AR (1999) Phase I study of an intraperitoneal liposomal cisplatin analog L-NDDP for treatment of peritoneal carcinomatosis. Proc Am Soc Clin Oncol 18:abstract no. 1405

Vichi P, Coin F, Renaud JP, Vermeulen W, Hoeijmakers JHJ, Moras D, Egly JM (1997) Cisplatin- and UV-damaged DNA lure the basal transcription factor TFIID/TBP. EMBO J 16:7444–7456

Voegeli R, Schumacher W, Engel J, Respondek J, Hilgard P (1990) D-19466, a new cyclobutane-platinum complex with antitumor activity. J Cancer Res Clin Oncol 116:439–442

Vollano JF, Al-Baker S, Dabrowiak JC, Schurig JE (1987) Comparative antitumor studies on platinum(II) and platinum(IV) complexes containing 1,2-diaminocyclohexane. J Med Chem 30:716–719

Wang K, Lu J, Li R (1996) The events that occur when cisplatin encounters cells. Coord Chem Rev 151:53–88

Welink J, Pechstein B, van der Vijgh WJF (1996) Determination of the two diastereoisomers of lobaplatin (D-19466) in plasma ultrafiltrate of cancer patients with a normal or an impaired kidney or liver function by high-performance liquid chromatography with ultraviolet detection. J Chromatography B 675:107–111

Wolf CR, Hayward IP, Lawrie SS, Buckton K, McIntyre MA, Adams DJ, Lewis AD, Scott ARR, Smyth JF (1987) Cellular heterogeneity and drug resistance in two ovarian adenocarcinoma cell lines derived from a single patient. Int J Cancer 39:695–702

Wong E, Giandomenico CM (1999) Current status of platinum-based antitumor drugs. Chem Rev 99:2451–2466

Woynarowski JM, Chapman WG, Napier C, Herzig MCS, Juniewicz P (1998) Sequence- and region-specificity of oxaliplatin adducts in naked and cellular DNA. Mol Pharmacol 54:770–777

Woynarowski JM, Faivre S, Herzig MCS, Arnett B, Chapman WG, Trevino AV, Raymond E, Chaney SG, Vaisman A, Varchenko M, Juniewicz PE (2000) Oxaliplatin-induced damage of cellular DNA. Mol Pharmacol 58:920–927

Yamada H, Uchida N, Maekawa R, Yoshioka T (2001) Sequence-dependent antitumor efficacy of combination chemotherapy with nedaplatin, a newly developed platinum, and paclitaxel. Cancer Lett 172:17–25

Yang LY, Li L, McLean D, Jiang H, Khokhar AR, Perez-Soler R (1998) Inefficient cellular repair of DNA adducts induced by cis-bis-neodecanoato-trans-R,R-1,2-diamminocyclohexane platinum(II). 10th NCI-EORTC Symp New Drugs in Cancer Therapy, p 137

Yoshida M, Khokhar AR, Siddik ZH (1994) Axial ligands and alicyclic ring size modulate the activity and biochemical pharmacology of ammine/cycloalkylamine-platinum(IV) complexes in tumor cells resistant to cis-diamminedichloroplatinum(II) or trans-1R,2R-1S,2S-diaminocyclohexanetetrachloroplatinum(IV). Cancer Res 54:4691–4697

Yoshioka T, Gamoh M, Shineha R, Ishibashi S, Shibata H, Suzuki T, Murakawa Y, Kato S, Shimodaira H, Kato S, Ishioka C, Kanamaru R (1999) A new combination chemotherapy with cis-diammine-glycolatoplatinum (Nedaplatin) and 5-fluorouracil for advanced esophageal cancers. Internal Med 38:844–848

Zákovská A, Nováková O, Balcarová Z, Bierbach U, Farrell N, Brabec V (1998) DNA interactions of antitumor trans-[PtCl2(NH3)(quinoline)]. Eur J Biochem 254:547–557

Zaludová R, Zákovská A, Kaspárkova J, Balcarová Z, Vrána O, Coluccia M, Natile G, Brabec V (1997) DNA modifications by antitumor trans-[PtCl2(E-iminoether)2]. Mol Pharmacol 52:354–361

Zamble DB, Lippard SJ (1999) The response of cellular proteins to cisplatin-damaged DNA. In: Lippert B (ed) Cisplatin. Chemistry and biochemistry of a leading anticancer drug. Verlag Helvetica Chimica Acta, Zürich; Wiley-VCH, Weinheim, pp 73–110

Zamble DB, Mu D, Reardon JT, Sancar A, Lippard SJ (1996) Repair of cisplatin-DNA adducts by the mammalian excision nuclease. Biochemistry 35:10004–10013

Zamboni WC, Gervais AC, Egorin MJ, Schellens JHM, Delauter BJ, Zuhowski EG, Pluim D, Hamburger DR, Working PK, Colbern G, Eiseman JL (2000) Disposition of platinum (Pt) in B16 murine melanoma tumours after administration of cisplatin or pegylated liposomal-cisplatin formulations (SPI-077 and SPI-077 B103). Proc 11th NCI EORTC AACR Symp:abstract no. 132

Zang DY, Lee KH, Lee JS, Lee JH, Kim WK, Kim SH, Kim WD, Kim DS, Kim JH, Kim BS, Cho YB, Kim DK, Kim KH (1999) Phase II trial of a novel platinum analog, SKI 2053R, in patients with previously untreated extensive-stage small-cell lung cancer. Am J Clin Oncol 22:495–498

Zeghari-Squalli N, Raymond E, Cvitkovic E, Goldwasser F (1999) Cellular pharmacology of the combination of the DNA topoisomerase I inhibitor SN-38 and the diaminocyclohexane platinum derivative oxaliplatin. Clin Cancer Res 5:1189–1196

Zehnulova J, Kasparkova J, Farrell N, Brabec V (2001) Conformation, recognition by high mobility group domain proteins, and nucleotide excision repair of DNA intrastrand cross-links of novel antitumor trinuclear platinum complex BBR3464. J Biol Chem 276:22191–22199

Zhang JG, Lindup WE (1993) Role of mitochondria in cisplatin-induced oxidative damage exhibited by renal cortical slices. Biochem Pharmacol 45:2215–2222

Zhang JG, Lindup WE (1994) Cisplatin nephrotoxicity: decreases in mitochondrial protein sulfhydryl concentration and calcium uptake by mitochondria from rat renal cortical slices. Biochem Pharmacol 47:1127–1135

Zhen W, Link CJ Jr, O'Connor PM, Reed E, Parker R, Howell SB, Bohr VA (1992) Increased gene-specific repair of cisplatin interstrand cross-links in cisplatin-resistant human ovarian cancer cell lines. Mol Cell Biol 12:3689–3698

Zimmermann W, Galanski M, Keppler BK (1999) New antitumor platinum-albumine conjugates for targeted drug delivery. 8th Int Symp Platinum and Other Metal Coordination Compounds in Cancer Chemotherapy:1.08

Zou Y, Van Houten B, Farrell N (1993) Ligand effects in platinum binding to DNA. A comparison of DNA binding properties for cis- and trans-[PtCl2(amine)2] (amine = NH3, pyridine). Biochemistry 32:9632–9638

Zou Y, Van Houten B, Farrell N (1994) Sequence specificity of DNA-DNA interstrand cross-link formation by cisplatin and dinuclear platinum complexes. Biochemistry 33:5404–5410

Rev Physiol Biochem Pharmacol (2003) 146:55–94
DOI 10.1007/s10254-002-0002-9

H.-G. Koch · M. Moser · M. Müller

Signal recognition particle-dependent protein targeting, universal to all kingdoms of life

Published online: 17 December 2002

Abstract The signal recognition particle (SRP) and its membrane-bound receptor represent a ubiquitous protein-targeting device utilized by organisms as different as bacteria and humans, archaea and plants. The unifying concept of SRP-dependent protein targeting is that SRP binds to signal sequences of newly synthesized proteins as they emerge from the ribosome. In eukaryotes this interaction arrests or retards translation elongation until SRP targets the ribosome-nascent chain complexes via the SRP receptor to the translocation channel. Such channels are present in the endoplasmic reticulum of eukaryotic cells, the thylakoids of chloroplasts, or the plasma membrane of prokaryotes. The minimal functional unit of SRP consists of a signal sequence-recognizing protein and a small RNA. The as yet most complex version is the mammalian SRP whose RNA, together with six proteinaceous subunits, undergo an intricate assembly process. The preferential substrates of SRP possess especially hydrophobic signal sequences. Interactions between SRP and its receptor, the ribosome, the signal sequence, and the target membrane are regulated by GTP hydrolysis. SRP-dependent protein targeting in bacteria and chloroplasts slightly deviate from the canonical mechanism found in eukaryotes. Pro- and eukaryotic cells harbour regulatory mechanisms to prevent a malfunction of the SRP pathway.

Abbreviations *EF-G* Elongation factor G · *ER* Endoplasmic reticulum · *Ffh* Fifty-four homologue · *GAP* GTPase activating protein · *GEF* Guanine nucleotide exchange factor · *HTH* Helix turn helix · *LHCP* Light-harvesting chlorophyll binding protein · *NAC* Nascent chain-associated complex · *ORF* Open reading frame · *PE* Phosphatidyl ethanolamine · *Pmf* Proton motive force · *RNC* Ribosome-nascent chain complexes · *RND* Resistance-nodulation cell division · *SRP* Signal recognition particle · *SR* SRP receptor · *TRAM* Translocating chain-associated membrane protein

H.-G. Koch (✉) · M. Moser · M. Müller
Institut für Biochemie und Molekularbiologie, Universität Freiburg,
Hermann-Herder-Straße 7, 79104 Freiburg, Germany
e-mail: hans-georg.koch@biochemie.uni-freiburg.de · Tel.: +49-761-2035250 ·
Fax: +49-761-2035274

Introduction

Protein transport events occurring at the endoplasmic reticulum (ER) of eukaryotic cells and the cytoplasmic membrane of prokaryotic organisms share many similarities. Resident proteins of both membranes span the lipid bilayer once or several times by α-helical stretches and their integration is usually mediated by uncleaved signal–anchor sequences. Proteins that are translocated across either membrane, collectively also termed secretory proteins, harbour cleavable N-terminal signal sequences. Prokaryotic and eukaryotic signal sequences have the same modular structure and are functionally exchangeable. Integration of membrane proteins and translocation of secretory proteins basically occur at the same sites (pores) within each membrane. In both types of membranes, these pores are composed of homologous components forming the Sec translocons. Parts of the Sec translocons are found populated by ribosomes, the membrane-bound ribosomes. Bacterial membrane and eukaryotic secretory proteins are targeted to the Sec translocons by the same molecular mechanism involving signal recognition particle (SRP) and its receptor (SRP receptor, SR).

Structure and assembly of the SRP

The functional core of SRP

The functional core of this ribonucleoprotein complex consists of the signal sequence binding subunit (SRP54 in eukaryotes and Ffh in prokaryotes) and the SRP RNA molecule (see Fig. 1). This core is conserved in all organisms, with the intriguing exception of chloroplasts, where the SRP lacks the RNA subunit.

Based on phylogenetic analyses, the SRP RNA can be divided into four different structural domains (I–IV) (Poritz et al. 1988; Larsen and Zwieb 1991), using the 300-nucleotide-long 7S RNA of the human SRP as reference (Fig. 1). Domain I comprises the 5' end of the molecule. Domain II is the main stem ending in two stem-loop structures which define domains III and IV. These four domains can be further subdivided into eight helices, one of which (helix 1) seems to be absent in eukaryotes. Helices 2–4 reside within or close to domain I and form the so-called *Alu* sequence of the 7S RNA. Helix 5 is located within domain II, helix 6 within domain III and helix 8 within domain IV. Helix 7 is located at the interface between domain III and IV and seems to be unique for the eukaryotic 7S RNA.

Of the four domains, only domain IV is conserved in all SRP-RNAs and this domain probably defines the minimal structural unit required for SRP function. The 114 nucleotide-long-4.5S RNA of the gram-negative bacterium *Escherichia coli* is homologous to the 7S RNA of eukaryotes, and basically consists only of domain IV (Poritz et al. 1988). The SRP RNAs of archaea and gram-positive bacteria more closely resemble the eukaryotic counterpart, but lack helix 7 at the interface of domains III and IV. In both prokaryotic groups, the 5' end and 3' end of the RNA molecule form the additional helix I (Kaine 1990), which is absent from eukaryotes.

SRP54/Ffh, the protein component of the functional core of SRP, comprises three domains, an amino-terminal N domain, a central GTPase domain (G domain) and a C-terminal M domain (Fig. 2A). The N domain is a four helix bundle (Fig. 2B) and forms together with the G domain a distinct structural element referred to as the NG domain (Freymann

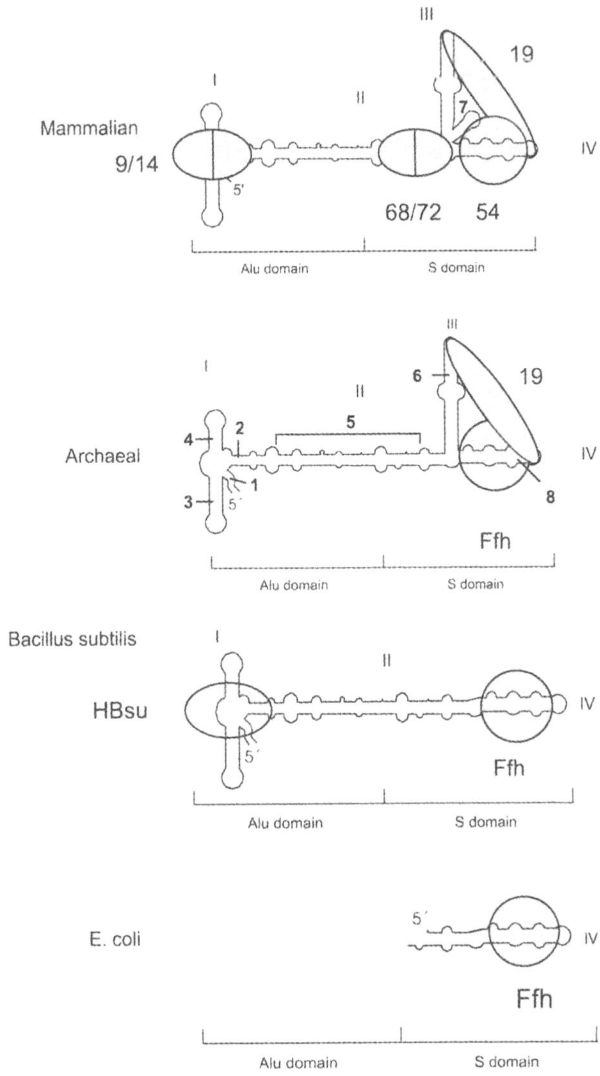

Fig. 1 The variable composition of SRP (modified from Eichler and Moll 2001). Shown are structure models of the SRP RNAs and the protein subunits (*in grey*). *Roman numerals* denote RNA domains and *small Arabic numerals* represent helices within the RNA molecules. *Big Arabic numerals* indicate the molecular masses of proteinaceous subunits. Most SRPs can be subdivided into an Alu domain and an S domain. Ffh is the SRP54 homologue of prokaryotic organisms. HBsu has been first characterized as a histone-like protein of *Bacillus subtilis*

et al. 1997). The G domain shows all the structural features of the Ras type superfamily of GTPases and contains four conserved sequence motives, which are involved in nucleotide binding (Bourne et al. 1991). The G domain has a β/α fold (Fig. 2B) with a five-stranded β sheet surrounded by α helices. A characteristic feature of the SRP subfamily of small GTPases is a unique insertion sequence within the G domain, the so-called I-box, or insertion box, element, which has an $\alpha\beta\alpha$ structure (Freymann et al. 1997). This insertion is also present in the G domain of the SRP receptor and has been implicated in both nucleotide exchange as well as in SRP–SR interactions (Freymann et al. 1997; Moser et al. 1997; see "The GTPase cycle during protein targeting"). The methionine-rich M domain is responsible for the interaction of SRP54/Ffh with both the SRP RNA (Römisch et al. 1990) and with signal sequences (Bernstein et al. 1989). RNA binding is mediated by a conserved helix-turn-helix (HTH) motif of the M domain which binds to the domain IV of the

SRP RNA (Kurita et al. 1996). The recently solved crystal structure of the complex between domain IV of the RNA and the M domain (Batey et al. 2000) indicates that the SRP RNA functions not only as scaffolding but forms, together with the flexible methionine side chains of the M domain the signal sequence binding site (see "The structural basis of signal sequence recognition").

SRP: variation on a theme

Although the basic concept of SRP-dependent protein targeting is conserved in all organisms, there is a surprising variation in the actual composition of the SRP in different organisms (Fig. 1).

The SRP of higher eukaryotes consists of six proteins (SRP9; SRP14; SRP19; SRP54; SRP68 and SRP72; named according to their apparent molecular masses) which are bound to the 7S RNA. It can be divided into two distinct domains of different functions (Gundelfinger et al. 1983; Siegel and Walter 1988a). The *Alu domain* comprises the subunits SRP9 and SRP14, bound as a heterodimer to the *Alu* sequence of the 7S RNA (Fig. 1). This domain is responsible for the retardation of translation, the so-called elongation arrest. The *S domain* contains the subunits SRP19, SRP54, SRP68 and SRP72 and forms the functional unit required for signal recognition and protein transport. SRP19 has been implicated in SRP assembly in both eukaryotes and archaea (see "Assembly of the SRP"). The SRP68 and SRP72 subunits are bound to the 7S RNA as a heterodimeric complex and are probably involved in the targeting function of the SRP by interacting with the SRP receptor (Siegel and Walter 1988a). It has also been proposed that SRP68/72 functions as guanine nucleotide dissociation stimulator (Althoff et al. 1994). The yeast Alu domain differs slightly from the mammalian domain. The 7S RNA lacks both helix 1 and helix 3 (cf. Fig. 1) and contains an SRP14 homodimer instead of the SRP9/14 heterodimer (Strub et al. 1999). The eukaryotic SRP forms a rod-shaped structure, 5–6 nm wide and 23–24 nm long, as revealed by electron microscopy (Andrews et al. 1985).

Although chloroplasts are of prokaryotic origin, the chloroplast SRP (cpSRP) differs markedly from prokaryotic or eukaryotic SRPs. The most prominent difference is the lack of an RNA component, which is exceptional among the known SRPs. Nevertheless, the RNA-binding HTH motif in the M domain (see above) is also conserved in cpSRP54 and *SRP RNA* genes were detected in some algal plastid genomes (Packer and Howe 2001). The second important difference is that the cpSRP54 subunit associates with a 43-kDa subunit (cpSRP43), which has no homologue in any other eukaryotic or prokaryotic SRP (Schuenemann et al. 1998). In chloroplasts, two pools of substrates are targeted to the thylakoids in an SRP-dependent manner, namely translation products of chloroplast ribosomes as well as cytosolically synthesized proteins after they have been transported into the stroma of the organelle. Accordingly, cpSRP54 is detected in a ribosome-bound and a soluble form, whereas cpSRP43 obviously exists only in a soluble form. This suggests that chloroplasts harbour two structurally and functionally different pools of cpSRP (Eichacker and Henry 2001), a ribosome-associated cpSRP54 for cotranslational targeting and a soluble cpSRP54–cpSRP43 complex for posttranslational targeting to the thylakoid (see also "SRP-dependent protein targeting in chloroplasts").

The exact stoichiometry of both subunits within the cpSRP is unclear. A complex consisting of one cpSRP54 and two cpSRP43 has been suggested (Nilsson et al. 1999; Groves et al. 2001). In vitro studies using immobilized peptide libraries have demonstrated that the positively charged C-terminal extension of cpSRP54 is essential for the interaction with cpSRP43 (Groves et al. 2001; Eichacker and Henry 2001). This extension is not conserved in other SRP GTPases and is presumably specific for chloroplast SRP (Eichacker and Henry 2001). Sequence analysis indicated that cpSRP43 from *Arabidopsis thaliana* contains typical motifs involved in protein–protein interaction (Klimyuk et al. 1999). Ankyrin repeats found at the N-terminus mediate interaction of cpSRP43 with its substrate LHCP, whereas chromatin-binding domains (chromo domains) located at the C-terminus are necessary for binding to the cpSRP54 subunit (Jonas-Straube et al. 2001). In contrast to chloroplasts, no proteins homologous to SRP subunits or to the SRP receptor are detectable in mitochondria.

Although homologues of SRP subunits have been detected in archaea, little is known about their cellular functions. SRP-like complexes have been characterized in different archaeal organisms such as *Archaeglobus fulgidus* (Moll et al. 1999; Bhuiyan et al. 2000) and *Pyrococcus furiosus* (Maeshima et al. 2001). The first crystal structure of an archaeal Ffh has recently been solved from *Acidianus ambivalens* (Moll et al. 1999; Montoya et al. 2000), indicating a structural organization very similar to bacteria and eukaryotes. The N, G, and M domains are conserved in the archaeal Ffh (Fig. 2A) and in vitro studies using recombinant Ffh from *Archaeglobus fulgidus* have revealed that archaeal SRP is able to interact with the signal sequence of preprolactin, a substrate of the eukaryotic SRP (Bhuiyan et al. 2000). In contrast to the bacterial SRP, the SRP19 subunit is present in all archaea characterized and sequenced so far (Eichler and Moll 2001; Fig. 1) and like in eukaryotes, SRP19 seems to be required for the assembly of the archaeal SRP (Diener and Wilson 2000).

The SRP of gram-positive bacteria has been mainly characterized in *Bacillus subtilis*. The *B. subtilis* SRP consists of an RNA (scRNA) (Struck et al. 1989) and two distinct proteins (Fig. 1): Ffh, the homologue of the eukaryotic SRP54 and the *E. coli* Ffh, and HBsu, a small histone-like protein (Honda et al. 1993; Nakamura et al. 1999b). The primary sequence of *B. subtilis* Ffh is rather similar to the *E. coli* Ffh sequence and it has the typical tripartite domain structure, with the M domain responsible for signal sequence binding and RNA interaction (Kurita et al. 1996; Bunai et al. 1996; Takamatsu et al. 1997). The 271-nucleotide-long scRNA of *B. subtilis* shows a secondary structure very similar to the eukaryotic 7S RNA (Fig. 1) and consists of three domains which correspond to domains I, II and IV of the 7S RNA (Struck et al. 1989; Nakamura et al. 1992). Thus, in contrast to the 4.5 S RNA of gram-negative bacteria, the *B. subtilis* scRNA contains an Alu-domain. Ffh is essential in both developmental stages of *B. subtilis*, the vegetative growth and also spore formation, which is accompanied by a high level of secretion. In *Streptococcus mutans*, a cariogenic bacterium present in dental plaques, the SRP pathway is involved in mediating acid tolerance (Gutierrez et al. 1999; Kremer et al. 2001). Surprisingly, the insertional inactivation of Ffh in *Streptococcus mutans* does not impair cell viability (Kremer et al. 2001).

HBsu interacts with the Alu domain of scRNA and it is suggested that HBsu folds into a supersecondary structure similar to the SRP9/14 heterodimer of eukaryotic SRP (Nakamura et al. 1999b). A speculative model is that HBsu might serve as a functional analogue of the eukaryotic SRP9/14 heterodimer, which is involved in translation arrest activity of SRP in eukaryotic cells. In *B. subtilis*, HBsu is essential for vegetative growth, whereas a truncated scRNA, lacking its Alu domain, is not. The Alu domain is, however, required for spore formation (Nakamura et al. 1999b).

The SRP of gram-negative bacteria is best characterized in *E. coli* and besides mammals and yeast, *E. coli* is the organism in which SRP-dependent protein targeting has thus far been investigated in most detail. Up to now, the *E. coli* SRP is the simplest version of

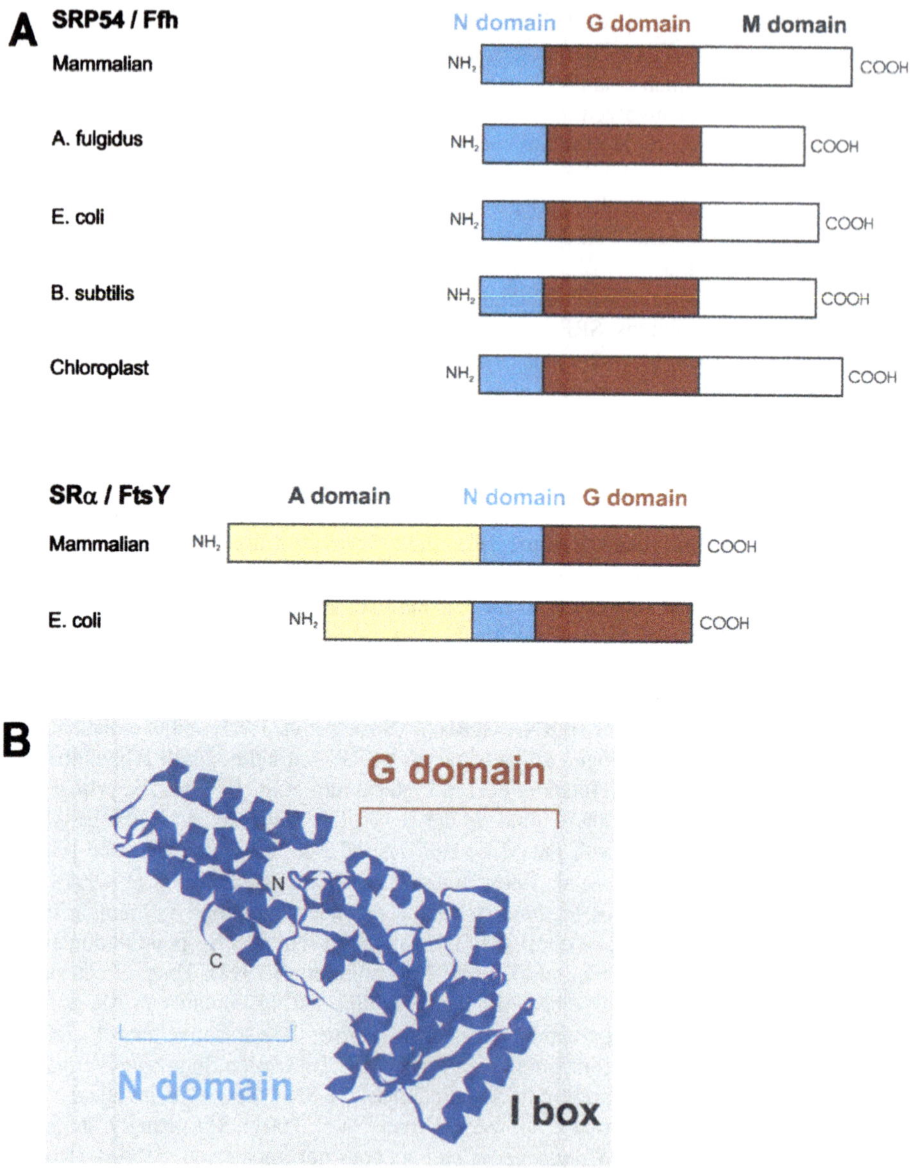

Fig. 2A, B Structural composition of SRP54/Ffh and SRα/FtsY. **A** Domain structure of proteins homologous to SRP54 and SRα. The N domain modulates the GTPase activity of the ras-like G domain. The M domain harbours the hydrophobic pocket required for substrate binding and interacts via a helix-turn-helix motif with the SRP RNA. The acidic A domain is involved in binding the SR to membranes. **B** Structure of the NG domain of Ffh from *Thermus aquaticus* (Freymann et al. 1997) taken from the Signal Recognition Particle Database (Gorodkin et al. 2001). The N domain is a four-helix bundle and forms, together with the G domain, a distinct structural element referred to as the NG domain. The G domain has a β/α fold and shows all the structural features of the ras type superfamily of GTPases. A characteristic feature of the SRP subfamily of small GTPases is a unique insertion sequence within the G domain, the so-called I-box, which has an $\alpha\beta\alpha$ structure

the SRP–ribonucleoprotein complex, consisting only of the core components Ffh, the homologue to SRP54 and 4.5S RNA, which corresponds mainly to domain IV of the eukaryotic 7S RNA (Fig. 1).

Assembly of the SRP

The assembly of the S domain of the eukaryotic SRP is a cooperative multistep process that takes place partially in the nucleolus and partially in the cytoplasm (Walter and Blobel 1983; Politz et al. 2000; Grosshans et al. 2001). In vitro reconstitution experiments have demonstrated that binding of SRP54 to the 7S RNA requires SRP19 (Walter and Blobel 1983). Using nuclease mapping and site-directed mutagenesis, two tetranucleotide loops within the RNA (Fig. 1, tips of helix 6 and helix 8) have been identified as primary binding sites for SRP19 (Siegel and Walter 1988b; Zwieb 1992). The archaeal SRP54, unlike the eukaryotic SRP54, has a significant affinity for the 7S RNA even in the absence of SRP19. This condition allowed studying the influence of SRP19 on the association of SRP54 with the 7S RNA. Chemical footprinting analyses with recombinant components from the archaeon *Archaeoglobus fulgidus* have thus suggested that binding of SRP19 to the tips of domains 3 and 4 causes a conformational change in the asymmetric bulge of domain 4, which then favours binding of SRP54 (Diener and Wilson 2000). Kinetic measurements using human components indicated that binding of SRP19 to the 7S RNA involves the formation of an unstable encounter complex preceding the assembly of the native complex, which then exhibits high affinity towards SRP54 (Rose and Weeks 2001). The crystal structure of an intermediate in the binary complex formation has recently been solved, indicating an initial contact between SRP19 and domain III (Wild et al. 2001). This is consistent with sequence homology searches, which suggest that SRP19 orthologues are found only in organisms whose SRP RNAs contain domain III, e.g. no SRP19 orthologues are found in bacterial species.

The SRP68/72 heterodimer formation occurs only in the presence of 7S RNA and occurs as a multistep process, in which binding of SRP68 to the 7S RNA seems to be a prerequisite for a stable SRP72–SRP68 interaction (Lütcke et al. 1993). The binding of SRP68 to the 7S RNA requires its N-terminal domain, while its C-terminal domain associates with the C-terminal domain of SRP72. Thus, SRP68 probably links the 7S RNA and SRP72 (Lütcke et al. 1993). During programmed cell death a 6-kDa C-terminal fragment of SRP72 is cleaved off, but this does not interfere with the protein transport activity (Utz et al. 1998).

The hierarchical assembly of the Alu domain of SRP starts with a heterodimerization of SRP9 and SRP14, which is a prerequisite for RNA binding (Strub and Walter 1990). Both proteins are structurally related and have a $\alpha\beta\beta\beta\alpha$-fold. The Alu domain in yeast contains an SRP14 homodimer instead of an SRP9/14 heterodimer (Strub et al. 1999). Based on structural as well as biochemical data, the primary recognition event is binding of SRP9/14 to the 5' end of the RNA. This initial contact induces the 3' end of the RNA to flip around and to bind to the 5' end-SRP9/14 complex, changing the open complex of the ribonucleoprotein to its final compact conformation (Weichenrieder et al. 2000, 2001).

Most of our knowledge on the assembly of the SRP is based on in vitro data with purified components. The assembly process in vivo, however, is presumably much more complex and has only lately been addressed. A recent study with fluorescent-labelled SRP RNA has demonstrated that upon microinjection into the nucleus of mammalian cells the RNA very rapidly localizes to the nucleoli. After this initial nucleolar localization, the

RNA leaves the nucleus and enters the cytoplasm (Jacobson and Pederson 1998). The SRP RNA is processed at the 3' end in mammalian cells and this step presumably requires an intact Alu domain, i.e. binding of the SRP9/14 heterodimer to the RNA (Chen et al. 1998). An intact Alu domain is also required for export of the SRP RNA into the cytoplasm (He et al. 1994; Chen et al. 1998). Thus, after dimerization of SRP9/14 in the cytoplasm, the heterodimer has to be transported into the nucleus to interact with the RNA. Indeed, a nucleolar localization not only of SRP14 but also of the SRP19, SRP68, and SRP72 orthologues from yeast and mammalian cells has recently been described (Politz et al. 2000; Grosshans et al. 2001). It seems therefore plausible that the assembly of the SRP, like the assembly of the ribosome, occurs to a large extent in the nucleolus. This pre-SRP complex, consisting most likely of SRP9/14, SRP19, SRP68/72 and 7S RNA in mammals and the corresponding homologues in yeast, then has to leave the nucleus, a process which in yeast requires the function of exportins and nucleoporins (Grosshans et al. 2001). The SRP54 subunit does not seem to be part of this pre-SRP complex, because it is mainly localized in the cytoplasm. Binding of SRP54 to the pre-SRP complex in the cytoplasm would then finally complete the assembly process of the SRP.

The SRP assembly in *E. coli* is naturally less complex, as from all that is known it involves only 4.5S RNA and Ffh. The physical interaction between both components had been initially demonstrated by the ability of anti-Ffh antibodies to immunoprecipitate the complex of Ffh and 4.5S RNA (Poritz et al. 1990; Ribes et al. 1990). The nucleotide sequence of the 4.5S RNA involved in Ffh binding is also conserved in the SRP54-binding nucleotide sequence of the eukaryotic 7S RNA (Wood et al. 1992). The in vivo stability of Ffh has been shown to be dependent on 4.5S RNA (Jensen and Pedersen 1994) and in vitro proteolysis experiments suggest that the structures of both M and NG domains of Ffh change upon 4.5S RNA binding (Zheng and Gierasch 1997). These changes are also accompanied by conformational changes within the 4.5S RNA (Lentzen et al. 1994). Recent structural data have provided evidence that it is not merely the stabilizing effect which led to the conservation of RNA as a component of the SRP, but that 4.5S RNA is probably directly involved in signal sequence recognition (Batey et al. 2000). Furthermore, the Ffh–FtsY interaction is thought to be modulated by 4.5S RNA (Peluso et al. 2000; Jagath et al. 2001; see "The GTPase cycle during protein targeting").

Substrates and substrate recognition

Recognition sequences

In practice, there are two ways to investigate whether or not a particular protein is a substrate of the SRP pathway. A direct molecular interaction is demonstrated by chemical cross-linking, which is usually performed on ribosome nascent chain complexes (RNCs), because with few exceptions only (see "Cotranslational protein targeting by SRP and elongation arrest"), substrate recognition by SRP occurs cotranslationally. Alternatively, the influence of SRP on the membrane transport of a particular protein is tested either by use of cell-free systems containing membrane vesicles or by an in vivo approach employing SRP mutant cells. Initially all secretory and integral membrane proteins analysed for transport into mammalian microsomes turned out to require SRP and SRP receptor. This was consistent with the demonstration of a cotranslational transport applicable to all these substrates. In parallel, their SRP-dependence was confirmed by cross-links between RNCs and SRP54 (Krieg et al. 1986; Wiedmann et al. 1987).

The situation turned into a more intricate scenario when the lower eukaryote yeast became amenable to similar experimental approaches. Secretory proteins were found to be translocated posttranslationally into yeast microsomes (Rothblatt and Meyer 1986), a process that required ATP (Waters and Blobel 1986) and cytosolic Hsp70s (Chirico et al., 1988; Deshaies et al. 1988). Although this posttranslational mode of translocation applies only to a subset of secretory proteins from yeast (Hansen and Walter 1988), the finding that yeast can grow in the absence of a functional SRP (Hann and Walter 1991; Stirling and Hewitt 1992) is consistent with a number of secretory proteins being translocated into the yeast ER independently of SRP.

Further analyses demonstrated that the selection between the SRP-dependent and SRP-independent targeting pathways of yeast is made on the basis of the hydrophobicity of the signal sequence, with those of higher average hydrophobicity being directed into the SRP pathway (Ng et al. 1996). In accordance with these findings, introduction of helix breakers into the core of the invertase signal sequence abolished SRP dependence (Rothe and Lehle 1998). Even a mammalian signal sequence was shown to become SRP-independent when its hydrophobicity was reduced (Hatsuzawa et al. 1997) whereas a bacterial secretory protein, LamB, could be converted into a proficient SRP substrate by increasing the number of leucine residues within its signal sequence (Zheng and Nicchitta 1999).

As detailed in "The SecA/SecB-dependent posttranslational protein targeting in *E. coli*", essentially all secretory proteins of bacteria are exported in a posttranslational process involving the activities of SecA and SecB. Only recently were inner membrane proteins unravelled as specific substrates of the *E. coli* SRP. The availability of cell-free assay systems with controllable levels of SecA/SecB and SRP/SR revealed that the integration of α-helical transmembrane domains was exclusively dependent on SRP/SR, with no involvement of SecA in this process (Koch et al. 1999; Scotti et al. 1999; for further details see "SRP-dependent cotranslational targeting in bacteria"). In contrast, some SRP/SR-dependent membrane proteins of *E. coli* were reported to also require SecA for assembly (Qi and Bernstein 1999; Valent et al. 1998). The controversial issue was addressed by recent studies demonstrating that distinct inner membrane proteins of *E. coli* in fact use both SRP/SR and SecA for their assembly, yet not arbitrarily but rather in a consecutive and complementary manner (Neumann-Haefelin et al. 2000). This class of inner membrane proteins of *E. coli* is characterized by the occurrence of long translocated domains either following a single signal anchor domain or forming an extended periplasmic loop within a multispanning membrane protein. Targeting and insertion of the signal anchor domain preceding such a translocated moiety is an SRP/SR-dependent process while the subsequent translocation step strictly requires SecA (Neumann-Haefelin et al. 2000). Thus, in bacteria such as *E. coli*, SRP/SR and SecA/SecB constitute two basically independent targeting pathways to the translocon of the inner membrane. This view is different from models suggesting an obligate role of SecA in all integration and translocation processes occurring at the Sec translocon (Driessen et al. 2001).

If bacteria in fact use parallel targeting pathways for secretory and membrane proteins, the question arises of how the cell discriminates between the two classes of proteins. Similar to what has been found for yeast, the bacterial SRP recognizes longer and more hydrophobic stretches of amino acids, as exemplified by the signal anchor sequences of inner membrane proteins (Valent et al. 1997; de Gier et al. 1998; Neumann-Haefelin et al. 2000; Lee and Bernstein 2001; Kim et al. 2001). In contrast, classic signal sequences of secretory proteins usually have fewer and less hydrophobic amino acids within their core region.

Nevertheless a number of secretory proteins of *E. coli* have also been claimed to be SRP substrates (Ribes et al. 1990; Poritz et al. 1990; Phillips and Silhavy 1992; Jensen et al. 1994; Luirink et al. 1994; Kim et al. 2001). Most of these studies employed depletion of the SRP/SR components from whole cells, which, however, inevitably reduces the levels of membrane proteins such as SecY (Koch et al. 1999). The observed decrease in translocation of a secretory protein under these conditions could therefore be caused primarily by a lack of active translocon rather than directly by a deficiency in SRP/SR.

Amongst those secretory proteins being compromised in their export by an SRP/SR depletion was repeatedly the periplasmic β-lactamase (Ribes et al. 1990; Poritz et al. 1990; Phillips and Silhavy 1992; Luirink et al. 1994). A comprehensive in vitro analysis of targeting and transport properties of β-lactamase did not, however, reveal any indication for SRP dependence (D. Beha, M. Müller, H.G. Koch, in preparation). The combined data available therefore suggest that the cleaved signal sequences of bacterial secretory proteins are not recognized by Ffh. As will be discussed in "Regulatory mechanisms compensating for and preventing impairment of, the SRP pathway", in addition to the lower hydrophobicity of these signal sequences, molecular chaperones contribute to their escape from the SRP.

The structural basis of signal sequence recognition

The recognition of eukaryotic signal sequences and prokaryotic signal anchor sequences by SRP is intricate because signal sequences are highly variable in sequence and have in common only a central hydrophobic core of about 10–20 amino acids which follows a usually positively charged N-terminus. The M domain of SRP54 (Fig. 2A) had been implicated early on in signal sequence recognition (Bernstein et al. 1989) and a direct interaction between the signal sequence and the M domain has been demonstrated by chemical cross-linking (Zopf et al. 1990). The M domain is characterized by an atypically high percentage of methionine residues (Zopf et al. 1990; Römisch et al. 1990) and due to the high flexibility of the hydrophobic side chain of methionine (Gellmann 1991), it was proposed that the methionine-rich domain constitutes a flexible, hydrophobic signal sequence binding pocket (Bernstein et al. 1989; Zopf et al.1990).

The recently solved crystal structure of Ffh from the thermophilic bacterium *Thermus aquaticus* provides compelling support for the methionine bristle hypothesis, although the number of methionine residues in the M domain of *T. aquaticus* Ffh is smaller than that of other M domains. The most prominent feature of the M domain is a large groove, comprised almost exclusively of hydrophobic amino acids (Keenan et al. 1998). This groove is about 25 Å long, 15 Å wide and 12 Å deep, which would be sufficient to accommodate approximately 20 amino acids in an α-helical conformation or about 16 amino acids in fully extended β-hairpin conformation. The groove is formed by three rather rigid α-helices and one very flexible loop (finger loop). The different positions of this loop in different crystals from *T. aquaticus* (Keenan et al. 1998) as well as from human SRP54 (Clemons et al. 1999) and *E. coli* Ffh (Batey et al. 2000) suggest that the finger loop is responsible for adjusting the size of the binding pocket to a given signal sequence (Bernstein 1998).

On the outside of the hydrophobic groove is located a conserved arginine-rich HTH motif, which is required for the interaction with the SRP RNA (Batey et al. 2000; Kurita et al. 1996). A comparison of the NMR structure of the free 4.5S RNA from *E. coli* in solution (Schmitz et al. 1999) with the recently solved crystal structure of the ribonucleoprotein core (domain IV of 4.5S RNA and the M domain of Ffh) of the SRP, reveals a significant conformational change of the RNA upon binding to the M domain (Batey et al. 2000). In contrast, the free M domain can almost completely be superimposed onto the M domain of the RNA–protein complex, indicating no major structural changes in the M domain upon RNA binding. In the complex, the hydrophobic core of the signal sequence binding pocket extends to the phosphodiester backbone of the RNA, creating a continuous surface for signal sequence interaction. Since the N-terminus of signal sequences contains usually positively charged residues, this surface provided by both protein and RNA could indicate that the interaction between SRP and signal sequences occurs via a combination of both hydrophobic and electrostatic interactions (Batey et al. 2000).

Cotranslational protein targeting by SRP and elongation arrest

Cotranslational protein targeting by SRP and elongation arrest in eukaryotes

The early biochemical analyses of the secretory pathway in eukaryotes during the 1960s had revealed that mRNAs coding for secretory proteins were bound to rough microsomes and potential mechanisms on how this was accomplished were suggested (reviewed in Blobel 1999). In the 1971 version of *The Signal Hypothesis*, Blobel and Sabatini (1971) postulated that binding of the mRNA to the ER is not via direct binding of the mRNA itself but that it occurs via the interaction of the nascent translation product with the ER membrane. A prerequisite of the suggested mechanism was an amino-terminal signal sequence that is recognized by a soluble factor, which then would bind the RNC to the ER membrane. Time course experiments with translating ribosomes detached from the ER membrane had suggested that the signal sequence is cleaved off cotranslationally and cotranslocationally by membrane-bound signal peptidase (Blobel and Dobberstein 1975a). Consequently, the posttranslational incubation of an in vitro synthesized secretory protein with microsomal membranes did not allow for translocation and signal sequence cleavage (Blobel and Dobberstein 1975b), suggesting a strict cotranslational process. Extensive biochemical analyses led to the identification of SRP as the soluble factor responsible for cotranslational targeting of the ribosome-associated nascent chains to the ER membrane (reviewed in Blobel 1999 and Keenan et al. 2001).

It is now evident that the cotranslational targeting proceeds in two distinct steps (Fig. 3): (a) the recognition step, in which SRP binds to the hydrophobic signal sequence as it emerges from the ribosome and (b) the association step, in which SRP-RNCs interact in a GTP-dependent manner with the SRP receptor and the translocon in the ER membrane. An intrinsic advantage of cotranslational protein targeting is that the coupling of translation and translocation should prevent misfolding of the nascent chain in the cytoplasm. A posttranslational mode, on the other hand, requires that completely synthesized proteins be kept in a translocation-competent conformation. But even in cotranslational targeting, the diffusion of SRP-bound RNC to the ER membrane has to be considerably faster than the speed of protein synthesis in order to avoid folding/misfolding. Nature has found an elegant way to cope with this problem, the so-called elongation arrest. This was first observed in a heterologous in vitro system, in which binding of canine SRP to wheat germ ribosome-associated chains caused a halt in translation (Walter and Blobel 1981). Later it was found that this arrest was transient and that it occurred at multiple sites during chain elongation (Lipp et al. 1987). A similar translation retardation was also observed in a homologous mammalian in vitro system (Wolin and Walter 1989) and also in yeast (Yaver et al. 1992; Mason et al. 2000), suggesting that elongation arrest is a conserved and

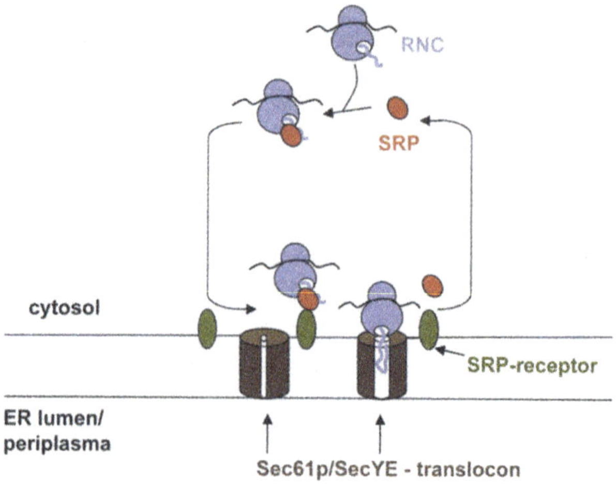

Fig. 3 SRP-dependent targeting of RNCs to the Sec61/SecYE translocon. The model similarly applies to the eukaryotic ER and the prokaryotic cytoplasmic membrane. SRP binds to the signal sequence as it emerges from the ribosome. Depicted are RNCs, ribosome nascent chain complexes, consisting of the small and large ribosomal subunits, the nascent chain (*blue wavy line*) and the mRNA (*black wavy line*). The SRP–RNC complex binds to the SRP receptor of the target membrane. SRP and SR dissociate and the RNC docks at the translocation channel (Sec61p/SecYE translocon) across which translocation of the nascent chain proceeds

functionally important feature of the eukaryotic SRP. This function of SRP increases the efficiency of protein translocation in vitro and in vivo, probably by enlarging the time window during which nascent chains can be targeted to the ER membrane in a translocation-competent conformation.

As mentioned earlier, the eukaryotic SRP can be divided into the S domain and the Alu domain. The Alu domain, consisting of the SRP9 and SRP14 subunits bound to the Alu-like sequence elements of the 7S RNA, is responsible for the elongation arrest. SRP subparticles assembled without SRP9 and SRP14 were fully active in protein translocation but were unable to induce elongation arrest (Siegel and Walter 1985, 1986). The removal of 20 C-terminally located amino acids of SRP14 exclusively impaired the elongation arrest activities without influencing protein targeting and ribosome binding and it has been suggested that the concurrent conformational change in the RNA is responsible for the loss of elongation arrest activity (Thomas et al. 1997). The flexibility and the dimensions of the SRP [approximately 6×24 nm, as estimated by electron microscopy (Andrews et al. 1985)] have led to the hypothesis that the S domain of SRP binds to the nascent chain near the ribosomal exit channel while the Alu domain binds near the acceptor site (A site) of the ribosome, thereby influencing translation (Bui und Strub 1999). The biochemical details on how the Alu domain impedes translation have not been elucidated so far. One simple explanation is that binding of the Alu domain interferes with the accessibility of the A site for elongation factors or tRNA. It is also possible that the Alu domain actively competes with elongation factors or tRNA, although there is no obvious structural similarity between elongation factors or tRNA and the Alu domain (Weichenrieder et al. 2000). The elongation arrest is released upon contact of the SRP-RNC with the SRP receptor in the ER membrane (Gilmore et al. 1982a; Meyer et al. 1982). Afterwards, the peptidyl transferase activity of the ribosome probably provides the major energy source for the translocation of secretory proteins through the Sec61 translocon.

SRP-dependent cotranslational targeting in bacteria

Until recently the prevailing view on protein transport in bacteria had been that secretory proteins, i.e. periplasmic and outer membrane proteins, depend on the so-called Sec system, which consists of the chaperone SecB, the motor protein SecA and the translocation channel in the inner membrane, composed of SecY, SecE and SecG as core components (reviewed in Müller et al. 2001). Small inner membrane proteins were originally considered to insert independently of the Sec system in a proton-motive-force (Pmf)-dependent reaction while other membrane proteins were thought to dependent on the Sec system as well (reviewed in Duong et al. 1997). Both the Sec-dependent and Sec-independent protein transport in bacteria were considered to be strictly posttranslational.

Based on the available genome sequences, it is now evident that SRP and its receptor are present in eukaryotes, bacteria and archaea, where they in general fulfil vital functions. In contrast, the SecA/SecB-dependent posttranslational membrane targeting appears to be unique to bacteria. No homologues of SecA and SecB have so far been identified in archaea and eukaryotes, with the exception of SecA in chloroplasts, which, however, are of prokaryotic origin. Thus, the SRP pathway probably represents a rather ancient mode of protein targeting. Assigning a specific function in protein transport to the bacterial SRP had been originally very controversial, in particular because various genetic screens for *E. coli* protein transport mutants had initially failed to identify SRP components (Schatz and Beckwith 1990). Redesigned genetic screens have only recently identified *E. coli* SRP mutants impaired in membrane protein assembly (Tian et al. 2000; Tian and Beckwith 2002). Pulse-chase experiments performed with conditional *E. coli* mutants of 4.5S RNA, Ffh and FtsY had suggested an involvement of these components in the export of a subset of bacterial secretory proteins (cf. "Recognition sequences"). This led to the idea that, similar to SecB, the bacterial SRP functions as a transport-specific chaperone rather than a protein-targeting factor (Wickner and Leonard 1996; Duong et al. 1997). On the other hand, replacing subunits of the eukaryotic SRP by the bacterial homologues led to partially active chimeric SRPs, pointing to a functional conservation between the bacterial and eukaryotic SRPs (Ribes et al. 1990; Bernstein et al. 1993).

During the last few years, the explicit function of the bacterial SRP was uncovered by biochemical and genetic approaches and these studies have changed our view on bacterial protein transport considerably. The initial observation was that the assembly of the inner membrane protein lactose permease was inhibited when *E. coli* cells were partially deprived of 4.5S RNA and Ffh (MacFarlane and Müller 1995). Similar findings were later obtained for other integral membrane proteins from *E. coli* (de Gier et al. 1996; Seluanov and Bibi 1997; Ulbrandt et al. 1997). These studies pointed to an involvement of SRP in the integration of bacterial inner membrane proteins, which would be consistent with the assumption that hydrophobic inner membrane proteins would preferentially be targeted in a cotranslational mode to avoid aggregation in an aqueous environment. The function of SRP in the integration of inner membrane proteins was directly demonstrated by a novel biochemical assay system, which made it possible to independently assay the activities of SecA/SecB and SRP/SR in protein transport in *E. coli* (Koch et al. 1999). In these analyses, it was shown that the integration of two multispanning membrane proteins, mannitol permease (MtlA) and SecY, into inside-out inner membrane vesicles of *E. coli* required all three purified SRP/SR components (4.5S RNA, Ffh and FtsY), but could not be driven by SecA, SecB and the Pmf. In contrast, in an identical experimental setup, these were the only components required for the translocation of a secretory protein, pOmpA, into membrane vesicles, a process in which the SRP components were completely inactive (Koch et al. 1999). These results suggested that in *E. coli*, SRP/SR and SecA/B constitute two distinct targeting devices for inner membrane and secreted proteins, respectively. Both targeting pathways converge at the SecY complex in the inner membrane (cf. "The bacterial SecYEG complex").

The salient feature of the eukaryotic SRP is that it mediates a cotranslational targeting to the ER membrane and the ability to induce a translational pause has been considered mandatory for cotranslationality. Although the elongation arrest activity of the eukaryotic SRP it is not a prerequisite for protein translocation into the ER (Siegel and Walter 1985), the lack of the SRP9 and SRP14 subunits in the bacterial SRP had been taken as evidence against a cotranslational protein targeting mediated by the bacterial SRP. As illustrated in Fig. 3, cotranslational targeting by SRP in eukaryotes proceeds in two distinct steps. SRP first binds to the signal sequence as it emerges from the ribosome and the complex of SRP and RNC then binds to the SRP receptor and is subsequently transferred to the translocation channel. By in vitro cross-linking experiments, this order of events has also been demonstrated for the bacterial SRP pathway. Membrane proteins were specifically recognized by the bacterial SRP via their hydrophobic signal anchor sequences and more importantly, recognition occurred only with ribosome-associated nascent chains (Valent et al. 1997; Beck et al. 2000), indicating a cotranslational interaction between the bacterial SRP and its substrates. The interaction between SRP and the ribosome-associated nascent chain was lost upon the addition of membrane vesicles and instead an interaction between the nascent chain and SecY became visible by chemical cross-linking (Valent et al. 1998; Beck et al. 2000). Again, the interaction between nascent chains of a membrane protein and SecY required the presence of the ribosome. Efficient interaction between the ribosome-bound nascent chain and SecY was only detectable in the presence of FtsY and GTP (Valent et al. 1998; Neumann-Haefelin et al. 2000). In contrast to ribosome-associated nascent chains of secretory proteins, those of membrane proteins were stably bound to the membrane, a process which was strictly dependent on the presence of SRP and FtsY (Neumann-Haefelin et al. 2000). These data demonstrate that SRP cotranslationally interacts with nascent chains of bacterial membrane proteins and initiates a cotranslational targeting to the SecY-translocon in the inner bacterial membrane.

The *B. subtilis* Ffh has been shown to bind to purified and denatured *E. coli* β-lactamase, provided that it contains a signal peptide of gram-negative or gram-positive origin (Bunai et al. 1996; Takamatsu et al. 1997). This interaction is also observed with mutant *B. subtilis* Ffh proteins incapable of binding the cognate scRNA, although scRNA is required for the proper in vivo function of SRP (Takamatsu et al. 1997). Similar in vitro studies employing purified components revealed binding of *Bacillus* SecA to the same precursors as Ffh and that Ffh enhances the affinity of SecA for the binary complex between SRP and substrate. Using a ligand affinity blotting technique, Ffh was shown to interact directly with SecA (Bunai et al. 1999). In vivo analyses revealed that depletion of both *B. subtillis* Ffh and SecA causes a significant reduction of secreted proteins (Bunai et al. 1999; Hirose et al. 2000). From these results the conclusion was drawn that *B. subtillis* Ffh and SecA interact to function cooperatively in protein export. Results of functional assays involving RNCs as natural substrates for *B. subtillis* SRP have not yet been reported. Furthermore it is currently unknown whether the observed reduction in extracellular proteins following depletion of Ffh is primarily due to a decreased number of active SecY translocons. In *E. coli* a coordinated activity of SRP and SecA has been demonstrated only for membrane proteins carrying large periplasmic loops (Neumann-Haefelin et al. 2000; Koch et al. 2002). Otherwise the SRP-dependent cotranslational protein targeting and the SecA/

SecB-dependent posttranslational targeting represent two independent, non-overlapping pathways in *E. coli* (Koch et al. 1999).

A cotranslational integration of a membrane protein has also been observed in archaea and the involvement of SRP has been postulated (Dale and Krebs 1999). Thus, the basic mechanism of SRP-dependent cotranslational protein targeting appears to be conserved between prokaryotes and eukaryotes.

Is elongation arrest dispensable in prokaryotes?

An important question in this respect is why the elongation arrest is a conserved feature in the eukaryotic SRP but obviously dispensable in bacteria and archaea. As mentioned in "Cotranslational protein targeting by SRP and elongation arrest in eukaryotes", the main function of elongation arrest is probably to enlarge the time window during which nascent chains can be targeted in a translocation competent conformation, i.e. before the nascent chain reaches a length that is prone to folding or aggregation. The mobility of RNCs in the cytosol and their ability to reach the membrane prior to folding or aggregation events is influenced by at least the diffusion coefficient, the diffusion distance and the speed of translation. Assuming that the diffusion coefficients for RNCs in the prokaryotic and eukaryotic cytoplasm are similar, these large complexes are estimated to diffuse at about 1 μm/s (Goodsell 1991; Elowitz et al. 1999; Schulz and Jorgensen 2001). The rate of translation in bacteria is estimated to be about 10 residues/s (Lorimer 1996) and considering the size of *E. coli* with 1×2 μm, the SRP-RNC should be able to reach the SRP receptor before the chain reaches a critical length.

Although prokaryotic cells are in most cases significantly smaller than eukaryotic cells, there is a significant size variation among the prokaryotes. It differs from 0.2×0.08 μm for *Thermodiscus sp.* up to 80×600 μm for *Epulopiscium fishelsoni* or up to a diameter of 750 μm for *Thiomargarita namibiensis* (Schulz and Jorgensen 2001), which is about 100 times the size of the yeast *S. cerevisiae*. Provided that the SRPs of these organisms also lack orthologues of SRP9/14, the dispensability of these two SRP subunits for the elongation arrest in prokaryotes cannot simply be attributed to a size phenomenon. One aspect that might also contribute to cotranslational targeting in the absence of elongation arrest is the observation that bacterial DNA is anchored to the cytoplasmic membrane during coupled transcription/translation of membrane proteins (Lynch and Wang 1993; Ma et al. 1994; Binenbaum et al. 1999). Tethering the respective genes to the cytoplasmic membrane further reduces the diffusion distance for RNCs in prokaryotes and would allow for cotranslational targeting without the need to arrest translation.

Alternatively, it is also conceivable that elongation arrest in bacteria is induced by an SRP9/14-independent mechanism. The 4.5S RNA is a potential candidate to couple cotranslational targeting with translation in bacteria. It has been shown that the estimated cellular concentration of 4.5S RNA is significantly higher than that of Ffh (Jensen and Pedersen 1994; Shibata et al. 1996) pointing to an additional, SRP-unrelated function of the 4.5S RNA. An involvement of 4.5S RNA in translation was concluded from the findings that a deficiency in 4.5S RNA is suppressed by mutations mapping to the elongation factor EF-G and amino acyl-tRNA synthetases (Brown 1987, 1989). The 4.5S RNA deficiency is also accompanied by an increased binding of EF-G to the ribosome (Nakamura et al. 1999a) and the 4.5S RNA has been shown to directly compete with 23S rRNA for binding to EF-G (Nakamura et al. 1999a). Ffh and EF-G recognize overlapping domains in the 4.5S RNA (Nakamura et al. 2001).

If a threshold level of 4.5S RNA were indeed required to maintain a normal rate of translation (Bourgaize and Fourniere 1987; Jensen et al. 1994) the possibility arises that Ffh by sequestering 4.5S RNA could retard translation. Affinity measurements have, however, demonstrated that half-maximal binding to 4.5S RNA in vitro occurs at 0.15 μM Ffh and 1.5 μM EF-G and accordingly almost all Ffh should be bound to 4.5S RNA in vivo, while only very little EF-G can form a stable complex with 4.5S RNA (Suzuma et al. 1999). The cellular concentrations of the three components have been estimated to be 1200 molecules 4.5S RNA/cell; 350 molecules Ffh/cell and 30,000 molecules EF-G/cell (Jensen and Pedersen 1994; Lorimer 1996). Given the variations in concentration of the three components and in their affinities for each other, a simple competition between Ffh and EF-G for 4.5S RNA as a mechanism to couple translation with cotranslational targeting does not appear very likely.

SRP-dependent protein targeting in chloroplasts

SRP-dependent proteins in chloroplasts do not only comprise chloroplast-encoded proteins such as the D1 protein, which is part of the water-splitting complex of photosystem II, but also nuclear-encoded proteins such as the LHCP family (light harvesting complexes) of proteins, which function as antenna pigments for photosystems I and II. These nuclear-encoded proteins are synthesized on cytosolic ribosomes and are transported into the stroma of chloroplasts before they are recognized by the chloroplast SRP. The cpSRP therefore has to be able to interact both cotranslationally as well as posttranslationally with its substrates.

The polytopic membrane protein D1, encoded by the plastid *psbA* gene, is synthesized in the chloroplast on membrane-bound ribosomes and seems to be cotranslationally integrated into the thylakoid membrane (Klein et al. 1988; Zhang et al. 1999). Immunoprecipitation and cross-linking studies have indicated that thylakoid-bound RNCs of D1 interact with cpSecY, the translocation channel in the thylakoid membrane (Zhang et al. 2001). The targeting of these RNCs to cpSecY involves their interaction with cpSRP54 but is independent of cpSRP43 and cpSecA (Nilsson et al. 1999). Ribosomes have been shown to pause at specific sites during translation of membrane-bound *psbA* mRNA, a phenomenon that was observed also in cyanobacteria expressing an orthologue of *psbA*. The underlying mechanism is, however, different from the elongation arrest activity of the mammalian SRP, probably involving a light-dependent regulation of translation elongation (Zhang et al. 1999, 2000; Tyystjärvi et al. 2001).

CpSRP when composed of the 54-kDa and 43-kDa subunits is able to mediate the post-translational integration of nuclear-encoded proteins of the LHCP family into the thylakoid membrane after they have been imported from the cytoplasm into the chloroplast stroma (Li et al. 1995). CpSRP interacts with LHCP and forms a soluble intermediate, termed transit complex. The insertion into the membrane requires GTP and cpFtsY (Hoffmann and Franklin 1994). Both cpSRP54 and cpSRP43 are involved in substrate binding (Groves et al. 2001) and are required for the posttranslational formation of the transit complex (Schuenemann et al. 1998). CpSRP43 interacts with a highly conserved sequence of LHCP, termed L18, located at the N-terminus of transmembrane helix three (deLille et al. 2000; Tu et al. 1999) while cpSRP54 binds to the transmembrane sequence three itself, which is an unusually long and hydrophobic transmembrane domain.

The different phenotypes of *Arabidopsis thaliana* mutants deficient in either cpSRP43 (Klimyuk et al. 1999) or cpSRP54 (Amin et al. 1999; Pilgrim et al. 1998) suggested that both proteins have no overlapping functions. In plastids lacking functional cpSRP54, proteins of the LHCP family as well as chloroplast-encoded and presumably cotranslationally targeted reaction centre proteins are affected. The first non-cotyledon leaves are yellow, indicating a defect in the assembly of a functional photosynthetic apparatus, but recover as the leaves mature and subsequent leaves develop normally. This suggests that alternative pathways can compensate for the loss of functional cpSRP. In contrast to the recovery of plants deficient in cpSRP54, cpSRP43 mutants develop pale green leaves throughout development, which suggests an improper integration of pigment-binding antenna proteins such as LHCP.

The SRP receptor and the GTPase cycle during protein targeting

The eukaryotic and bacterial SRP receptors

The SRP-mediated elongation arrest in eukaryotes is abolished upon the addition of microsomal membranes, which led to the proposal of a membrane-bound SRP receptor (SR) (Gilmore et al. 1982a). The ability to release the translational arrest had been used to purify SR from microsomal membranes (Gilmore et al. 1982b; Meyer et al. 1982) and to identify the heterodimeric SR. The larger SRα subunit is peripherally associated with the ER membrane via the smaller SRβ subunit, which is an integral membrane protein with one transmembrane segment. Both subunits were shown to bind GTP (Connolly and Gilmore 1989; Miller et al. 1995) and to be exclusively located in the ER membrane (Gilmore et al. 1982a). Binding of SRα to SRβ requires both hydrophobic and hydrophilic sequences located in the N-terminal domain of SRα (Young et al. 1995). The assembly process seems to involve a cotranslational targeting of SRα to the ER membrane, which is independent of SRP. Instead, a pause site within the mRNA, caused by a stem-loop structure similar to ribosomal frameshift structures, allows the folding of the N-terminal domain, which then interacts with SRβ before translation resumes (Young and Andrews 1996). A cotranslational targeting to the membrane has also been postulated for FtsY, the bacterial SRα orthologue (see "Contacting the Sec61 complex in the ER membrane"), although whether this targeting is SRP-dependent has not been addressed.

FtsY, the bacterial SR had been originally identified as a putative cell division protein, encoded in one operon together with two additional genes, *ftsE* and *ftsX* (Crickmore and Salmond 1986; Gill and Salmond 1987). Only later was the homology between SRα and FtsY recognized (Bernstein et al. 1989; Römisch et al. 1989) and its involvement in protein targeting was postulated (Luirink et al. 1994). Like Ffh, FtsY has a tripartite structure consisting of the A, N, and G domains (Fig. 2A). The crystal structures of the NG domains of the bacterial Ffh (Freymann et al. 1997) and FtsY (Montoya et al. 1997) exhibit a very similar conformation, comprising an N-terminal four-helix bundle tightly associated with the Ras-like GTPase domain (cf. Fig. 2B). FtsY is a hydrophilic protein that is partially localized in the cytoplasm and partially associated with the cytoplasmic membrane (Luirink et al. 1994), although a membrane-anchoring protein homologous to SRβ has not been identified in bacteria. Whether the soluble FtsY has a biological function is still an unsolved issue. Although the A domain is required for proper function, it can be replaced functionally by an unrelated membrane anchor, suggesting that FtsY has to be membrane-attached for its function (Zelazny et al. 1997). On the other hand, the function of the yeast SR is not significantly impaired by deleting the transmembrane domain of SRβ, suggesting that a cytosolic SR is functional in protein targeting (Ogg et al. 1998).

Several studies have addressed the question of how FtsY binds to the membrane. In vivo experiments with individually expressed FtsY domains have indicated that the A, the NG, and the G domains can bind to the membrane independently of each other (de Leeuw et al. 1997). Binding presumably occurs through direct interaction between FtsY and anionic phospholipids and involves two lipid binding sites, one of which is located in the NG domain (de Leeuw et al. 2000). In a separate study, neither the A domain nor the NG domain were found to be sufficient for membrane binding, which rather required the complete AN domain (Millman and Andrews 1999) and the zwitterionic phospholipid phosphatidyl ethanolamine (PE) (Millman et al. 2001). In the absence of PE, binding of FtsY to the membrane was shown to be trypsin-sensitive, which could indicate the requirement for a membrane protein to tether FtsY to the membrane (Millman et al. 2001). An attractive candidate for this membrane protein would be SecY, which indeed had been shown to be trypsin-sensitive (Swidersky et al. 1992). The ambiguity on how FtsY binds to the membrane has obviously been solved in several gram-positive bacteria, where the primary sequences of FtsY homologues reveal the presence of a sufficiently hydrophobic N-terminal domain to serve as a transmembrane anchor, although this has not yet been verified experimentally (Bibi et al. 2001).

The GTPase cycle during protein targeting

The GTPases encompass a protein family of highly conserved molecular switches responsible for regulating many complex functions such as protein synthesis, cell cycling and protein trafficking (Caldon et al. 2001). Despite their diverse functions, all GTPases are believed to have evolved from a single common precursor (Bourne 1995).

Protein targeting by SRP in eukaryotes involves three different GTPases (SRP54, SRα and SRβ); two are involved in bacteria (Ffh and FtsY) and the available genome sequences suggest that archaea probably also have two GTPases involved (Ffh and FtsY). Sequence comparison as well as structural information has revealed that SRP54, SRα, Ffh and FtsY probably constitute a new subfamily of small Ras-like GTPases (Bourne et al. 1990) with a relatively low affinity for nucleotides. In contrast, SRβ probably belongs to the Arf subfamily of GTPases, a family of small GTPases which seem to be absent in bacteria (Caldon et al. 2001). In comparison to the SRP54-subfamily, SRβ has a higher affinity for nucleotides (Miller et al. 1995; Bacher et al. 1999).

The general mechanism of GTPases is described in the molecular switch model (Bourne et al. 1991; Millman and Andrews 1997). During binding and hydrolysis of GTP, the enzyme passes through three different conformational states: GTP-bound, GDP-bound and empty. Initially, the GTPase is in the empty state and inactive, upon binding of GTP the enzyme is activated by a conformational change, which only then allows the interaction with its target molecule. Simultaneously with the GTPase-target interaction, GTP is hydrolyzed and the active state of the enzyme is turned off. After release of GDP, the enzyme returns to the empty state. This empty state, however, is considered to be a transient intermediate during the exchange of GDP for GTP. Both the intrinsically low rate of hydrolysis and nucleotide exchange are regulated by GTPase activating proteins (GAPs) and guanine nucleotide exchange factors (GEFs).

The GTPase cycles of SRP and SR obviously do not exactly follow this general model but instead have several striking particularities, which led to the concerted switch model for the SRP GTPases (Millman and Andrews 1997). The most prominent feature is that binding of SRP54 to the signal sequence and the subsequent targeting of the ribosome to the SRP receptor (Fig. 3) precedes GTP binding by SRP54 and SRα and thus, the empty states of SRP54 and SRα play a central role in protein targeting (Rapiejko and Gilmore 1997). The stability of these empty states is also reflected by the fact that the GTPase domains of both Ffh and FtsY have been crystallized in the empty state (Freymann et al. 1997; Montoya et al. 1997). These structures reveal that many of the amino acid side chains predicted to interact with the β and γ phosphates of GTP are shielded by a system of salt bridges and hydrogen bonds, which is most likely the explanation for the stability of the empty form. The interaction of SRP54/Ffh with SRα/FtsY leads to GTP-binding and to a reciprocal stimulation of the GTPase activities (Powers and Walter 1995). Thus, both proteins obviously act as GAPs for each other, which seem to be unique among GTPases. The recently solved structure of Ffh bound with the nonhydrolyzable GTP analogue GMP-PNP suggests a mechanism by which activation of the Ffh-GTPase is regulated at the level of entry of a preloaded GTP (primed state) into the active site (Padmanabhan and Freymann 2001). It is conceivable that the interaction of SRP with SR promotes the concerted transition of both proteins from the primed state into the active state.

The requirement for an additional priming step for both Ffh and FtsY preceding GTP binding has recently been proposed (Lu et al. 2001). This is based on the observation that mutating a conserved glycine at the interface between the N and G domain of Ffh or FtsY severely impaired their ability to interact with each other. This interface has been implicated in intradomain communication (Newitt and Bernstein 1997; Freymann et al 1999) and it is suggested that the above-mentioned mutation interferes with communicating substrate binding by the M domain to the NG domain (Lu et al. 2001), which is considered to prime SRP for its interaction with SR. A similar priming step is suggested to be imposed onto FtsY upon binding to lipids or to the translocation channel in the membrane.

The SRP-SR interaction proceeds primarily via their respective NG domains (Zopf et al. 1993; Peluso et al. 2000), but is further modulated by the RNA bound to the M domain of SRP. It has been proposed that complex formation between Ffh and FtsY is facilitated by 4.5S RNA in a catalytic fashion (Peluso et al. 2000). As described in "Assembly of SRP", binding of 4.5S RNA to the M domain of SRP induces a conformational change in the NG domain (Zheng and Gierasch 1997) that might be mandatory for efficient interaction with FtsY. In agreement with this notion, mutations within the 4.5S RNA that neither affect Ffh binding nor the interaction of SRP with ribosome-associated nascent chains, impair the ability of SRP to interact with FtsY (Jagath et al. 2001). Based on fluorescence measurements, a reversible conformational change is also imposed on FtsY during its interaction with SRP (Jagath et al. 2000).

Although GTP hydrolysis by both SRP54/Ffh and SRα/FtsY is stimulated upon contact with each other, this is probably not sufficient to coordinate signal sequence release from SRP54/Ffh with the accessibility of the translocon. As described earlier in this section, binding of SRP54 to the signal sequence emerging from the ribosome stimulates GTP binding. Likewise, GTP-binding by SRα is stimulated by adding purified Sec61 (Bacher et al. 1996) pointing to the translocon as an additional regulator of the GTPase cycle. This is verified by experiments using proteoliposomes, which show that dissociation of SRP54 from the signal sequence is blocked in the absence of Sec61 (Song et al. 2000). An additional layer of complexity to the GTPase cycle during protein targeting is conferred by SRβ, which is the least understood GTPase in the SRP pathway of eukaryotes. It serves as a membrane anchor for SRα, but the transmembrane domain, in contrast to its GTPase domain, is not required for its function (Ogg et al. 1998). The interaction of SRβ with SRα is nucleotide-dependent (Legate et al. 2000) and seems to require SRβ in its GTP-bound state. SRβ in its GTP-bound state also interacts with the ribosome, which stimulates its

GTPase activity (Bacher et al. 1999). Thus, the ribosome probably serves as a GAP for SRβ. A direct contact between SRβ and a 21-kDa ribosomal protein has recently been identified by chemical cross-linking (Fulga et al. 2001). This contact, however, is only detectable in the empty or in the GDP-bound state of SRβ. GTP-bound SRβ is presumably required for signal sequence release from SRP and for the coordination with the presence of the translocon (Fulga et al. 2001). This could impose a certain proofreading activity in that two different contacts are required for a functional targeting: the contact between the ribosome and SRβ and the contact between SRα and SRP.

As mentioned earlier, bacteria lack an orthologue of SRβ and FtsY, the SRα orthologue exists in a soluble and membrane-bound form (Luirink et al. 1994). The GTPase activity of FtsY is stimulated upon contact with Ffh (Powers and Walter 1995) and upon contact with phospholipids (de Leeuw et al. 2000). There is also evidence that FtsY interacts with membrane proteins and thus the interaction of FtsY with the bacterial membrane might occur as a two-step process. Probably the initial binding proceeds via interactions with phospholipids and is followed by interactions with a membrane protein (Millman et al. 2001). In vitro cross-linking experiments have demonstrated that soluble FtsY is not sufficient to induce dissociation of SRP from RNCs, but rather requires the context of the inner membrane (Valent et al. 1998; Neumann-Haefelin et al., unpublished data). Thus, a premature release of the ribosome-associated chains from SRP in bacteria is prevented even though a bona fide membrane receptor for FtsY has not been identified. Considering that there is only a single target membrane in bacteria, the intrinsic ability of FtsY to bind to the membrane might be sufficient to ensure that proteins are specifically transferred to the bacterial SecYE translocon.

For returning to the empty state, many GTPases depend on an external GEF to dissociate GDP. Biochemical evidence has indicated that nucleotide exchange in SRP54 and SRα does not depend on externally added factors (Miller et al. 1993, 1994). Based on kinetic analyses, it has been proposed that FtsY and Ffh have a built-in nucleotide exchange activity, located in the so-called I-box or Insertion box domain (IBD) (Moser et al. 1997; Jagath et al. 1998). This I-box is a unique structural motif which is inserted within the G domain (Fig. 2B) and seems to be characteristic of the SRP54-GTPase subfamily. Due to the minimal structural rearrangements within the I-box upon binding of Mg^{2+}-GDP, it has been questioned, however, whether the I-box is the primary exchange factor in the SRP GTPase (Freymann et al. 1999). For SRβ, whether GDP–GTP exchange is catalyzed intrinsically or whether it depends on an external exchange factor has not yet been addressed.

In summary, binding and hydrolysis of GTP by the SRP GTPases appears to be a strictly regulated process which involves three GTPases (two in bacteria) and at least the ribosome and the translocation channel. In a model reconciling the available data, SRP54 in its empty state binds to the signal sequence as it emerges from the ribosome. GTP-binding to SRP54 would be induced upon contact with the ribosome or the signal sequence. This would lead to the above-mentioned primed state of SRP54 reflecting a conformation that does not yet allow GTP to access the catalytic centre. Likewise, contact of SR with the (empty) translocon would induce GTP-binding by SRβ, which only then would lead to a binding of SRα to SRβ. A conformational change caused by this interaction mediates binding of GTP to SRα without concomitant hydrolysis. Only the contact between SRα and SRP54 via their respective NG domains would achieve access of GTP to the catalytic domains. The concurrent GTP hydrolysis would result in the release of the signal sequence from SRP and the dissociation of SRP from SR. This in turn facilitates the transfer of the nascent chain to the translocation channel and enables SRP to initiate a new targeting cycle. The same principle applies to the SRP-GTPase cycle in bacteria. Here, in the light of only one targeting membrane and the lack of SRβ, binding of FtsY to the membrane and/or contacting the translocation channel would induce GTP binding to FtsY and thus priming the bacterial SRP receptor for its interaction with the SRP-RNC. Although many important details during the GTPase cycle of SRP-mediated protein targeting have been experimentally addressed, the details on how signal sequence release is coordinated with its accessibility to the translocation channel remain poorly understood. An additional layer of complexity stems from observations that components of the translocation complex in the endoplasmic reticulum are regulated by phosphorylation, a process which involves a Ca^{2+}-dependent protein kinase (Gruss et al. 1999).

Interaction of the ribosome nascent chain complex with the translocation channel in the ER membrane and the bacterial inner membrane

Upon interaction of SRP with the signal sequence, the SRP-RNC is directed to the SRP receptor and subsequently to the protein conducting channel in the membrane (Fig. 3). The major constituent of the translocation pore in the ER membrane is the heterotrimeric Sec61 protein complex (Görlich and Rapoport 1993). This complex is highly conserved; the Sec61α, Sec61β, and Sec61γ subunits from mammals are homologous to the yeast Sec61, Sbh1, and Sss1 proteins, respectively. Moreover, Sec61α and Sec61γ are homologous to the prokaryotic SecY and SecE, pointing to a mechanism of protein translocation that is highly conserved (for detailed reviews on the Sec61p and SecYEG complexes see Rapoport et al. 1996 and Müller et al. 2001).

Contacting the Sec61 complex in the ER membrane

The Sec61 complex is required for the translocation into the ER of all substrates so far tested. Sec61α, the largest subunit of the complex, can be cross-linked to ribosome-associated nascent chains, indicating that they are in close proximity to the complex during translocation (Mothes et al. 1994). Proteoliposomes containing only three components, the SRP-receptor, the Sec61 complex and the TRAM protein (translocating chain-associated membrane protein), are competent to translocate all proteins tested so far (Görlich and Rapoport 1993). For some proteins, however, TRAM seems to be dispensable and whether a protein is TRAM-dependent or TRAM-independent seems to relate to diverse structural features of its signal sequence (Voigt et al. 1996). Furthermore, no TRAM homologue has been detected in the yeast genome, which is an additional indication for TRAM not being an essential component of the Sec61 translocon.

Purified Sec61 complexes form cylindrical oligomers with a diameter of 85 Å and a central pore of about 20 Å (Hanein et al. 1996). Oligomer formation is stimulated upon ribosome contact and upon contact with the Sec62/63p complex, which is required for posttranslational transport across the ER membrane (see "Posttranslational protein targeting to the Sec61 complex in eukaryotes"). The cryo electron microscopic structure of a 80S ribosome–Sec61 complex has been determined both in the presence and absence of a signal sequence-containing nascent polypeptide chain. The three-dimensional reconstruction of the ribosome–Sec61 complex in the absence of a signal sequence indicates a perfect alignment between the central pore of the Sec61 complex and the exit tunnel of the ribosome (Beckmann et al. 1997). Thus, during cotranslational protein transport, the nascent chain probably moves directly from the aqueous channel within the ribosome to an aqueous pore within the membrane. Furthermore, this analysis suggested one major attachment site between the Sec61 complex and the large ribosomal subunit leaving a lateral opening of about 15 Å. This lack of a tight seal between the ribosome and the Sec61 complex is surprising in light of collisional quenching experiments, which have demonstrated that the ribosome provides a permeability barrier on the cytoplasmic side of the Sec61 channel (Crowley et al. 1994). Similarly, the Sec61 channel is sealed on the lumenal side by the Hsp70 chaperone BiP (Hamman et al. 1998). Therefore, the 15-Å gap had been attributed to the absence of the signal sequence in the original cryo electron microscopic study (Beckmann et al. 1997). However, in recent cryo electron microscopic studies with both mammalian and yeast ribosome–Sec61 complexes in the presence of signal sequences, the lateral gap is still visible (Menetret et al. 2000; Beckmann et al. 2001). This is of double importance. First, it indicates that the permeability barrier on the cytoplasmic side of the channel can be maintained without a tight ribosome–Sec61 contact, which might be the result of dynamic conformational changes within the channel itself (Beckmann et al. 2001). Secondly, the presence of a lateral opening is in agreement with protease protection assays showing that emerging nascent chains of a membrane protein with extended cytosolic loops are accessible to added protease (Hegde and Lingappa 1996). This is not the result of ribosome detachment from the translocon during the synthesis of cytosolic domains because cross-linking results (Mothes et al. 1997) have demonstrated that both the lumenal and the cytosolic domains of a membrane protein are synthesized while the ribosome is in contact with the membrane. Furthermore, specialized secretory proteins have been described whose translocation into the ER pauses at discrete sites, which leads to a transient exposure of polypeptide domains to the cytosol. This process is regulated by the above-mentioned TRAM (Hegde et al. 1998).

In yeast, a second copy of Sec61p, termed Sshlp (Sec sixty-one homologue), has been identified, which is localized in the ER and is about 30% identical to Sec61p (Finke et al. 1996). Sshlp associates with Sbh2p, a Sbhlp homologue, and Ssslp, forming a complex that is able to bind ribosomes with similar affinity as the Sec61p complex (Robb and Brown 2001; Prinz et al. 2000). Interestingly, the Sshlp complex does not bind the Sec62/63 complex, which is essential for the posttranslational transport in yeast (see "Posttranslational protein targeting to the Sec61 complex in eukaryotes"), suggesting that it is only active in cotranslational protein transport (Wilkinson et al. 2001). It has been postulated that the Sshlp translocon might serve as a backup translocon during times of high secretion load (Robb and Brown 2001).

The Sec61 complex in eukaryotes as well as the SecYEG complex in bacteria (see "The bacterial SecYEG complex") do not only function as a receptor for RNCs, but also bind non-translating ribosomes (Görlich et al. 1992; Kalies et al. 1994; Prinz et al. 2000). SRP, however, is able to mediate efficient and productive targeting of signal sequence-containing RNCs to the Sec61 complex even in the presence of a vast excess of non-translating ribosomes (Raden and Gilmore 1998; Neuhof et al. 1998). Thus, binding of SRP to the signal sequence and its subsequent interaction with the SRP receptor obviously conveys onto SRP-RNCs a selective advantage in membrane targeting over non-translating ribosomes or over ribosomes translating non-SRP substrates.

Consistent with the idea that SRP-RNCs do not necessarily encounter unoccupied Sec61 complexes, it has recently been shown that large ribosomal subunits remain in stable association with the ER membrane after run-off translation (Seiser and Nicchitta 2000; Potter and Nicchitta 2000). This finding raises the possibility that ribosome exchange between cytosolic and membrane-bound ribosomes includes only the small ribosomal subunit. Interestingly, membrane-bound ribosomes engaged in the synthesis of cytosolic proteins detached from the ER membrane and completed translation in the cytosol (Potter and Nicchitta 2000). This led to the proposal that elongation of translation reduces the affinity of membrane-bound ribosomes for the ER membrane. Only in the presence of a signal sequence and the interacting SRP/SR is the ribosome-membrane contact maintained (Potter et al. 2001).

The existence of membrane-bound ribosomes has also been demonstrated in *E. coli* (Randall and Hardy 1983; Prinz et al. 2000). During the SRP-dependent targeting of bacterial membrane proteins, the RNCs are in stable contact with the SecY translocon (Valent et al. 1998; Beck et al. 2000; Neumann-Haefelin et al. 2000). This indicated that also in bacteria translating ribosomes are delivered to the membrane via the SRP-SR interaction. An alternative pathway for ribosome targeting only depending on FtsY has recently been suggested. In vivo studies of the cellular distribution of ribosomes in *E.coli* led to the observation that the depletion of Ffh did not interfere with membrane binding of ribosomes, whereas depleting FtsY significantly reduced the amount of membrane-bound ribosomes (Herskowitz and Bibi 2000). It has therefore been proposed that a cotranslational targeting of FtsY (Herskowitz et al. 2001) represents a mechanism by which ribosomes become tethered to the membrane.

The bacterial SecYEG complex

As in eukaryotes, the minimal functional unit required for protein translocation across the bacterial inner membrane turned out to be a heterotrimeric protein complex, termed SecYEG. Only SecY and SecE are required for cell viability in *E. coli*, and proteoliposomes reconstituted with just SecY and SecE are capable of translocating proteins, albeit at low efficiency (Akimaru et al. 1991).

The bacterial SecYEG complex has been primarily characterized with respect to its function as translocation channel for the SecA/B-dependent posttranslational protein transport (reviewed in Mori and Ito 2001). Using different experimental approaches, it has now become evident that SecY and SecE are also required for the SRP-dependent cotranslational targeting. By chemical- and site-specific cross-linking studies, nascent chains of SRP-dependent substrates have been shown to contact SecY and SecE during integration (Valent et al. 1998; Beck et al. 2000; Neumann-Haefelin et al. 2000; Beck et al. 2001). Jamming the SecY channel significantly reduces the integration of SRP-dependent membrane proteins (Koch et al. 1999) and reducing the cellular level of SecY and SecE equally impairs SecA-dependent and SRP-dependent protein transport (Koch and Müller 2000). Surprisingly, in a *secG* deletion mutant, the integration of SRP-dependent proteins still occurs at wild type level under conditions that completely abolish translocation of SecA-dependent proteins (Koch and Müller 2000). This suggests that the function of SecG is limited to the SecA-dependent posttranslational transport. This notion is supported by the observation that SecG has so far been found only in organisms which also contain SecA.

Recently the involvement of YidC, a 60-kDa protein of the *E. coli* inner membrane, in the integration of SRP-dependent substrates has been recognized. YidC seems to be functionally and structurally closely associated with the SecYEG translocon (Scotti et al. 2000) and is involved in the lateral transfer of transmembrane domains from the translocon into the lipid bilayer. This is suggested by site-specific cross-linking experiments, in which the transmembrane domain of an SRP-dependent membrane protein has been shown to first contact SecY and SecE and then YidC before it is released into the lipid phase (Beck et al. 2001; Urbanus et al. 2001). In addition, the first transmembrane domain of a polytopic membrane protein was shown to be retained by YidC during insertion of subsequent transmembrane helices (Beck et al. 2001). This finding suggests an important function of YidC in the formation of helix bundles of polytopic membrane proteins prior to their release into the membrane lipids. Like YidC, its orthologues Oxa1p in the inner mitochondrial membrane and Alb3 in the thylakoid membrane, are specifically involved in membrane protein assembly (Hell et al. 1998; Moore et al. 2000). In contrast to its mitochondrial orthologue Oxa1p, YidC itself is an SRP substrate and is integrated through the SecYE channel into the membrane (Koch et al. 2002; Urbanus et al. 2002). Oxa1p, on the other hand, is part of a Sec-independent integration complex in the inner mitochondrial membrane through which membrane proteins, like the nuclear-encoded Oxa1p itself, are integrated from the matrix side into the inner membrane (Herrmann et al. 1997; Hell et al. 2001).

The SecYEG complex transiently associates with a second membrane protein complex, the SecDFYajC complex (Duong and Wickner 1997). The SecDF proteins constitute a special family within the RND superfamily of proteins (resistance-nodulation-cell division), which comprises a huge family of proteins, all of which work as multifunctional efflux pumps (Tseng et al. 1999). The exact role of the SecDFYajC complex in protein transport is not clear, but has been mainly associated with the posttranslational SecA/B-dependent protein transport (see "The SecA/SecB-dependent posttranslational protein targeting in *E. coli*"). Whether the transport of SRP-dependent substrates is influenced in *secDF* mutants has yet to be addressed. Strikingly, SecDF homologues have also been identified in archaea, which do not contain SecA, pointing to a function of SecDF not restricted to an interaction with SecA-dependent substrates.

The *B. subtilis* SecYE complex forms ring-like structures in detergent solution that are very similar to those of the Sec61p complex (Meyer et al. 1998). As for the Sec61 complex, the outer diameter is estimated to be about 85 Å and the inner pore size about 15–20 Å. Based on sedimentation data, it was proposed that about three copies of SecYE form the channel (Meyer et al. 1998). Electron microscopy of the SecYEG complex from *E. coli* indicates a larger dimension of about 100 Å and a central cavity of about 50 Å (Manting et al. 2000). It has been proposed that the active translocase is composed of four SecYEG complexes and only assembled upon contact with SecA, the motor protein for posttranslational protein transport in bacteria (Manting et al. 2000). The need of SecA for the recruitment of SecY, SecE, and SecG to form an active translocon is difficult to reconcile with the SecA-independent integration of SRP substrates through SecYE. Thus, other factors than SecA should also be capable of inducing oligomerization of the translocon; one possible candidate is the ribosome, as observed for the assembly of the Sec61p complex in eukaryotes (Hanein et al. 1996). A different experimental setup has questioned the oligomeric state of the active translocon and let to the proposal that the active form of the translocase is a monomeric SecYEG (Yahr and Wickner 2000). A first low-resolution crystal structure of *E.coli* SecYEG at 9 Å and analytical ultracentrifugation experiments support the presence of more than one copy of SecYEG in the channel. In these analyses, however, the transition between the monomeric and tetrameric form of SecYEG was not dependent on either SecA or the ribosome (Collinson et al. 2001).

There is now general agreement that the bacterial SRP pathway delivers proteins to the SecYE channel; one example, however, has been described in which an SRP substrate is integrated in the absence of SecYE (Cristobal et al. 1999). The integration of a largely truncated version of the *E. coli* membrane protein ProW, consisting of only one transmembrane domain and a 100-amino-acid-long N-terminal periplasmic loop has been shown to be severely impaired by reducing the cellular concentrations of 4.5S RNA, but not by reducing the level of SecYE. Whether this is a general phenomenon for small membrane proteins with a large periplasmic N-tail or limited to the construct used in the above-mentioned study has not been addressed so far.

SRP-independent protein targeting to the Sec61 translocon of the ER and to the SecYEG translocon of the bacterial inner membrane

Posttranslational protein targeting to the Sec61 complex in eukaryotes

In addition to the cotranslational mode of translocation, proteins can also be transported in an SRP-independent, posttranslational mode across the ER membrane. This transport pathway has been mainly characterized in yeast, but is present in higher eukaryotes as well. In yeast, the hydrophobicity of the signal sequence seems to be the decisive feature with respect to which of the two mechanisms is operative. Precursor proteins with long hydrophobic core regions within their signal sequence show a more pronounced SRP dependency (see "Recognition sequences"). In mammalian cells, the length of the precursor protein seems to be the critical feature and it has been shown that only polypeptides of less than 70 amino acids are transported posttranslationally (Müller and Zimmermann 1987). In contrast to the cotranslational mode in which the coupling of protein synthesis and transport prevents folding or misfolding of the polypeptide, in the posttranslational mode, premature folding and aggregation have to be prevented by cytosolic chaperones, mainly of the Hsp70 family and to some extent also by others (Eggers et al. 1997; Zimmermann 1998; Plath and Rapoport 2000).

The posttranslational protein transport in yeast differs from the cotranslational mode also at the level of the ER membrane, where in addition to the Sec61 subcomplex ("Contacting the Sec61 complex in the ER membrane"), the so-called Sec62/63 subcomplex is involved. This subcomplex comprises the proteins Sec62p, Sec63p, Sec71p and Sec72p (Deshaies et al. 1991; Panzner et al. 1995). Posttranslational translocation of proteins into the yeast ER proceeds in two distinct steps (reviewed in Rapoport et al. 1999 and Stirling 1999). The substrate first binds to the Sec complex via its signal sequence. In this step, precursors are first targeted to the Sec62/63 subcomplex and then delivered to the Sec61 subcomplex. Thus, the Sec62/63 subcomplex might serve as a membrane receptor for the SRP-independent, posttranslational transport in yeast. For the subsequent translocation reaction through the Sec61 complex (step 2), a lumenal Hsp70, termed Kar2p in yeast and BiP in mammalian cells, is required. Recruitment of Kar2p proceeds via Sec63, which contains a lumenal J domain, typical for cochaperones interacting with Hsp70s (Sadler et al. 1989). Sequential binding and release of the substrate by Hsp70s is believed to drive the posttranslational transport across the Sec61 channel (Matlack et al. 1999). The discovery of homologues of Sec62p and Sec63p in mammalian ER (Meyer et al. 2000; Tyedmers et al. 2000) suggested that they might also be involved in the translocation of cotranslationally targeted secretory proteins. This was in fact recently demonstrated using novel *sec63* and *kar2* mutants of yeast, which exhibited defects in both SRP-dependent and SRP-independent protein translocation (Young et al. 2001).

The SecA/SecB-dependent posttranslational protein targeting in *E. coli*

In general, translocation of secretory proteins, i.e. proteins residing in the periplasm or the outer membrane of gram-negative bacteria, across the SecYEG channel occurs posttranslationally. These secretory proteins are synthesized as precursor with an N-terminal, cleavable signal sequence and as in eukaryotes, these proteins have to be retained in a translocation competent conformation by cytosolic chaperones. SecB serves as an export-specific chaperone for many secreted proteins in gram-negative bacteria, primarily by binding to the mature part of the exported protein. SecB is a homotetrameric protein consisting of 17-kDa monomers that keep preproteins in a loosely folded conformation for export (Randall and Hardy 1995). The recently available crystal structure of SecB indicates the presence of two distinct peptide binding sites within the tetramer. Site 1 appears to selectively recognize hydrophobic and aromatic regions of polypeptides, while the second site is supposed to preferentially bind hydrophobic and basic residues (Xu et al. 2000; Driessen 2001), consistent with the preferential binding of SecB to sequence motifs enriched in aromatic and basic residues (Knoblauch et al. 1999). In addition to SecB, other general chaperones are probably also involved in the maintenance of transport competence, in particular the DnaK system (Hsp70) and the GroEL/GroES system (Hsp60) (Wild et al. 1996; Kusukawa et al. 1989). One particular important chaperone is the ribosome-bound peptidyl-prolyl isomerase trigger factor, which is probably the first protein to stably interact with ribosome-associated nascent chains of cytosolic and secretory proteins (Hesterkamp et al. 1996; Scholz et al. 1997; Valent et al. 1997; Beck et al. 2000) and which in addition appear to be involved in routing proteins into the SecA/B-dependent pathway (see "Regulatory mechanisms compensating for and preventing impairment of, the SRP pathway").

Besides its chaperone function, SecB has also targeting function (Ernst et al. 1994), which is mediated by its affinity for SecA, the motor protein of posttranslational transport (Breukink et al. 1995; Fekkes et al. 1997). SecA is an essential homodimeric protein consisting of 102-kDa monomers, each comprising one ATPase domain with a high-affinity ATP binding site (Mitchell and Oliver 1993) and two regulatory domains (Sianidis et al. 2001). SecA exists in both soluble and membrane-bound forms and its membrane association is mainly mediated by the interaction with SecY, which involves the carboxy terminus of SecA (Snyders et al. 1997). SecA binds to both the signal sequence as well as to the mature part of a secretory protein (Akita et al. 1990; Triplett et al. 2001; Wang et al. 2000). Upon ATP binding to SecA, SecB dissociates from the SecA–SecY–preprotein complex (Fekkes et al. 1997). In an ATP-dependent reaction, SecA inserts and de-inserts into the SecYEG channel, thereby pushing the preprotein across the membrane (Economou 2000). The insertion and de-insertion cycle of SecA is coupled to a conformational change of SecG, which changes its orientation in the membrane (Nishiyama et al. 1996). During translocation, the signal sequence of the protein is cleaved off by the periplasmic signal peptidase. The ATP hydrolysis by SecA is not the only energy source for the posttranslational transport; during later stages of the translocation the Pmf is also involved, although the exact function has not been defined (Manting and Driessen 2000).

As mentioned in "The bacterial SecYEG complex", the SecYEG complex transiently associates with the SecDFYajC complex. SecD and SecF were shown to stabilize the inserted form of SecA (Kim et al. 1994; Economou et al. 1995) and a secretory protein has been shown to be in contact with SecD during translocation (Beck et al. 2000). Overexpressing YajC suppresses a dominant negative phenotype of a particular *secY* mutant, suggesting an interaction between YajC and SecY (Taura et al. 1994). Whether the SecDFYajC complex is specific for posttranslational transport remains to be analysed.

Regulatory mechanisms compensating for and preventing impairment of the SRP pathway

Cellular responses towards reduced levels of SRP/SRP receptor

The intrinsic advantage of a cotranslational mode of protein transport is that the newly synthesized polypeptides do not require chaperoning to stay in a transport-competent form. Accordingly, in organisms harbouring alternative co- and posttranslational targeting pathways to the Sec translocon, the cotranslational one is required for hydrophobic membrane proteins that are prone to aggregation. Consistent with a detrimental effect on cell physiology that the accumulation of aggregation-sensitive hydrophobic proteins exerts, reduced cellular concentrations of SRP components lead to the induction of a heat-shock response. This was first described for *E. coli* strains with reduced 4.5S RNA levels (Bourgaize et al. 1990). A recent analysis of an Ffh depletion strain of *E. coli* has suggested that the target of the heat-shock response is up-regulation of proteases rather than of chaperones (Bernstein and Hyndman 2001). Thus, *E. coli* tries to prevent the accumulation of potentially toxic aggregates in the cytoplasm by increasing the cellular concentrations of proteases.

In *S. cerevisiae*, deletion of any component of the SRP pathway results in exceedingly sick cells, but in contrast to virtually all other known organisms, these cells are viable (Hann and Walter 1991). Yeast responds in two distinct phases to SRP depletion. First, cytoplasmic Hsp70 levels are drastically increased concurrently with a reduction in the growth rate. During the second phase, the Hsp90 and Hsp104 levels are increased, together with the lumenal Hsp70 BiP (Arnold and Wittrup 1994). A more extended analysis revealed an even more global response of *S. cerevisiae* to the loss of SRP in that more than 700 open reading frames (ORFs) are either induced or repressed. Among them are all ORFs encoding heat-shock proteins and chaperones and many ORFs coding for proteins of the translation machinery and of the protein degradation machinery (Mutka and Walter 2001). Thus, despite its ability to posttranslationally transport proteins into the ER, yeast requires strategies to cope with the loss of the SRP pathway such as reducing the rate of protein synthesis concomitantly with increasing the cellular concentrations of chaperones and proteases.

In chloroplasts, the level of ClpC, a protein related to the 90-kDa class of heat-shock proteins (Hsp90) is elevated in cpSRP54 and cpSRP43 mutants (Amin et al. 1999) and this may also partially compensate for the defect of a functional cpSRP.

Measures to prevent an overload of the SRP pathway

As detailed in the preceding section, an overload of the SRP pathway risking the concept of cotranslational protein targeting and transport is a detrimental event for living cells. For this reason, the translocon always has to be accessible for SRP-RNCs.

Guaranteeing unrestricted access of the SRP-RNCs to the translocon is particularly critical in bacteria because of the limited number of SecY molecules. There are about 300 copies of SecY per cell (Matsuyama et al. 1992), through which more than 10^6 SecA-dependent periplasmic and outer membrane proteins are exported per generation (Pugsley 1993). This corresponds to about 100 secreted polypeptides per minute and translocation channel. Considering this immense number of postranslationally transported proteins, bacteria need a very efficient scanning process to ensure preferential access of SRP substrates to the translocon. This is especially important, since the tendency to aggregate is intrinsically high for the specific SRP substrates of bacteria, i.e. hydrophobic membrane proteins.

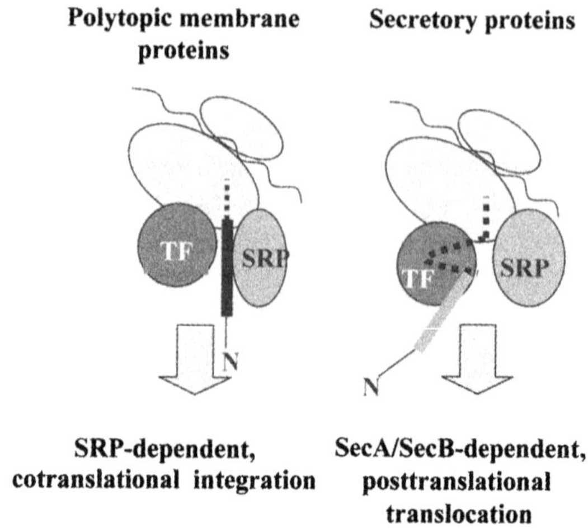

Fig. 4 Discrimination between nascent chains of membrane and secretory proteins in *E. coli*. The signal anchor sequence of a membrane protein is depicted in *black* whereas the signal sequence of a secretory protein is shown in *grey*. Trigger factor (*TF*) prevents the interaction of *SRP* with the signal sequence of a secretory protein

The mechanisms by which the cell discriminates between the SecA/B-dependent and SRP-dependent substrates have recently been addressed by a chemical and site specific cross-linking approach (Fig. 4). Membrane proteins are specifically recognized by SRP via their hydrophobic membrane spanning segments (cf. "Recognition sequences"). This recognition was shown to occur only in the presence of the ribosome (Valent et al. 1998; Beck et al. 2000), which is in agreement with the cotranslational mode of targeting. In an identical experimental setup, ribosome-associated nascent chains of the SecA-dependent protein pOmpA were found to be exclusively cross-linked to trigger factor, a ribosome-associated chaperone (Beck et al. 2000). In the absence of trigger factor, however, both SecA and, even more surprisingly, SRP were found to interact with ribosome-associated nascent chains of OmpA. These interactions were prevented by adding purified trigger factor prior to cross-linking (Beck et al. 2000).

These results suggest that at the stage of the ribosome, polytopic membrane proteins of *E. coli* are selected for cotranslational targeting by interacting with SRP, whereas the interaction with trigger factor prevents secreted proteins from entering the cotranslational SRP pathway. Although the primary binding sites for trigger factor are located within the mature part of OmpA, subsequently to its binding to the mature part trigger factor is also able to interact with the signal sequence of OmpA (Eisner et al., unpublished data), thereby probably shielding the signal sequence against unwanted interactions with SRP. If trigger factor in fact plays a critical role in clearing the SecYE translocon from prematurely, i.e. cotranslationally, targeted RNCs of secretory proteins, it would be expected to be an essential protein. This, however, was found not to be the case (Guthrie and Wickner 1990). The discrepancy was recently solved by demonstrating that the *E. coli* Hsp70 chaperone DnaK, which becomes essential in the absence of trigger factor (Deuerling et al. 1999; Teter et al. 1999), can substitute for trigger factor in preventing an SRP-mediated targeting of OmpA-RNCs (Eisner et al. unpublished). Chaperones such as trigger factor and DnaK therefore appear to fulfil important regulatory functions by forcing secretory proteins of bacteria to bypass an SRP-dependent cotranslational targeting to the Sec translocon.

Are there eukaryotic chaperones that function analogously to trigger factor? The eukaryotic nascent-chain-associated complex (NAC), a heterodimeric cytosolic protein, has been suggested to bind to all empty ribosomes and nascent polypeptide chains emerging from translating ribosomes (Wiedmann et al. 1994). NAC is located very close to the exit site on the large ribosomal subunit, in concert with which it forms part of the protease-protective environment of a nascent polypeptide chain (Wang et al. 1995). In vitro in the absence of NAC, also RNCs carrying cytosolic proteins were found to be targeted to the ER, which even resulted in translocation, albeit with low efficiency. This mistargeting of nascent cytosolic polypeptides, which occurred independently of SRP (Lauring et al. 1995), could be prevented by NAC. Only in case of an emerging signal sequence would NAC be replaced by SRP, which would then lead to targeting of the SRP-RNCs to the Sec61 complex. In further studies, however, conflicting results were reported, showing that NAC does not prevent the non-selective interaction between ribosomes and Sec61 (Neuhof et al. 1998; Raden and Gilmore 1998).

A completely different function of NAC has been suggested by recent studies in which NAC of yeast was found be involved the targeting and translocation of precursor proteins into mitochondria (George et al. 1998; Fünfschilling and Rospert 1999). Irrespective of the controversial findings on the function of NAC, the similarities between NAC and trigger factor are only superficial. Whereas both ribosome-associated factors bind to nascent chains early on during synthesis and can prevent their inappropriate membrane-targeting, the respective targets are totally different. In eukaryotes, NAC would be directed towards non-secreted proteins and interfere with their unwanted targeting to the ER, a process that proceeds independently of SRP. On the contrary, the bacterial trigger factor would prevent secretory proteins from being incorrectly recognized by SRP. Thus the involvement of trigger factor and DnaK in avoiding an overload and subsequent impairment of the bacterial SRP pathway has so far remained unprecedented.

Acknowledgement Work of the authors' laboratory was supported by the Deutsche Forschungsgemeinschaft, the Sonderforschungsbereich 388, the European Union (grant QLK3-CT-1999–00917), the Fonds der Chemischen Industrie, and the Forschungskommission of the University Hospital Freiburg.

Note added in proof Recent studies have identified the ribosomal protein L23 as a binding site for trigger factor (Kramer et al. (2002) Nature 419:171–174) and its eukaryotic homologue L23A as binding site for SRP (Pool et al. (2002) Science 297:1345–1348)

References

Akimaru J, Matsuyama S, Tokuda H, Mizushima S (1991) Reconstitution of a protein translocation system containing purified SecY, SecE, and SecA from Escherichia coli. Proc Natl Acad Sci U S A 88:6545–6549

Akita M, Sasaki S, Matsuyama S, Mizushima S (1990) SecA interacts with secretory proteins by recognizing the positive charge at the amino terminus of the signal peptide in Escherichia coli. J Biol Chem 265:8164–8169

Althoff S, Selinger D, Wise JA (1994) Molecular evolution of SRP cycle components: functional implications. Nucl Acid Res 22:1933–1947

Amin P, Sy DA, Pilgrim ML, Parry DH, Nussaume L, Hoffman NE (1999) Arabidopsis mutants lacking the 43- and 54-kilodalton subunits of the chloroplast signal recognition particle have distinct phenotypes. Plant Physiol 121:61–70

Andrews DW, Walter P, Ottensmeyer FP (1985) Structure of the signal recognition particle by electron microscopy. Proc Natl Acad Sci U S A 82:785–789

Arnold CE, Wittrup KD (1994) The stress response to loss of signal recognition particle function in Saccharomyces cerevisiae. J Biol Chem 269:30412–30418

Bacher G, Lütcke H, Jungnickel B, Rapaport TA, Dobberstein B (1996) Regulation by the ribosome of the GTPase of the signal-recognition particle during protein targeting. Nature 381:248–251

Bacher G, Pool M, Dobberstein B (1999) The ribosome regulates the GTPase of the β-subunit of the signal recognition particle receptor. J Cell Biol 146:723–730

Batey RT, Rambo RP, Lucast L, Rha B, Doudna JA (2000) Crystal structure of the ribonucleoprotein core of the signal recognition particle. Science 287:1232–1239

Beck K, Wu LF, Brunner J, Müller M (2000) Discrimination between SRP-and SecA/SecB-dependent substrates involves selective recognition of nascent chains by SRP and trigger factor. EMBO J 19:134–143

Beck K, Eisner G, Trescher D, Dalbey RE, Brunner J, Müller M (2001) YidC, an assembly site for polytopic Escherichia coli membrane proteins located in immediate proximity to the SecYE translocon and lipids. EMBO Rep 2:709–714

Beckmann R, Bubeck D, Grassucci R, Penczek P, Verschoor A, Blobel G, Frank J (1997) Alignment of conduits for the nascent polypeptide chain in the ribosome-Sec61 complex. Science 278:2123–2126

Beckmann R, Spahn CM, Eswar N, Helmers J, Penczek PA, Sali A, Frank J, Blobel G (2001) Architecture of the protein-conducting channel associated with the translating 80S ribosome. Cell 107:361–372

Bernstein HD (1998) Protein targeting: getting into the groove. Current Biol 8:R715–R718

Bernstein HD, Hyndman JB (2001) Physiological basis for conservation of the signal recognition particle targeting pathway in Escherichia coli. J Bacteriol 183:2187–2197

Bernstein HD, Poritz MA, Strub K, Hoben PJ, Brenner S, Walter P (1989) Model for signal sequence recognition from amino-acid sequence of 54 K subunit of signal recognition particle. Nature 340:482–486

Bernstein HD, Zopf D, Freymann DM, Walter P (1993) Functional substitution of the signal recognition particle 54-kDa subunit by its Escherichia coli homolog. Proc Natl Acad Sci U S A 90:5229–5233

Bhuiyan SH, Gowda K, Hotokezaka H, Zwieb C (2000) Assembly of archaeal signal recognition particle from recombinant components Nucleic Acids Res 15:1365–1373

Bibi E, Herskovits AA, Bochkareva ES, Zelazny A (2001) Putative integral membrane SRP receptors. TIBS 26:15–16

Binenbaum Z, Parola AH, Zaritsky A, Fishov I (1999) Transcription- and translation-dependent changes in membrane dynamics in bacteria: testing the transertion model for domain formation. Mol Microbiol 32:1173–1182

Blobel G (1999) Protein targeting. Biosci Rep 20:303–344

Blobel G, Dobberstein B (1975a) Transfer of proteins across membranes: I. Presence of proteolytically processed and unprocessed nascent immunoglobin light chains on membrane-bound ribosomes of murine myeloma. J Cell Biol 67:835–851

Blobel G, Dobberstein B (1975b) Transfer of proteins across membranes: II. Reconstitution of functional rough microsomes from heterologous components. J Cell Biol 67:852–862

Blobel G, Sabatini D (1971) Dissociation of mammalian polyribosomes into subunits by puromycin. Proc Natl Acad Sci U S A 68:390–394

Bourgaize DB, Fournier MJ (1987) Initiation of translation is impaired in E. coli cells deficient in 4.5S RNA. Nature 325:281–284

Bourgaize DB, Phillips TA, VanBogelen RA, Jones PG, Neidhardt FC, Fournier MJ (1990) Loss of 4.5S RNA induces the heat shock response and lambda prophage in Escherichia coli. J Bacteriol 172:1151–1154

Bourne HR (1995) GTPases: a family of molecular switches and clocks. Philosoph Trans Royal Soc London B Biol Sci 349:283–289

Bourne HR, Sanders DA, McCormick F (1990) The GTPase superfamily: a conserved switch of diverse cell functions. Nature 348:125–132

Bourne HR Sanders DA, McCormick F (1991) The GTPase superfamily: conserved structure and molecular mechanism. Nature 349:117–127

Breukink E, Nouwen N, van Raalte A, Mizushima S, Tommassen J, de Kruijff B (1995) The C terminus of SecA is involved in both lipid binding and SecB binding. J Biol Chem 270:7902–7907

Brown S (1987) Mutations in the gene for EF-G reduce the requirement for 4.5S RNA in the growth of E. coli. Cell 49:825–833

Brown S (1989) Time of action of 4.5S RNA in Escherichia coli translation. J Mol Biol 209:79–90

Bui N, Strub K (1999) New insights into the signal recognition and elongation arrest activities of the signal recognition particle. Biol Chem 380:135–145

Bunai K, Takamatsu H, Horinaka T, Oguro A, Nakamura K, Yamane K (1996) Bacillus subtilis Ffh, a homologue of mammalian SRP54, can intrinsically bind to the precursors of secretory proteins. Biochem Biophys Res Commun 227:762–767

Bunai K, Yamada K, Hayashi K, Nakamura K, Yamane K (1999) Enhancing effect of Bacillus subtilis Ffh, a homologue of the SRP54 subunit of the mammalian signal recognition particle, on the binding of SecA to precursors of secretory proteins in vitro. J Biochem 125:151–159

Caldon CE, Yoong P, March PE (2001) Evolution of a molecular switch: universal bacterial GTPases regulate ribosome function. Mol Microbiol 41:289–297

Chen Y, Sinha K, Perumal K, Gu J, Reddy R (1998) Accurate 3' end processing and adenylation of human signal recognition particle RNA and Alu RNA in vitro. J Biol Chem 273:35023–35031

Chirico WJ, Waters MG, Blobel G (1988) 70 K heat shock related proteins stimulate protein translocation into microsomes. Nature 332:805–810

Clemons WM, Gowda K, Black SD, Zwieb C, Ramakrishnan V (1999) Crystal structure of the conserved subdomain of human protein SRP54M at 2.1 Å resolution: evidence for the mechanism for signal peptide binding. J Mol Biol 292:697–705

Collinson I, Breyton C, Duong F, Tziatzios C, Schubert D, Or E, Rapoport T, Kuhlbrandt W (2001) Projection structure and oligomeric properties of a bacterial core protein translocase. EMBO J 20:2462–2471

Connolly T, Gilmore R (1989) The signal recognition particle receptor mediates the GTP-dependent displacement of SRP from the signal sequence of the nascent polypeptide. Cell 57:599–610

Crickmore N, Salmond GP (1986) The Escherichia coli heat shock regulatory gene is immediately downstream of a cell division operon: the fam mutation is allelic with rpoH. Mol Gen Genet 203:535–539

Cristobal S, Scotti P, Luirink J, von Heijne G, de Gier JW (1999) The signal recognition particle-targeting pathway does not necessarily deliver proteins to the sec-translocase in Escherichia coli. J Biol Chem 274:20068–20070

Crowley KS, Liao S, Worrell VE, Reinhart GD, Johnson AE (1994) Secretory proteins move through the endoplasmic reticulum membrane via an aqueous, gated pore. Cell 78:461–471

Dale H, Krebs MP (1999) Membrane insertion kinetics of a protein domain in vivo. J Biol Chem 274:22693–22698

De Gier JWL, Mansournia P, Valent QA, Phillips GJ, Luirink J, von Heijne G (1996) Assembly of a cytoplasmic membrane protein in Escherichia coli is dependent on the signal recognition particle. FEBS Lett 399:307–309

De Gier JW, Scotti PA, Saaf A, Valent QA, Kuhn A, Luirink J, von Heijne G (1998) Differential use of the signal recognition particle translocase targeting pathway for inner membrane protein assembly in Escherichia coli. Proc Natl Acad Sci U S A 95:14646–14651

De Leeuw E, Poland D, Mol O, Sinning I, ten Hagen-Jongman CM, Oudega B, Luirink J (1997) Membrane association of FtsY, the E. coli SRP receptor. FEBS Lett 416:225–229

De Leeuw E, te Kaat K, Moser C, Menestrina G, Demel R, de Kruijff B, Oudega B, Luirink J, Sinning I (2000) Anionic phospholipids are involved in membrane association of FtsY and stimulate its GTPase activity. EMBO J 19:531–541

DeLille J, Peterson EC, Johnson T, Moore M, Kight A, Henry R (2000) A novel precursor recognition element facilitates posttranslational binding to the signal recognition particle in chloroplasts. Proc Natl Acad Sci U S A 97:1926–1931

Deshaies RJ, Koch BD, Werner-Washburne M, Craig EA, Schekman R (1988) A subfamily of stress proteins facilitates translocation of secretory and mitochondrial precursor polypeptides. Nature 332:800–805

Deshaies RJ, Sanders SL, Feldheim DA, Schekman R (1991) Assembly of yeast Sec proteins involved in translocation into the endoplasmic reticulum into a membrane-bound multisubunit complex. Nature 349:806–808

Deuerling E, Schulze-Specking A, Tomoyasu T, Mogk A, Bukau B (1999) Trigger factor and DnaK cooperate in folding of newly synthesized proteins. Nature 400:693–696

Diener JL, Wilson C (2000) Role of SRP19 in assembly of the Archaeoglobus fulgidus signal recognition particle. Biochemistry 39:12862–12874

Driessen AJ (2001) SecB, a molecular chaperone with two faces. Trends Microbiol 9:193–196

Driessen AJ, Manting EH, van der Does C (2001) The structural basis of protein targeting and translocation in bacteria. Nat Struct Biol 8:492–498

Duong F, Wickner W (1997) The SecDFYajC domain of preprotein translocase controls preprotein movement by regulating SecA membrane cycling. EMBO J 16:4871–4879

Duong F, Eichler J, Price A, Leonard MR, Wickner W (1997) Biogenesis of the Gram-negative bacterial envelope. Cell 91:567–573

Economou A (2000) Bacterial protein translocase: a unique molecular machine with an army of substrates. FEBS Lett 2000 476:18–21

Economou A, Pogliano JA, Beckwith J, Oliver DB, Wickner W (1995) SecA membrane cycling at SecYEG is driven by distinct ATP binding and hydrolysis events and is regulated by SecD and SecF. Cell 1995 83:1171–1181

Eggers DK, Welch WJ, Hansen WJ (1997) Complexes between nascent polypeptides and their molecular chaperones in the cytosol of mammalian cells. Mol Biol Cell 8:1559–1573

Eichacker LA, Henry R. Function of a chloroplast SRP in thylakoid protein export (2001) Biochim Biophys Acta 12:120–134
Eichler J, Moll R (2001) The signal recognition particle of archaea. Trends Microbiol 9:130–136
Elowitz MB, Surette MG, Wolf PE, Stock JB, Leibler S (1999) Protein mobility in the cytoplasm of Escherichia coli. J Bacteriol 181:197–203
Ernst F, Hoffschulte HK, Thome-Kromer B, Swidersky UE, Werner PK, Muller M (1994) Precursor-specific requirements for SecA, SecB, and Δ μH+ during protein export of Escherichia coli. J Biol Chem 269:12840–12845
Fekkes P, van der Does C, Driessen AJ (1997) The molecular chaperone SecB is released from the carboxy-terminus of SecA during initiation of precursor protein translocation. EMBO J 16:6105–6113
Finke K, Plath K, Panzner S, Prehn S, Rapoport TA, Hartmann E, Sommer T (1996) A second trimeric complex containing homologs of the Sec61p complex functions in protein transport across the ER membrane of S. cerevisiae. EMBO J 15:1482–1494
Freymann DM, Keenan RJ, Stroud RM, Walter P (1997) Structure of the conserved GTPase domain of the signal recognition particle. Nature 385:361–364
Freymann DM, Keenan RJ, Stroud RM, Walter P (1999) Functional changes in the structure of the SRP GTPase on binding GDP and Mg2+GDP. Nature Struct Biol 6:793–801
Fulga TA, Sinning I, Dobberstein B, Pool MR (2001) SRβ coordinates signal sequence release from SRP with ribosome binding to the translocon. EMBO J 20:2338–2247
Fünfschilling U, Rospert S (1999) Nascent polypeptide-associated complex stimulates protein import into yeast mitochondria. Mol Biol Cell 10:3289–3299
Gellman S (1991) On the role of methionine residues in the sequence-independent recognition of nonpolar protein surfaces. Biochemistry 30:6633–6636
George R, Beddoe T, Landl K, Lithgow T (1998) The yeast nascent polypeptide-associated complex initiates protein targeting to mitochondria in vivo. Proc Natl Acad Sci U S A 95:2296–2301
Gill DR, Salmond GP (1987) The Escherichia coli cell division proteins FtsY, FtsE and FtsX are inner membrane-associated. Mol Gen Genet 210:504–508
Gilmore R, Walter P, Blobel G (1982a) Protein translocation across the endoplasmic reticulum. I. Detection in the microsomal membrane of a receptor for the signal recognition particle. J Cell Biol 95:463–469
Gilmore R, Walter P, Blobel G (1982b) Protein translocation across the endoplasmic reticulum. II. Isolation and characterization of the signal recognition particle receptor. J Cell Biol 95:470–477
Goodsell DS (1991) Inside a living cell. TIBS 16:203–206
Görlich D, Rapoport TA (1993) Protein translocation into proteoliposomes reconstituted from purified components of the endoplasmic reticulum membrane. Cell 75:615–630
Görlich D, Prehn S, Hartmann E, Kalies KU, Rapoport TA (1992) A mammalian homolog of SEC61p and SECYp is associated with ribosomes and nascent polypeptides during translocation. Cell 71:489–503
Gorodkin J, Knudsen B, Zwieb C, Samuelsson T (2001) SRPDB (Signal Recognition Particle Database). Nucleic Acids Res 29:169–170
Grosshans H, Deinert K, Hurt E, Simos G (2001) Biogenesis of the signal recognition particle (SRP) involves import of SRP proteins into the nucleolus, assembly with SRP-RNA, and Xpo1p-mediated export. J Cell Biol 153:745–761
Groves MR, Mant A, Kuhn A, Koch J, Dubel S, Robinson C, Sinning I (2001) Functional characterization of recombinant chloroplast signal recognition particle. J Biol Chem. 276:27778–27786
Gruss OJ, Feick P, Frank R, Dobberstein B (1999) Phosphorylation of components of the ER translocation site. Eur J Biochem 260:785–793
Gundelfinger ED, Krause E, Melli M, Dobberstein B (1983) The organization of the 7SL RNA in the signal recognition particle. Nucl Acid Res 11:7363–7374
Guthrie B, Wickner W (1990) Trigger factor depletion or overproduction causes defective cell division but does not block protein export. J Bacteriol 172:5555–5562
Gutierrez JA, Crowley PJ, Cvitkovitch DG, Brady LJ, Hamilton IR, Hillman JD, Bleiweis AS(1999) Streptococcus mutans ffh, a gene encoding a homologue of the 54 kDa subunit of the signal recognition particle, is involved in resistance to acid stress. Microbiology 145:357–366
Hamman BD, Hendershot LM, Johnson AE (1998) BiP maintains the permeability barrier of the ER membrane by sealing the lumenal end of the translocon pore before and early in translocation. Cell 92:747–758
Hanein D, Matlack KE, Jungnickel B, Plath K, Kalies KU, Miller KR, Rapoport TA, Akey CW (1996) Oligomeric rings of the Sec61p complex induced by ligands required for protein translocation. Cell 87:721–732
Hann BC, Walter P (1991) The signal recognition particle in S. cerevisiae. Cell 67:131–144

Hansen W, Walter P (1988) Prepro-carboxypeptidase Y and a truncated form of pre-invertase, but not full-length pre-invertase, can be posttranslationally translocated across microsomal vesicle membranes from Saccharomyces cerevisiae. J Cell Biol 106:1075–1081

Hatsuzawa K, Tagaya M, Mizushima S (1997) The hydrophobic region of signal peptides is a determinant for SRP recognition and protein translocation across the ER membrane. J Biochem 121:270–277

He XP, Bataille N, Fried HM (1994) Nuclear export of signal recognition particle RNA is a facilitated process that involves the Alu sequence domain. J Cell Sci 107:903–912

Hegde RS, Lingappa VR (1996) Sequence-specific alteration of the ribosome-membrane junction exposes nascent secretory proteins to the cytosol. Cell 85:217–228

Hegde RS, Voigt S, Rapoport TA, Lingappa VR (1998) TRAM regulates the exposure of nascent secretory proteins to the cytosol during translocation into the endoplasmic reticulum. Cell 92:621–631

Hell K, Herrmann JM, Pratje E, Neupert W, Stuart RA (1998) Oxa1p, an essential component of the N-tail protein export machinery in mitochondria. Proc Natl Acad Sci U S A 95:2250–2255

Hell K, Neupert W, Stuart RA (2001) Oxa1p acts as a general membrane insertion machinery for proteins encoded by mitochondrial DNA. EMBO J 20:1281–1288

Herrmann JM, Neupert W, Stuart RA (1997) Insertion into the mitochondrial inner membrane of a polytopic protein, the nuclear-encoded Oxa1p. EMBO J 16:2217–2226

Herskovitz AA, Bibi E (2000) Association of Escherichia coli ribosomes with the inner membrane requires the signal recognition particle receptor but is independent of the signal recognition particle. Proc. Natl Acad Sci U S A 97:4621–4626

Herskowitz AA, Seluanov A, Rajsbaum R, ten Hagen-Jongman CM, Henrichs T, Bochkareva ES, Phillips GJ, Probst FJ, Nakae T, Ehrmann M, Luirink J, Bibi E (2001) Evidence for coupling of membrane targeting and function of the signal recognition particle (SRP) receptor FtsY. EMBO Rep 2:1040–1046

Hesterkamp T, Hauser S, Lutcke H, Bukau B (1996) Escherichia coli trigger factor is a prolyl isomerase that associates with nascent polypeptide chains. Proc Natl Acad Sci U S A 93:4437–4441

Hirose I, Sano K, Shioda I, Kumano M, Nakamura K, Yamane K (2000) Proteome analysis of Bacillus subtilis extracellular proteins: a two-dimensional protein electrophoretic study. Microbiology 146:65–75

Hoffman NE, Franklin AE (1994) Evidence for a stromal GTP requirement for the integration of a chlorophyll a/b-binding polypeptide into thylakoid membranes. Plant Physiol 105:295–304

Honda K, Nakamura K, Nishiguchi M, Yamane K (1993) Cloning and characterization of a Bacillus subtilis gene encoding a homolog of the 54-kilodalton subunit of mammalian signal recognition particle and Escherichia coli Ffh. J Bacteriol 175:4885–4894

Jacobson MR, Pederson T (1998) Localization of signal recognition particle RNA in the nucleolus of mammalian cells. Proc Natl Acad Sci U S A 95:7981–7986

Jagath JR, Rodnina MV, Lentzen G, Wintermeyer W (1998) Interaction of guanine nucleotides with the signal recognition particle from Escherichia coli. Biochemistry 37:15408–15413

Jagath JR, Rodnina MV, Wintermeyer W (2000) Conformational changes in the bacterial SRP receptor FtsY upon binding of guanine nucleotides and SRP. J Mol Biol 295:745–753

Jagath JR, Matassova NB, de Leeuw E, Warnecke JM, Lentzen G, Rodnina MV, Luirink J, Wintermeyer W (2001) Important role of the tetraloop region of 4.5S RNA in SRP binding to its receptor FtsY. RNA 7:293–301

Jensen C, Pedersen S (1994) Concentrations of 4.5S RNA and Ffh protein in Escherichia coli: the stability of Ffh protein is dependent on the concentration of 4.5S RNA. J Bacteriol 176:7148–7145

Jensen CG, Brown S, Pedersen S (1994) Effect of 4.5S RNA depletion on Escherichia coli protein synthesis and secretion. J Bacteriol 176:2502–2506

Jonas-Straube E, Hutin C, Hoffman NE, Schunemann D (2001) Functional analysis of the protein-interacting domains of chloroplast SRP43. J Biol Chem 276:24654–24660

Kaine BP (1990) Structure of the archaebacterial 7S RNA molecule. Mol Gen Genet 221:315–321

Kalies KU, Görlich D, Rapoport TA (1994) Binding of ribosomes to the rough endoplasmic reticulum mediated by the Sec61p-complex. J Cell Biol 126:925–934

Keenan R, Freymann DM, Walter P, Stroud RM (1998) Crystal structure of the signal sequence binding subunit of the signal recognition particle. Cell 94:181–191

Keenan RJ, Freymann DM, Stroud RM, Walter P (2001) The signal recognition particle. Annu Rev Biochem 70:755–775

Kim J, Rusch S, Luirink J, Kendall DA (2001) Is Ffh required for export of secretory proteins? FEBS Lett 505:245–248

Kim YJ, Rajapandi T, Oliver D (1994) SecA protein is exposed to the periplasmic surface of the E. coli inner membrane in its active state. Cell 78:845–853

Klein RR, Mason HS, Mullet JE Light-regulated translation of chloroplast proteins (1988) I. Transcripts of psaA-psaB, psbA, and rbcL are associated with polysomes in dark-grown and illuminated barley seedlings. J Cell Biol 106:289–301

Klimyuk VI, Persello-Cartieaux F, Havaux M, Contard-David P, Schuenemann D, Meiherhoff K, Gouet P, Jones JD, Hoffman NE, Nussaume L (1999) A chromodomain protein encoded by the arabidopsis CAO gene is a plant-specific component of the chloroplast signal recognition particle pathway that is involved in LHCP targeting. Plant Cell 11:87–99

Knoblauch NT, Rudiger S, Schonfeld HJ, Driessen AJ, Schneider-Mergener J, Bukau B (1999) Substrate specificity of the SecB chaperone. J Biol Chem 274:34219–34225

Koch HG, Müller M (2000) Dissecting the translocase and integrase functions of the Escherichia coli SecYEG translocon. J Cell Biol 150:689–694

Koch HG, Hengelage T, Neumann-Haefelin C, MacFarlane J, Hoffschulte HK, Schimz KL, Mechler B, Müller M (1999) In vitro studies with purified components reveal signal recognition particle (SRP) and SecA/SecB as constituents of two independent protein targeting pathways of Escherichia coli. Mol Biol Cell 10:2163–2173

Koch HG, Moser M, Schimz KL, Müller M (2002) The integration of YidC into the cytoplasmic membrane of E. coli requires the signal recognition particle, SecA and SecYEG. J Biol Chem 277:5715–5718

Kremer BH, van der Kraan M, Crowley PJ, Hamilton IR, Brady LJ, Bleiweis AS (2001) Characterization of the sat operon in Streptococcus mutans: evidence for a role of Ffh in acid tolerance. J Bacteriol 183:2543–2552

Krieg UC, Walter P, Johnson AE (1986) Photocrosslinking of the signal sequence of nascent preprolactin to the 54-kilodalton polypeptide of the signal recognition particle. Proc Natl Acad Sci U S A 83:8604–8608

Kurita K, Honda K, Suzuma S, Takamatsu H, Nakamura K, Yamane K (1996) Identification of a region of Bacillus subtilis Ffh, a homologue of mammalian SRP54 protein, that is essential for binding to small cytoplasmic RNA. J Biol Chem 271:13140–13146

Kusukawa N, Yura T, Ueguchi C, Akiyama Y, Ito K (1989) Effects of mutations in heat-shock genes groES and groEL on protein export in Escherichia coli. EMBO J 8:3517–3521

Larsen N, Zwieb C (1991) SRP-RNA sequence alignment and secondary structure. Nucl Acid Res 19:209–215

Lauring B, Sakai H, Kreibich G, Wiedmann M (1995) Nascent polypeptide-associated complex protein prevents mistargeting of nascent chains to the endoplasmic reticulum. Proc Natl Acad Sci U S A 92:5411–5415

Lee HC, Bernstein HD (2001) The targeting pathway of Escherichia coli presecretory and integral membrane proteins is specified by the hydrophobicity of the targeting signal. Proc Natl Acad Sci U S A 98:3471–3476

Legate KR, Falcone D, Andrews DW (2000) Nucleotide dependent binding of the GTPase domain of the signal recognition particle receptor β-subunit to the α-subunit. J Biol Chem 275:27439–27446

Lentzen G, Dobberstein B, Wintermeyer W (1994) Formation of SRP-like particle induces a conformational change in E. coli 4.5S RNA. FEBS Lett 348:233–238

Li X, Henry R, Yuan J, Cline K, Hoffman NE (1995) A chloroplast homologue of the signal recognition particle subunit SRP54 is involved in the posttranslational integration of a protein into thylakoid membranes. Proc Natl Acad Sci U S A 92:3789–3793

Lipp J, Dobberstein B, Haeuptle MT (1987) Signal recognition particle arrests elongation of nascent secretory and membrane proteins at multiple sites in a transient manner. J Biol Chem 262:1680–1684

Lorimer GH (1996) A quantitative assessment of the role of the chaperon in proteins in protein folding in vivo. FASEB J 10:5–9

Lu Y, Qi HY, Hyndman JB, Ulbrandt ND, Teplyakov A, Tomasevic N, Bernstein HD (2001) Evidence for a novel GTPase priming step in the SRP protein targeting pathway. EMBO J 20:6724–6734

Luirink J, ten Hagen-Jongman CM, van der Weijden CC, Oudega B, High S, Dobberstein B, Kusters R (1994) An alternative protein targeting pathway in Escherichia coli: studies on the role of FtsY. EMBO J 13:2289–2296

Lütcke H, Prehn S, Asford A, Remus M, Frank R, Dobberstein B (1993) Assembly of the 68- and 72-kD proteins of signal recognition particle with 7S RNA. J Cell Biol 121:977–985

Lynch AS, Wang JC (1993) Anchoring of DNA to bacterial cytoplasmic membrane through cotranscriptional synthesis of polypeptides encoding membrane proteins or proteins for export: a mechanism of plasmid hypernegative supercoiling in mutants deficient in DNA topoisomerase I. J Bacteriol 175:1645–1655

Ma D, Cook DN, Pon NG, Hearst JE (1994) Efficient anchoring of RNA polymerase in Escherichia coli during coupled transcription-translation of genes encoding integral inner membrane polypeptides. J Biol Chem 269:15362–15370
MacFarlane J, Müller M (1995) The functional integration of a polytopic membrane protein of Escherichia coli is dependent on the bacterial signal recognition particle. Eur J Biochem 233:766–771
Maeshima H, Okuno E, Aimi T, Morinaga T, Itoh T (2001) An archaeal protein homologous to mammalian SRP54 and bacterial Ffh recognizes a highly conserved region of SRP RNA. FEBS Lett 507:336–340
Manting EH, Driessen AJ (2000) Escherichia coli translocase: the unravelling of a molecular machine. Mol Microbiol 37:226–238
Manting EH, van Der Does C, Remigy H, Engel A, Driessen AJ (2000) SecYEG assembles into a tetramer to form the active protein translocation channel. EMBO J 19:852–861
Mason N, Ciufo LF, Brown JD (2000) Elongation arrest is a physiologically important function of signal recognition particle. EMBO J 19:4164–4174
Matlack KE, Misselwitz B, Plath K, Rapoport TA (1999) BiP acts as a molecular ratchet during posttranslational transport of prepro-alpha factor across the ER membrane. Cell 97:553–564
Matsuyama S, Fujita Y, Sagara K, Mizushima S (1992) Overproduction, purification and characterization of SecD and SecF, integral membrane components of the protein translocation machinery of Escherichia coli. Biochim Biophys Acta 1122:77–84
Menetret JF, Neuhof A, Morgan DG, Plath K, Radermacher M, Rapoport TA, Akey CW (2000) The structure of ribosome-channel complexes engaged in protein translocation. Mol Cell 6:1219–1232
Meyer DI, Krause E, Dobberstein B (1982) Secretory protein translocation across membranes-the role of the "docking protein". Nature 297:647–650
Meyer HA, Grau H, Kraft R, Kostka S, Prehn S, Kalies KU, Hartmann E (2000) Mammalian Sec61 is associated with Sec62 and Sec63. J Biol Chem 275:14550–14557
Meyer TH, Menetret JF, Breitling R, Miller KR, Akey CW, Rapoport TA (1998) The bacterial SecY/E translocation complex forms channel-like structures similar to those of the eukaryotic Sec61p complex. J Mol Biol 1285:1789–1800
Miller JD, Wilhelm H, Gierasch L, Gilmore R, Walter P (1993) GTP binding and hydrolysis by the signal recognition particle during initiation of protein translocation. Nature 366:351–354
Miller JD, Bernstein HD, Walter P (1994) Interaction of E. coli Ffh/4.5S ribonucleoprotein and FtsY mimics that of mammalian signal recognition particle and its receptor. Nature 367:657–659
Miller JD, Tajima S, Lauffer L, Walter P (1995) The β subunit of the signal recognition particle receptor is a transmembrane GTPase that anchors the α-subunit, a peripheral GTPase, to the endoplasmic reticulum membrane. J Cell Biol 128:273–282
Millman JS, Andrews DW (1997) Switching the model: a concerted mechanism for GTPases in protein targeting. Cell 89:673–676
Millman JS, Andrews DW (1999) A site-specific, membrane-dependent cleavage event defines the membrane binding domain of FtsY. J Biol Chem 274:33227–33234
Millman JS, Qi HY, Vulcu F, Bernstein HD, Andrews DW (2001) FtsY binds to the Escherichia coli inner membrane via interactions with phosphatidylethanolamine and membrane proteins. J Biol Chem 276:25982–25989
Mitchell C, Oliver D (1993) Two distinct ATP-binding domains are needed to promote protein export by Escherichia coli SecA ATPase. Mol Microbiol 10:483–497
Moll R, Schmidtke S, Schafer G (1999) Domain structure, GTP-hydrolyzing activity and 7S RNA binding of Acidianus ambivalens ffh-homologous protein suggest an SRP-like complex in archaea. Eur J Biochem 259:441–448
Montoya G, Svensson C, Luirink J, Sinning I (1997) Crystal structure of the NG domain from the signal-recognition particle receptor FtsY. Nature 385:365–368
Montoya G, Kaat K, Moll R, Schafer G, Sinning I (2000) The crystal structure of the conserved GTPase of SRP54 from the archaeon Acidianus ambivalens and its comparison with related structures suggests a model for the SRP-SRP receptor complex. Structure Fold Des 15:515–525
Moore M, Harrison MS, Peterson EC, Henry R (2000) Chloroplast Oxa1p homolog albino3 is required for post-translational integration of the light harvesting chlorophyll-binding protein into thylakoid membranes. J Biol Chem 275:1529–1532
Mori H, Ito K (2001) The Sec protein-translocation pathway. Trends Microbiol 9:494–500
Moser C, Mol O, Goody RS, Sinning I (1997) The signal recognition particle receptor of Escherichia coli (FtsY) has a nucleotide exchange factor built into the GTPase domain. Proc Natl Acad Sci U S A 94:11339–11344
Mothes W, Prehn S, Rapoport TA (1994) Systematic probing of the environment of a translocating secretory protein during translocation through the ER membrane. EMBO J 13:3973–3982

Mothes W, Heinrich SU, Graf R, Nilsson I, von Heijne G, Brunner J, Rapoport TA (1997) Molecular mechanism of membrane protein integration into the endoplasmic reticulum. Cell 89:523–533
Müller G, Zimmermann R (1987) Import of honeybee prepromelittin into the endoplasmic reticulum: structural basis for independence of SRP and docking protein. EMBO J 6:2099–2107
Müller M, Koch HG, Beck J, Schäfer U (2001) Protein traffic in bacteria: multiple routes from the ribosome to and across the membrane. Progr Nucl Acid Res Mol Biol 66:107–157
Mutka SC, Walter P (2001) Multifaceted physiological response allows yeast to adapt to the loss of the signal recognition particle-dependent protein-targeting pathway. Mol Biol Cell 12:577–588
Nakamura K, Imai Y, Nakamura A, Yamane K (1992) Small cytoplasmic RNA of Bacillus subtilis: functional relationship with human signal recognition particle 7S RNA and Escherichia coli 4.5S RNA. J Bacteriol 174:2185–2192
Nakamura K, Fujii Y, Shibata T, Yamane K (1999a) Depletion of Escherichia coli 4.5S RNA leads to an increase in the amount of protein elongation factor EF-G associated with ribosomes. Eur J Biochem 259:543–550
Nakamura K, Yahagi S, Yamazaki T, Yamane K (1999b) Bacillus subtilis histone-like protein, HBsu, is an integral component of a SRP-like particle that can bind the Alu domain of small cytoplasmic RNA. J Biol Chem 274:13569–13576
Nakamura K, Miyamoto H, Suzuma S, Sakamato T, Kawai G, Yamane K (2001) Minimal functional structure of Escherichia coli 4.5S RNA required for binding elongation factor EF-G. J Biol Chem 276:22844–22849
Neuhof A, Rolls MM, Jungnickel B, Kalies KU, Rapoport TA (1998) Binding of signal recognition particle gives ribosome/nascent chain complexes a competitive advantage in endoplasmic reticulum membrane interaction. Mol Biol Cell 9:103–115
Neumann-Haefelin C, Schäfer U, Müller M, Koch HG (2000) SRP-dependent co-translational targeting and SecA-dependent translocation analyzed as individual steps in the export of a bacterial protein. EMBO J 19:6419–6426
Newitt JA, Bernstein HD (1997) The N domain of the signal recognition particle 54 kDa subunit promotes efficient signal sequence binding. Eur J Biochem 245:720–729
Ng DT, Brown JD, Walter P (1996) Signal sequences specify the targeting route to the endoplasmic reticulum membrane. J Cell Biol 134:269–278
Nilsson R, Brunner J, Hoffman NE, van Wijk KJ (1999) Interactions of ribosome nascent chain complexes of the chloroplast-encoded D1 thylakoid membrane protein with cpSRP54. EMBO J 18:733–742
Nishiyama K, Suzuki T, Tokuda H (1996) Inversion of the membrane topology of SecG coupled with SecA-dependent preprotein translocation. Cell 85:71–81
Ogg SC, Barz WP, Walter P (1998) A functional GTPase domain, but not its transmembrane domain, is required for function of the SRP receptor β-subunit. J Cell Biol 142:341–354
Packer JC, Howe CJ (2001) Algal plastid genomes encode homologues of the SRP-associated RNA. Mol Microbiol 27:508–510
Padmanabhan S, Freymann DM (2001) The conformation of bound GMPPNP suggests a mechanism for gating the active site of the SRP GTPase. Structure 9:859–867
Panzner S, Dreier L, Hartmann E, Kostka S, Rapoport TA (1995) Posttranslational protein transport in yeast reconstituted with a purified complex of Sec proteins and Kar2p. Cell 81:561–570
Peluso P, Herschlag D, Nock S, Freymann DM, Johnson AE, Walter P (2000) Role of 4.5S RNA in assembly of the bacterial signal recognition particle with its receptor. Science 288:1640–1643
Phillips GJ, Silhavy TJ (1992) The E. coli ffh gene is necessary for viability and efficient protein export. Nature 359:744–746
Pilgrim ML, van Wijk KJ, Parry DH, Sy DA, Hoffman NE (1998) Expression of a dominant negative form of cpSRP54 inhibits chloroplast biogenesis in Arabidopsis. Plant J 13:177–186
Plath K, Rapoport TA (2000) Spontaneous release of cytosolic proteins from posttranslational substrates before their transport into the endoplasmic reticulum. J Cell Biol 151:167–178
Politz JC, Yarovoi S, Kilroy SM, Gowda K, Zwieb C, Pederson T (2000) Signal recognition particle components in the nucleolus. Proc Natl Acad Sci U S A 97:55–60
Poritz MA, Strub K, Walter P (1988) Human SRP RNA and E. coli 4.5S RNA contain a highly homologous structural domain. Cell 55:4–6
Poritz MA, Bernstein HD, Strub K, Zopf D, Wilhelm H, Walter P (1990) An E. coli ribonucleoprotein containing 4.5S RNA resembles mammalian signal recognition particle. Science 250:1111–1117
Potter MD, Nicchitta CV (2000) Regulation of ribosome detachment from the mammalian endoplasmic reticulum membrane. J Biol Chem 275:33828–33835
Potter MD, Seiser RM, Nicchitta CV (2001) Ribosome exchange revisited: a mechanism for translation-coupled ribosome detachment from the ER membrane. Trends Cell Biol 11:112–115

Powers T, Walter P (1995) Reciprocal stimulation of GTP hydrolysis by two directly interacting GTPases. Science 269:1422–1424

Prinz A, Hartmann E, Kalies KU (2000) Sec61p is the main ribosome receptor in the endoplasmic reticulum of Saccharomyces cerevisiae. Biol Chem 381:1025–1029

Pugsley AP (1993) The complete general secretory pathway in Gram-negative bacteria. Microbiol Rev 57:50–108

Qi HY, Bernstein HD (1999) SecA is required for the insertion of inner membrane proteins targeted by the Escherichia coli signal recognition particle. J Biol Chem 274:8993–8997

Raden D, Gilmore R (1998) Signal recognition particle-dependent targeting of ribosomes to the rough endoplasmic reticulum in the absence and presence of the nascent polypeptide-associated complex. Mol Biol Cell 9:117–130

Randall LL, Hardy SJ (1983) Preparation of free and membrane-bound polysomes from Escherichia coli. Meth Enzymol 97:70–76

Randall LL, Hardy SJ (1995) High selectivity with low specificity: how SecB has solved the paradox of chaperone binding. TIBS 20:65–69

Rapiejko PJ, Gilmore R (1997) Empty site forms of the SRP54 and SRα GTPases mediate targeting of ribosome-nascent chain complexes to the endoplasmic reticulum. Cell 89:703–713.

Rapoport TA, Jungnickel B, Kutay U (1996) Protein transport across the eukaryotic endoplasmic reticulum and bacterial inner membranes. Annu Rev Biochem 65:271–303

Rapoport TA, Matlack KE, Plath K, Misselwitz B, Staeck O (1999) Posttranslational protein translocation across the membrane of the endoplasmic reticulum. Biol Chem 380:1143–1150

Ribes V, Römisch K, Giner A, Dobberstein B, Tollervey D (1990) E. coli 4.5S RNA is part of a ribonucleoprotein particle that has properties related to signal recognition particle. Cell 63:591–600

Robb A, Brown JD (2001) Protein transport: two translocons are better than one. Mol Cell 8:484–486

Römisch K, Webb J, Herz J, Prehn S, Frank R, Vingron M, Dobberstein B (1989) Homology of the 54 K protein of the signal recognition particle, docking protein, and two E. coli proteins with putative GTP-binding domains. Nature 340:478–482

Römisch K, Webb J, Lingelbach K, Gausepohl H, Dobberstein B (1990) The 54 kDa protein of signal recognition particle contains a methionine rich RNA binding domain. J Cell Biol 111:1793–1802

Rose MA, Weeks KM (2001) Visualizing induced fit in early assembly of the human signal recognition particle. Nature Struct Biol 8:515–520

Rothblatt JA, Meyer DI (1986) Secretion in yeast: reconstitution of the translocation and glycosylation of alpha-factor and invertase in a homologous cell-free system Cell 44:619–628

Rothe C, Lehle L (1998) Sorting of invertase signal peptide mutants in yeast dependent and independent on the signal-recognition particle. Eur J Biochem 252:16–24

Sadler I, Chiang A, Kurihara T, Rothblatt J, Way J, Silver P (1989) A yeast gene important for protein assembly into the endoplasmic reticulum and the nucleus has homology to DnaJ, an Escherichia coli heat shock protein. J Cell Biol 109:2665–2675

Schatz PJ, Beckwith J (1990) Genetic analysis of protein export in Escherichia coli. Ann Rev Genet 24:215–248

Schmitz U, James TL, Lukavsky P, Walter P (1999) Structure of the most conserved internal loop in SRP RNA. Nature Struct Biol 6:634–638

Scholz C, Stoller T, Zarnt T, Fischer G, Schmid FX (1997) Cooperation of enzymatic and chaperone functions of trigger factor in the catalysis of protein folding. EMBO J 16:54–58

Schuenemann D, Gupta S, Persello-Cartieaux F, Klimyuk VI VI, Jones JDG, Nussaume L, Hoffman NE (1998) A novel signal recognition particle targets light-harvesting proteins to the thylakoid membranes Proc Natl Acad Sci U S A 95:10312–10316

Schulz HN, Jorgenson BB (2001) Big Bacteria. Ann Rev Microbiol 55:105–137

Scotti PA, Valent QA, Manting EH, Urbanus ML, Driessen AJ, Oudega B, Luirink J (1999) SecA is not required for signal recognition particle-mediated targeting and initial membrane insertion of a nascent inner membrane protein. J Biol Chem 274:29883–29888

Scotti PA, Urbanus ML, Brunner J, de Gier JW, von Heijne G, van der Does C, Driessen AJ, Oudega B, Luirink J (2000) YidC, the Escherichia coli homologue of mitochondrial Oxa1p, is a component of the Sec translocase. EMBO J 19:542–549

Seiser RM, Nicchitta CV (2000) The fate of membrane-bound ribosomes following the termination of protein synthesis. J Biol Chem 275:33820–33827

Seluanov A, Bibi E (1997) FtsY, the prokaryotic signal recognition particle receptor homologue, is essential for biogenesis of membrane proteins (1997). J Biol Chem 272:2053–2055

Shiabata T, Fujii Y, Nakamura Y, Nakamura K, Yamane K (1996) Identification of protein synthesis elongation factor G as a 4.5S RNA-binding protein in Escherichia coli. J Biol Chem 271:13162–13168

Sianidis G, Karamanou S, Vrontou E, Boulias K, Repanas K, Kyrpides N, Politou AS, Economou A (2001) Cross-talk between catalytic and regulatory elements in a DEAD motor domain is essential for SecA function. EMBO J 20:961–970

Siegel V, Walter P (1985) Elongation arrest is not a prerequisite for secretory protein translocation across the microsomal membrane. J Cell Biol 100:1913–1921

Siegel V, Walter P (1986) Removal of the Alu structural domain from signal recognition particle leaves its protein translocation activity intact. Nature 320:81–84

Siegel V, Walter P (1988a) Each of the activities of signal recognition particle (SRP) is contained within a distinct domain: analysis of biochemical mutants of SRP. Cell 52:39–49

Siegel V, Walter P (1988b) Binding sites of the 19 kDa and 68/72 kDa signal recognition particle (SRP) proteins on SRP RNA as determined in protein-RNA footprinting. Proc Natl Acad Sci U S A 85:4315–4319

Snyders S, Ramamurthy V, Oliver D (1997) Identification of a region of interaction between Escherichia coli SecA and SecY proteins. J Biol Chem 272:11302–11306

Song W, Raden D, Mandon EC, Gilmore R (2000) Role of Sec61α in the regulated transfer of the ribosome-nascent chain complex from the signal recognition particle to the translocation channel. Cell 100:333–343

Stirling CJ (1999) Protein targeting to the endoplasmic reticulum in yeast. Microbiology 145:991–998

Stirling CJ, Hewitt EW (1992) The S. cerevisiae SEC65 gene encodes a component of yeast signal recognition particle with homology to human SRP19. Nature 356:534–537

Strub K, Walter P (1990) Assembly of the Alu domain of the signal recognition particle (SRP): dimerization of the two protein components is required for efficient binding to SRP RNA. Mol Cell Biol 10:777–784

Strub K, Fornallaz M, Bui N (1999) The Alu domain homolog of the yeast signal recognition particle consists of an SRP14p homodimer and a yeast-specific RNA structure. RNA 10:1333–1347

Struck JC, Hartmann RK, Toschka HY, Erdmann VA (1989) Transcription and processing of Bacillus subtilis small cytoplasmic RNA. Mol Gen Genet 215:478–482

Suzuma S, Hayashi K, Nakamura K, Yamane K (1999) Analysis of Escherichia coli 4.5S RNA binding affinity to Ffh and EF-G. FEMS Microbiol Lett 180:271–277

Swidersky UE, Rienhöfer-Schwer A, Werner PK, Ernst F, Benson SA, Hoffschulte HK, Müller M (1992) Biochemical analysis of the biogenesis and function of the Escherichia coli export factor SecY. Europ J Biochem 207:803–811

Takamatsu H, Bunai K, Horinaka T, Oguro A, Nakamura K, Watabe K, Yamane K (1997) Identification of a region required for binding to presecretory protein in Bacillus subtilis Ffh, a homologue of the 54-kDa subunit of mammalian signal recognition particle. Eur J Biochem 248:575–582

Taura T, Akiyama Y, Ito K (1994) Genetic analysis of SecY: additional export-defective mutations and factors affecting their phenotypes. Mol Gen Genet 243:261–269

Teter SA, Houry WA, Ang D, Tradler T, Rockabrand D, Fischer G, Blum P, Georgopoulos C, Hartl FU (1999) Polypeptide flux through bacterial Hsp70: DnaK cooperates with trigger factor in chaperoning nascent chains. Cell 97:755–765

Thomas Y, Bui N, Strub K (1997) A truncation in the 14 kDa protein of the signal recognition particle leads to tertiary structure changes in the RNA and abolishes the elongation arrest activity of the particle. Nucl Acid Res 25:1920–1929

Tian H, Beckwith J (2002) Genetic screen yields mutations in genes encoding all known components of the Escherichia coli signal recognition particle pathway. J Bacteriol 184:111–118

Tian H, Boyd D, Beckwith J (2000) A mutant hunt for defects in membrane protein assembly yields mutations affecting signal recognition particle and Sec machinery. Proc Natl Acad Sci U S A 97:4730–4735

Triplett TL, Sgrignoli AR, Gao FB, Yang YB, Tai PC, Gierasch LM (2001) Functional signal peptides bind a soluble N-terminal fragment of SecA and inhibit its ATPase activity. J Biol Chem 276:19648–19655

Tseng TT, Gratwick KS, Kollman J, Park D, Nies DH, Goffeau A, Saier MH Jr (1999) The RND permease superfamily: an ancient, ubiquitous and diverse family that includes human disease and development proteins. J Mol Microbiol Biotechnol 1:107–125

Tu CJ, Schuenemann D, Hoffman NE (1999) Chloroplast FtsY, chloroplast signal recognition particle, and GTP are required to reconstitute the soluble phase of light-harvesting chlorophyll protein transport into thylakoid membranes. J Biol Chem 274:27219–27224

Tyedmers J, Lerner M, Bies C, Dudek J, Skowronek MH, Haas IG, Heim N, Nastainczyk W, Volkmer J, Zimmermann R (2000) Homologs of the yeast Sec complex subunits Sec62p and Sec63p are abundant proteins in dog pancreas microsomes. Proc Natl Acad Sci U S A 97:7214–7219

Tyystjärvi T, Herranen M, Aro EM (2001) Regulation of translation elongation in cyanobacteria: membrane targeting of the ribosome nascent-chain complex controls the synthesis of D1 protein. Molec Microbiol 40:476–484

Ulbrandt ND, Newitt JA, Bernstein HD (1997) The E. coli signal recognition particle is required for the insertion of a subset of inner membrane proteins. Cell 88:187–196

Urbanus ML, Scotti PA, Froderberg L, Saaf A, de Gier JW, Brunner J, Samuelson JC, Dalbey RE, Oudega B, Luirink J (2001) Sec-dependent membrane protein insertion: sequential interaction of nascent FtsQ with SecY and YidC. EMBO Rep 2:524–529

Urbanus ML, Froderberg L, Drew D, Bjork P, de Gier JW, Brunner J, Oudega B, Luirink J (2002) Targeting, insertion and localization of E. coli YidC. J Biol Chem 277:12718–12723

Utz PJ, Hottelet M, Le TM, Kim SJ, Geiger ME, van Venrooij WJ, Anderson P (1998) The 72 kDa component of signal recognition particle is cleaved during apoptosis. J Biol Chem 273:35362–35370

Valent QA, de Gier JWL, von Heijne G, Kendall DA, ten Hagen-Jongman CM, Oudega B, Luirnk J (1997) Nascent membrane and presecretory proteins synthesized in Escherichia coli associate with signal recognition particle and trigger factor. Mol Microbiol 25:53–64

Valent QA, Scotti PA, High S, de Gier JWL, von Heijne G, Lentzen G, Wintermeyer W, Oudega B, Luirink J (1998) The Escherichia coli SRP and SecB targeting pathways converge at the translocon. EMBO J 17:2504–2512

Voigt S, Jungnickel B, Hartmann E, Rapoport TA (1996) Signal sequence-dependent function of the TRAM protein during early phases of protein transport across the endoplasmic reticulum membrane. J Cell Biol 134:25–35

Walter P, Blobel G (1981) Translocation of proteins across the endoplasmic reticulum. II. Signal recognition protein (SRP) mediates the selective binding to microsomal membranes of in vitro assembled polysomes synthesizing secretory protein. III. Signal recognition protein (SRP) causes signal sequence-dependent and site-specific arrest of chain elongation that is released by microsomal membranes. J Cell Biol 91:551–561

Walter P, Blobel G (1983) Disassembly and reconstitution of signal recognition particle. Cell 34:525–533

Wang L, Miller A, Kendall DA (2000) Signal peptide determinants of SecA binding and stimulation of ATPase activity. J Biol Chem 275:10154–10159

Wang S, Sakai H, Wiedmann M (1995) NAC covers ribosome-associated nascent chains thereby forming a protective environment for regions of nascent chains just emerging from the peptidyl transferase center. J Cell Biol 130:519–528

Waters MG, Blobel G (1986) Secretory protein translocation in a yeast cell-free system can occur posttranslationally and requires ATP hydrolysis. J Cell Biol 102:1543–1550

Weichenrieder O, Wild K, Strub K, Cusack S (2000) Structure and assembly of the Alu domain of the mammalian signal recognition particle. Nature 408:167–173

Weichenrieder O, Stehlin C, Kapp U, Birse DE, Timmins PA, Strub K, Cusack S (2001) Hierarchical assembly of the Alu domain of the mammalian signal recognition particle. RNA 7:731–740

Wickner W, Leonard MR (1996) Escherichia coli preprotein translocase. J Biol Chem 271:29514–29516

Wiedmann M, Kurzchalia TV, Hartmann E, Rapoport TA (1987) A signal sequence receptor in the endoplasmic reticulum membrane. Nature 328:830–833

Wiedmann B, Sakai H, Davis TA, Wiedmann M (1994) A protein complex required for signal-sequence-specific sorting and translocation. Nature 370:434–440

Wild J, Rossmeissl P, Walter WA, Gross CA (1996) Involvement of the DnaK-DnaJ-GrpE chaperone team in protein secretion in Escherichia coli. J Bacteriol 178:3608–3613

Wild K, Sinning I, Cusack S (2001) Crystal structure of an early protein-RNA assembly complex of the signal recognition particle. Science 294:598–601

Wilkinson BM, Tyson JR, Stirling CJ (2001) Sshlp determines the translocation and dislocation capacities of the yeast endoplasmic reticulum. Dev Cell 1:401–409

Wolin SL, Walter P (1989) Signal recognition particle mediates a transient elongation arrest of preprolactin in reticulocyte lysate. J Cell Biol 109:2617–2622

Wood H, Luirink J, Tollervey D (1992) Evolutionary conserved nucleotides within the E. coli 4.5S RNA are required for association with P48 in vitro and for optimal function in vivo. Nucl Acid Res 20:5919–5925

Xu Z, Knafels JD, Yoshino K (2000) Crystal structure of the bacterial protein export chaperone secB. Nat Struct Biol 7:1172–1177

Yahr TL, Wickner WT (2000) Evaluating the oligomeric state of SecYEG in preprotein translocase. EMBO J 19:4393–4401

Yaver DS, Matoba S, Ogrydziak DM (1992) A mutation in the signal recognition particle 7S RNA of the yeast Yarrowia lipolytica preferentially affects synthesis of the alkaline extracellular protease: in vivo evidence for translational arrest. J Cell Biol 116:605–616

Young BP, Craven RA, Reid PJ, Willer M, Stirling CJ (2001) Sec63p and Kar2p are required for the translocation of SRP-dependent precursors into the yeast endoplasmic reticulum in vivo. EMBO J 20:262–271

Young JC, Andrews DW (1996) The signal recognition particle receptor α-subunit assembles co-translationally on the endoplasmic reticulum membrane during mRNA-encoded translation pause in vitro. EMBO J 15:172–181

Young JC, Ursini J, Legate KR, Miller JD, Walter P, Andrews DW (1995) An amino-terminal domain containing hydrophobic and hydrophilic sequences binds the signal recognition particle receptor α subunit to the β subunit on the endoplasmic reticulum membrane. J Biol Chem 270:15650–15657

Zelazny A, Seluanov A, Cooper A, Bibi E (1997) The NG domain of the prokaryotic signal recognition particle receptor, FtsY, is fully functional when fused to an unrelated integral membrane polypeptide. Proc Natl Acad Sci U S A 94:6025–6029

Zhang L, Paakkarinen V, van Wijk KJ, Aro EM (1999) Co-translational assembly of the D1 protein into photosystem II. J Biol Chem 274:16062–16067

Zhang L, Paakkarinen V, van Wijk KJ, Aro EM (2000) Biogenesis of the chloroplast-encoded D1 protein: regulation of translation elongation, insertion, and assembly into photosystem II. Plant Cell 12:1769–1782

Zhang L, Paakkarinen V, Suorsa M, Aro EM (2001) A SecY homologue is involved in chloroplast-encoded D1 protein biogenesis. J Biol Chem 276:37809–37814

Zheng N, Gierasch LM (1997) Domain interaction in E. coli SRP: stabilization of M domain by RNA is required for effective signal sequence modulation of NG domain. Mol Cell 1:79–87

Zheng T, Nicchitta CV (1999) Structural determinants for signal sequence function in the mammalian endoplasmic reticulum. J Biol Chem 274:36623–36630

Zimmermann R (1998) The role of molecular chaperones in protein transport into the mammalian endoplasmic reticulum. Biol Chem 379:275–282

Zopf D, Bernstein HD, Johnson AE, Walter P (1990) The methionine-rich domain of the 54 kD protein subunit of the signal recognition particle contains an RNA binding site and can be crosslinked to a signal sequence. EMBO J 9:4511–4517

Zopf D, Bernstein HD, Walter P (1993) GTPase domain of the 54 k subunit of the mammalian signal recognition particle is required for protein translocation but not for signal sequence binding. J Cell Biol 120:1113–1121

Zwieb C (1992) Recognition of a tetranucleotide loop of signal recognition particle RNA by protein SRP 19. J Biol Chem 267:15650–15656

Rev Physiol Biochem Pharmacol (2003) 146:95–158
DOI 10.1007/s10254-002-0003-8

B. C. Burckhardt · G. Burckhardt

Transport of organic anions across the basolateral membrane of proximal tubule cells

Published online: 30 January 2003

Abstract Renal proximal tubules secrete diverse organic anions (OA) including widely prescribed anionic drugs. Here, we review the molecular properties of cloned transporters involved in uptake of OA from blood into proximal tubule cells and provide extensive lists of substrates handled by these transport systems. Where tested, transporters have been immunolocalized to the basolateral cell membrane. The sulfate anion transporter 1 (sat-1) cloned from human, rat and mouse, transported oxalate and sulfate. Drugs found earlier to interact with sulfate transport in vivo have not yet been tested with sat-1. The Na^+-dicarboxylate cotransporter 3 (NaDC-3) was cloned from human, rat, mouse and flounder, and transported three Na^+ with one divalent di- or tricarboxylate, such as citric acid cycle intermediates and the heavy metal chelator 2,3-dimercaptosuccinate (succimer). The organic anion transporter 1 (OAT1) cloned from several species was shown to exchange extracellular OA against intracellular α-ketoglutarate. OAT1 translocated, e.g., anti-inflammatory drugs, antiviral drugs, β-lactam antibiotics, loop diuretics, ochratoxin A, and *p*-aminohippurate. Several OA, including probenecid, inhibited OAT1. Human, rat and mouse OAT2 transported selected anti-inflammatory and antiviral drugs, methotrexate, ochratoxin A, and, with high affinities, prostaglandins E_2 and $F_{2\alpha}$. OAT3 cloned from human, rat and mouse showed a substrate specificity overlapping with that of OAT1. In addition, OAT3 interacted with sulfated steroid hormones such as estrone-3-sulfate. The driving forces for OAT2 and OAT3, the relative contributions of all OA transporters to, and the impact of transporter regulation by protein kinases on renal drug excretion in vivo must be determined in future experiments.

Introduction

Renal proximal tubules perform two major tasks: (1) reabsorption of water, salts and organic compounds, which should not be lost with the urine, and (2) secretion of waste products and potentially toxic xenobiotics to ensure their effective disposal with urine. For

B. C. Burckhardt · G. Burckhardt (✉)
Abteilung Vegetative Physiologie und Pathophysiologie, Zentrum Physiologie,
Georg-August-Universität Göttingen, Humboldtallee 23, 37073 Göttingen, Germany
e-mail: gburckh@gwdg.de · Tel.: +49-551-395881 · Fax: +49-551-395883

transcellular secretion processes, compounds must cross the basolateral and the luminal cell membranes consecutively. Since the kidneys handle mainly hydrophilic substances, sufficiently fast translocation across cell membranes requires the presence of transporter proteins. Many transporters have been cloned to date. Their primary structure is known and their properties have been studied individually after heterologous expression in a variety of cell lines. In this review, we focus on the cloned organic anion transporters, sat-1, NaDC-3, OAT1, OAT2, and OAT3, located at the basolateral membrane of renal proximal tubule cells. Earlier results obtained at the levels of intact kidneys, isolated tubules, or membrane vesicles are included, if needed, for better understanding of the properties of the cloned transporters.

Renal proximal tubule cells transport various endogenous and exogenous organic anions. Endogenous organic anions include aliphatic mono- (e.g., lactate, pyruvate, acetoacetate, β-hydroxybutyrate), di- and tricarboxylates (oxalate and Krebs cycle intermediates such as citrate, α-ketoglutarate, succinate), anionic vitamins (e.g., ascorbate, folate), anionic hormones (prostaglandins), second messengers (cAMP, cGMP), bile salts (cholate, glyco- or taurocholate), urate, and a series of anionic metabolites of neurotransmitters (e.g., homovanillate, vanillin mandelate, 5-hydroxyindole acetate). The list of exogenous organic anions transported across proximal tubular epithelium includes ACE inhibitors (captoprilate, enalaprilat and others), angiotensin receptor blockers (losartan and related compounds), β-lactam antibiotics (penicillins, cephalosporins), antiviral drugs (e.g., acyclovir, amantadine, azidothymidine), nonsteroidal anti-inflammatory drugs (e.g., acetylsalicylate, diclofenac, ibuprofen), diuretics (bumetanide, furosemide, thiazides), sedatives (barbiturates) and a number of test agents such as p-aminohippurate (PAH), phenol red and some X-ray contrast agents. These examples underline the physiological and pharmacological importance of proximal tubules in the handling of organic anions of diverse chemical structures.

Whereas many endogenous organic anions are reabsorbed, most exogenous organic anions as well as endogenous waste products undergo net secretion in proximal tubules. The secretion process begins with the uptake of organic anions from the blood into the cell across the basolateral membrane and ends with their exit across the luminal membrane. The uptake of the negatively charged compounds occurs against the inside negative electrical membrane potential difference and in some cases in addition against an intracellular-to-extracellular organic anion concentration difference. Such "uphill" transport requires the input of energy, which can be derived from the cotransport with sodium ions or from the exchange against an intracellularly accumulated organic anion such as α-ketoglutarate. Since a large variety of organic anions must be translocated, a number of transporters with broad specificity are needed. Here we briefly summarize the molecular properties of the cloned transporters and provide alphabetical lists of compounds that have been tested so far (up to May 2002) for sat-1, NaDC-3, OAT1, OAT2, and OAT3.

In the lists, we indicate whether or not substrates were translocated or induced currents in cells expressing one of the above-mentioned transporters. Whenever uptake of a radiolabeled substrate into expressing cells was statistically significantly greater than uptake into nonexpressing control cells, we show a plus sign (+) in the tables, even if stimulation of uptake by carrier expression was smaller than twofold. Low degrees of stimulation are, however, discussed in later sections. Furthermore, the lists show compounds which were tested as possible inhibitors of the uptake of radiolabeled model substrates. A minus sign (–) indicates a lack of inhibition, and a plus sign (+) a statistically significant inhibition. We chose not to indicate the degree of inhibition, because the experiments have been conducted with various inhibitor concentration over label concentration ratios. When available, we include the apparent K_m values for uptake, as well as IC_{50} of K_i values for inhibition.

In the final sections of this review, we summarize the interaction of selected drug groups, antibiotics, antiviral drugs, diuretics, and nonsteroidal anti-inflammatory drugs, with the cloned transporters to demonstrate that drugs may utilize more than a single organic anion transporter in the basolateral cell membrane of proximal tubule cells.

The sulfate-anion antiporter (sat-1)

Physiological roles of sulfate and renal sulfate handling

As recently reviewed (Beck and Silve 2001; Markovich 2001; Morris and Murer 2001), the divalent, inorganic anion sulfate, SO_4^{2-}, plays an important role in the biosynthesis of proteoglycans and, hence, in the formation of the extracellular matrix. Sulfate is also required for conjugation reactions in the liver, for heparin synthesis, and for the activity of some gastrointestinal hormones such as gastrin or cholecystokinin. For sulfation reactions, sulfate must be taken up into the cells. Disorders in sulfate transport processes lead to severe diseases such as diastrophic dysplasia (Hästbacka et al. 1994) and congenital chloride diarrhea syndrome (Moseley et al. 1999). The affected transporters, DTDST (diastrophic dysplasia sulfate transporter) in diastrophic dysplasia and DRA (down-regulated in adenoma) are structurally related to the hepatic and renal sulfate transporter, sat-1 (Markovich 2001). Mutations in another related transporter, Pendrin, are responsible for the Pendred syndrome (Everett et al. 1997).

The kidneys play an important role in the maintenance of a plasma sulfate level at about 0.3 mM in adult humans (Cole and Scriver 1980). Sulfate is freely filtered in the glomeruli and actively reabsorbed in the proximal tubules (Hierholzer et al. 1960). The uptake of sulfate from the tubule lumen into the cells is coupled with Na^+, as was demonstrated earlier in rat kidney microperfusion studies (David and Ullrich 1992; Ullrich et al. 1980) and in experiments with brush-border membrane vesicles from rat (Lücke et al. 1980) and rabbit kidneys (Schneider et al. 1984; Turner 1984) (see Fig. 1). The luminal (brush-border) sulfate transporter, NaSi-1 or SLC13A1, has been cloned from rat, mouse and man (Beck and Markovich 2000; Lee et al. 1999, 2000; Markovich et al. 1993) . The NaSi-1 orthologs share homologies with the Na^+-coupled dicarboxylate cotransporters, NaDCs (see section on "The sodium-coupled dicarboxylate cotransporter, NaDC-3"), but only the human NaSi-1 was inhibited by the NaDC substrates succinate and citrate (Lee et al. 2000). Prototypical substrates of NaSi-1 are sulfate, thiosulfate, and a number of tetrahedral oxyanions such as selenate and tungstate (Lee et al. 2000; Markovich et al. 1994; Markovich and Regeer 1999).

The exit of sulfate across the basolateral membrane involves an anion exchanger (see Fig. 1), as has been suggested from studies on intact rat kidney (Ullrich et al. 1984b), isolated rabbit proximal tubules (Brazy and Dennis 1981), and was demonstrated in a number of experiments on renal basolateral membrane vesicles (BLMV). A pH difference (intravesicular pH to extravesicular pH) accelerated the uptake of sulfate into BLMV from flounder (Renfro and Pritchard 1982), rat (Hagenbuch et al. 1985; Löw et al. 1984; Pritchard and Renfro 1983; Shimada and Burckhardt 1986) and bovine kidneys (Talor et al. 1987), suggesting the presence of an SO_4^{2-}/OH^- exchanger. A bicarbonate concentration

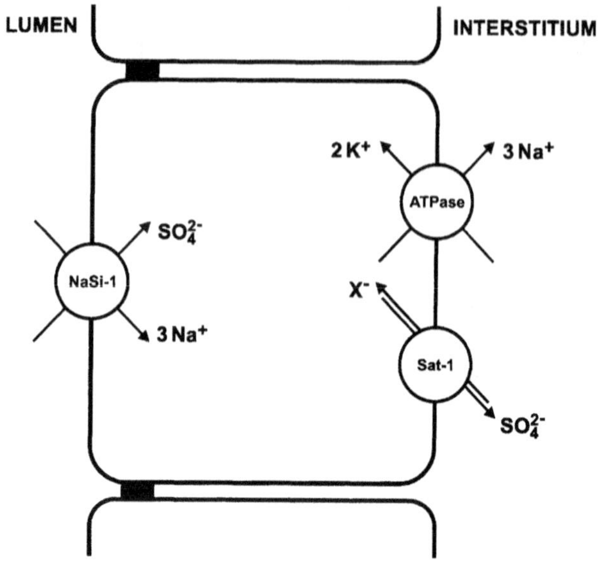

Fig. 1 Absorption of sulfate in renal proximal tubules. Sulfate uptake from the lumen across the luminal (brush-border) membrane occurs by the Na^+-sulfate cotransporter, NaSi-1. Sulfate exit across the basolateral membrane into the interstitium involves the sulfate-anion exchanger, sat-1. X^- may be OH^-, HCO_3^-, but not Cl^-. The Na^+,K^+-ATPase in the basolateral membrane energizes sulfate uptake through NaSi-1 by pumping Na^+ ions out of the cell

difference (intravesicular-to-extravesicular) also *trans*-stimulated sulfate uptake in rat (Hagenbuch et al. 1985; Pritchard and Renfro 1983), rabbit (Kuo and Aronson 1988), chicken (Renfro et al. 1987), and bovine renal BLMV (Talor et al. 1987), indicating SO_4^{2-}/HCO_3^- exchange as another mode of the antiporter. The antiporter also performed SO_4^{2-}/SO_4^{2-}, SO_4^{2-}/SSO_3^{2-} exchange (Löw et al. 1984; Pritchard and Renfro 1983; Shimada and Burckhardt 1986), SO_4^{2-}/oxalate exchange (Kuo and Aronson 1988), and was inhibited by the stilbenes, 4,4'-di-isothiocyanatostilbene-2,2'-disulfonate (DIDS) or 4-acetamido-4'-isothiocyanatostilbene-2,2'-disulfonate (SITS) (Bästlein and Burckhardt 1986; Grinstein et al. 1980; Hagenbuch et al. 1985; Kuo and Aronson 1988; Pritchard and Renfro 1983; Renfro et al. 1987; Renfro and Pritchard 1982; Talor et al. 1987). Chloride ions did not interfere with sulfate transport in rat and rabbit renal BLMV (Hagenbuch et al. 1985; Kuo and Aronson 1988; Pritchard and Renfro 1983), suggesting that SO_4^{2-}/Cl^- exchange does not present a mode of antiport.

Substrates of the sulfate/anion antiporter in the basolateral membrane

The substrate specificity of the SO_4^{2-}/anion antiporter was tested in vesicles and to a much greater extent in the intact rat kidney. In rabbit renal BLMV, $^{35}SO_4^{2-}/HCO_3^-$ exchange was inhibited by unlabeled sulfate, thiosulfate, oxalate, bumetanide, furosemide, α-cyanohydroxycinnamate, and probenecid, but not by phosphate, nitrate, chloride, acetate, lactate, pyruvate, malonate, succinate, α-ketoglutarate, citrate, urate, and PAH (Pritchard and Renfro 1983). A systematic survey of putative substrates of the sulfate transporter in the basolateral membrane of rat renal proximal tubules in situ was performed in Ullrich's laboratory. Besides the tetrahedral oxyanions thiosulfate, selenate and molybdate (Ullrich et al. 1984b), some sulfonates/aminosulfonates (Ullrich et al. 1985a), disulfonates and di- and tricarboxylates (Ullrich et al. 1985b), salicylate analogs (Ullrich et al. 1985c), sulfamoyl compounds, diphenylamine-carboxylates (Ullrich et al. 1985d), substituted benzene analogs (Ullrich et al. 1988), and sulfate esters (Ullrich et al. 1990) inhibited the uptake of labeled sulfate from the capillaries into the proximal tubule cells of rat kidney in situ. Based on the apparent K_i values found in situ, substrates have been classified into five categories (Ullrich 1997; Ullrich et al. 1985a). Category I includes tetrahedral inorganic anions as well as divalent organic anions such as oxalate, disulfonates (SITS, DIDS, disulfonic dyes), benzene-dicarboxylates, benzene-sulfocarboxylates, and diphosphonates with two negative charges at a proper distance. Monovalent sulfate esters belong to category II, and monovalent sulfonates with a neighboring OH or NH group (e.g., 2-hydroxyethane sulfonate, NAP-taurine) to category III. In category IV are sulfonates and phosphonates with hydrophobic moieties (e.g., phenol red and related dyes), and category V comprises aromatic carboxylates with neighboring OH or NH groups and additional electronegative side groups on the aromatic ring. Interestingly, the sulfate transporter interacted both with bulky charges such as SO_4^{2-} and $R\text{-}SO_3^-$, as well as with planar carboxyl groups. In the bulky mode, the transporter accepted high affinity compounds having two negative charges at distances of 4–7 Å, whereas in the planar mode the two charges had to be closer (3–4 Å), as found in oxalate or maleate (Fritzsch et al. 1989). A hydrophobic domain close to the negative charge(s) increased the affinity, suggesting that the transporter binds substrates by hydrophic interaction as well as by interaction with negatively charged groups.

Among endogenous substrates interacting with the basolateral sulfate transporter (SO_4^{2-}/anion antiporter) were oxalate (app. K_i 1.2 mM; Ullrich et al. 1984b), sulfated steroid hormones (corticosterone sulfate, cortisol sulfate, estriol sulfate; (Ullrich et al. 1990, 1991c) and some bile salts (cholate, chenodeoxycholate, glycocholate; Ullrich 1998). With respect to pharmacology, it is important to note that the sulfamoyl- and phenoxy-diuretics acetazolamide, ethacrynate, furosemide, hydrochlorothiazide, piretanide and tienilate (Fritzsch et al. 1989), the diphosphonates alendronate, clodronate, etodronate and pamidronate (Ullrich et al. 1997), the antibiotics cefodizime, cefsulodin, ceftriaxone and ticarcillin (Ullrich et al. 1989a), and the anti-inflammatory drugs diclofenac, fenoprofen, flufenamate, indomethacin, ketoprofen, niflumate and paracetamol sulfate (Ullrich et al. 1991a, 1993c, 1994) interacted with the sulfate transport in intact rat kidney. In most cases, the apparent K_i values were greater than 1 mM, indicating a low affinity of the sulfate transporter for these substrates. If, however, the V_{max} were also high, the sulfate transporter could contribute to the renal excretion of diuretics, diphosphonates, anti-inflammatory drugs and some antibiotics. At present, however, drug translocation by the sulfate transporter has not yet been tested.

Cloning, gene, and tissue distribution of sat-1

For the molecular properties of sat-1, see Table 1. The SO_4^{2-}/HCO_3^- antiporter was first cloned from rat liver and called sulfate anion transporter-1 or sat-1 (Bissig et al. 1994). A sat-1 clone with an identical sequence was also isolated from rat brain (Lee et al. 1999) and another clone, differing in only two bases, from rat kidneys (Karniski et al. 1998). The cloning of the human and mouse sat-1 was reported in a review (Markovich 2001), but full details have not yet been published. The mouse sat-1 gene (*Sat1*, *slc26a1*) was localized to chromosome 5, where it partially overlaps with the gene of the murine α-L-iduronidase located on the opposite strand (Clarke et al. 1994). The human *SAT-1* gene, *SLC26A1*, is on chromosome *4p16.3* (Markovich 2001). Northern blot analysis revealed strong bands of rat sat-1 (rsat-1) in liver and kidneys. After prolonged exposure, weak bands in brain and

Table 1 Properties of the cloned sat-1 (SLC26A1)

Cloned from	Human, rat, mouse
Gene	Human: chromosome *4p16.3*
Expression	Liver, kidneys >>brain, muscle
Localization	Kidneys: proximal tubules (S1, S2>S3), basolateral membrane; brain: hippocampus, cerebellum
Protein	702–704 amino acids; 12 TM helices; 2–3 *N*-glycosylation sites; several sites for protein kinase C, casein kinase II; 1 site for tyrosine kinase
Transport	Transport of sulfate; SO_4^{2-}/OH^- (HCO_3^-) antiport not yet investigated with cloned sat-1
Regulation	Down-regulation by protein kinase C in oocytes[a]

[a] Data not shown but mentioned in Markovich 2001

muscle became visible; small and large intestine, heart and lungs did not show any signals (Bissig et al. 1994). In situ hybridization with rat brain tissues indicated the presence of rsat-1 mRNA in the hippocampus and in the cerebellum (Lee et al. 1999). Antibodies against rsat-1 stained the basolateral membrane of the proximal tubule cells of segments S1 and S2 and, less intensely, of segment S3 (Karniski et al. 1998).

Sat-1 protein

As predicted from the open reading frame, rsat-1 protein is composed of 703 amino acids (Bissig et al. 1994). Mouse and human sat-1 have 704 and 702 amino acids, respectively (Markovich 2001) (see Table 1). Secondary structure predictions revealed 12 transmembrane helices (M1–M12) with the N- and C-termini located at the cytosolic side. The loops between M3 and M4, and between M11 and M12, carry three potential *N*-glycosylation sites, supporting the topological model which places these loops at the extracellular side of the membrane (Bissig et al. 1994). In Western blots the sat-1 expressed in sf9 cells showed an apparent molecular weight of 68 kDa, whereas the anti-sat-1 antibodies reacted with 81-kDa bands in blots from rabbit and rat renal basolateral membranes (Karniski et al. 1998). The larger band in native membranes appeared to reflect the fully glycosylated sat-1. With rsat-1, several consensus sequences were found for phosphorylation by protein kinase C, casein kinase II, as well as one tyrosine kinase site (Markovich 2001). Their role in regulation of rsat-1 activity remains to be elucidated.

Functional characterization of sat-1 substrates

For an overview of substrates, see Table 2. When the rat liver sat-1 was expressed in oocytes, $^{35}SO_4^{2-}$ uptake was stimulated over noninjected oocytes. Sulfate uptake was not enhanced by the presence of sodium, indicating that sat-1 is a Na^+-independent transporter. Transport was saturable, showing an apparent K_m of 0.14 mM sulfate and was inhibited in the presence of 1 mM DIDS or 5 mM oxalate. Succinate and cholate had no effect (Bissig et al. 1994) (see Table 2). The renal rsat-1 was expressed in sf9 cells and showed uptake of labeled sulfate and oxalate. Unlabeled sulfate inhibited the uptake of labeled oxalate and vice versa, indicating that both divalent anions are translocated by rsat-1 (Bissig et al. 1994). Moreover, sulfate uptake by renal rsat-1 was inhibited by thiosulfate, probenecid and phenol red (Markovich and Regeer 1999). Thiosulfate, DIDS, oxalate, probenecid,

Table 2 Substrate specificity of the cloned sat-1

Compound	Species	Expression system	Transport or K_m (μM)	Inhibition or IC_{50} (μM)	References
Cholate	Rat	Oocyte		–	Bissig et al. 1994; Lee et al. 1999; Markovich et al. 1994
4,4'-di-Isothiocyanatostilbene-2,2'-disulfonate (DIDS)	Rat	Oocyte		30[1]; 150[2]	[1]Bissig et al. 1994; [2]Lee et al. 1999
Oxalate	Rat	sf9 cells	+	+	Karniski et al. 1998
		Oocyte		+	Bissig et al. 1994; Lee et al. 1999; Markovich et al. 1994
Phenol red	Rat	Oocyte		+	Lee et al. 1999; Markovich et al. 1994
Probenecid	Rat	Oocyte		+	Lee et al. 1999; Markovich et al. 1994
Succinate	Rat	Oocyte		–	Bissig et al. 1994; Lee et al. 1999; Markovich et al. 1994
Sulfate	Rat	Oocyte	136[1]	+[2]	[1]Bissig et al. 1994; [2]Karniski et al. 1998
		sf9 cells	+	+	Karniski et al. 1998
Thiosulfate	Rat	Oocyte		+	Lee et al. 1999; Markovich et al. 1994

and phenol red, but not succinate and cholate, inhibited rat brain sat-1 expressed in oocytes (Lee et al. 1999).

A survey of Table 2 shows that only a few organic anions were tested as possible inhibitors and substrates of sat-1. Given the interaction of the sulfate transporter in the intact kidney with a considerable series of organic compounds including drugs (see "Substrates of the sulfate/anion antiporter in the basolateral membrane"), it remains to be determined which of these substrates interacts with, and is transported by, the isolated sat-1. An open question is also whether expressed sat-1 does indeed function as an antiporter, as suggested from earlier vesicle data. To our knowledge, the *trans*-effects of OH^- and HCO_3^- have not been tested on the isolated sat-1.

The sodium-dicarboxylate symporter (NaDC-3)

Physiological role of di- and tricarboxylates and renal di- and tricarboxylate handling

With exception of oxalate, endogenous di- and tricarboxylates such as oxaloacetate, fumarate, succinate, α-ketoglutarate, isocitrate and citrate are intermediates of the citric acid cycle (Krebs cycle) and constitute central substrates of energy metabolism in mitochondria. In human plasma, citrate levels are in the range of 50–300 μM (Simpson 1983). The concentration for α-ketoglutarate was 8.6 μM in children (Rocchiccioli et al. 1984). In rats, plasma α-ketoglutarate concentration was 12 μM (Martin et al. 1989). Uptake of di- and tricarboxylates across cell membranes requires the input of energy, because it occurs

Fig. 2 Absorption of di- and tricarboxylates in renal proximal tubules. Uptake of dicarboxylates (*DC*$^{2-}$) and divalent tricarboxylates into proximal tubule cells occurs from both the lumen and the blood/interstitium by Na^+-coupled transport. The cotransporters, NaDC-1 and NaDC-3, are structurally related, but not identical. Both systems are energized by the electrochemical potential difference of Na^+ maintained by the Na^+,K^+-ATPase

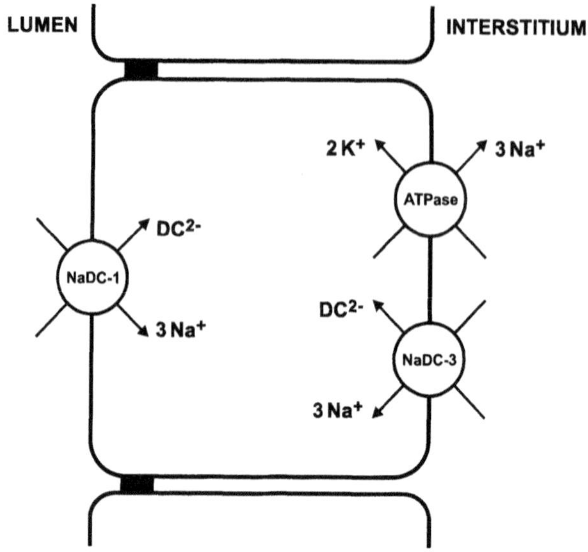

against an inside negative electrical potential difference and against intracellular-to-extracellular concentration differences. Rat proximal tubule cells, for instance, show an electrical potential difference of more than -70 mV (Burckhardt BC et al. 1984). The intracellular α-ketoglutarate concentration reached 636 µM in rat kidney cortex slices at 10 µM bath α-ketoglutarate (Pritchard 1995). To overcome the energetic barrier of -70 mV and a 96-fold concentration difference, a symport of one α-ketoglutarate with more than one Na^+ ion is required.

Krebs cycle intermediates are absorbed from both the primary filtrate and the capillaries into proximal tubule cells (see Fig. 2), leading to a net extraction of citrate, α-ketoglutarate and succinate from the circulation (Martin et al. 1989; Simpson 1983). The di- and tricarboxylates taken up into the cells serve as metabolic fuels: addition of succinate or α-ketoglutarate to suspended rat renal proximal tubules increased oxygen consumption and gluconeogenesis (Guder et al. 1971), and extracellular citrate provided 10% and 15% of the energy needs of human and dog kidneys, respectively (Baruch et al. 1975; Nieth and Schollmeyer 1966). In addition, the intracellular-to-extracellular α-ketoglutarate concentration difference serves as a driving force for the organic anion/α-ketoglutarate antiporter, OAT1, located in the basolateral membrane of proximal tubule cells (see below). Studies on the intact rat kidney (Sheridan et al. 1983) as well as on vesicles isolated from luminal membrane (brush-border membrane vesicles, BBMV) (Wright EM et al. 1982; Wright SH et al. 1980, 1982) revealed the presence of an Na^+-coupled transport system for di- and tricarboxylates. The basic characteristics of the symporter are a 3 Na^+:1 dicarboxylate^{2-} stoichiometry, inhibition by Li^+, comparably low affinity for succinate (higher than 0.15 mM), preferred interaction with four-carbon dicarboxylates with the carboxylate groups in the *trans* position, and insensitivity of succinate transport to pH. Citrate is accepted in its divalent form. The Na^+-dicarboxylate cotransporters (NaDC-1) present in the luminal membrane of renal proximal tubule cells have been cloned from rabbit (Pajor 1995), human (Pajor 1996), rat (Chen et al. 1998; Sekine et al. 1998a) and mouse kidneys (Pajor and Sun 2000). Their characteristics perfectly matched those known from earlier studies on BBMV (for reviews see Pajor 1999, 2000).

Substrates of the Na^+-dependent dicarboxylate transporter in the basolateral membrane

The uptake of Krebs cycle intermediates from the capillaries across the basolateral membrane into proximal tubule cells is also a Na^+-dependent process, as demonstrated in the intact rat kidney (Ullrich et al. 1984a) as well as in studies with basolateral membrane vesicles (BLMV) isolated from rat (Burckhardt G 1984) and rabbit (Jørgensen et al. 1983; Sheikh et al. 1982) kidneys (see also Fig. 2). Similar to the luminal transporter, dicarboxylate transport in BLMV was Na^+-dependent, electrogenic, Li^+-sensitive, and was inhibited most strongly by four- or five-carbon dicarboxylates, with a preference for the *trans* configuration of the carboxylate groups. A larger series of aliphatic and aromatic di- and tricarboxylates was tested in the intact rat kidney (Ullrich et al. 1984a). It turned out that the distance between the negative charges is critical: for proper interaction the charges had to be at least 5 Å apart from each other (Fritzsch et al. 1989). A substitution of aliphatic dicarboxylates in the α-position alone as, for example, in α-ketoglutarate, did not change the affinity toward the transporter, but a substitution in the α and β position, as, for example, in 2,3-dimercaptosuccinate decreased the affinity. Altogether, the structural requirements for interaction with the luminal and the basolateral dicarboxylate transporter were similar, with the exception of 2,3-dimethylsuccinate, 2,3-dimercaptosuccinate, *cis*-aconitate and a few other dicarboxylates, which were accepted only by the dicarboxylate transporter in the basolateral membrane (Ullrich et al. 1984a). Moreover, the pH sensitivity of succinate transport differed, showing no pH dependence in BBMV and a pH optimum at around pH 7.5 in BLMV from rat kidney (Burckhardt G 1984). Finally, the affinity of the basolateral Na^+-dicarboxylate cotransporters for succinate was higher than that of the luminal transporter, indicating that the NaDC-transporter proteins are related, but not identical. Indeed, molecular cloning revealed a separate branch of the cotransporter family, the NaDC-3 orthologs.

Cloning, gene, and tissue distribution of NaDC-3

For a survey of molecular properties, see Table 3. The NaDC-3 has been cloned from human placenta (hNaDC-3; Wang et al. 2000), rat placenta (rNaDC-3; Kekuda et al. 1999), rat kidneys (SDCT-2; Chen et al. 1999), mouse brain (mNaDC-3; Pajor et al. 2001), and winter flounder kidney (fNaDC-3; Steffgen et al. 1999) (see also Table 3). Rat SDCT-2 and rNaDC-3 have been given different names, but are otherwise identical. The human gene (*SLC22A3*) is over 80 kb long, comprised of 13 exons and 12 introns, and has been located to chromosome *20q12–13.1* (Wang et al. 2000).

In Northern blot studies on rat tissues, rNaDC-3 transcripts were detected in the kidneys with an additional, faint band in brain. Heart, stomach, small intestine and skeletal muscles did not show any signals (Kekuda et al. 1999). RT-PCR revealed an expression of NaDC-3 also in rat placenta, the source for isolation of the clone. With an SDCT2 probe, signals were found in kidneys >>liver >brain; heart, spleen, lung, skeletal muscle, and testis were negative (Chen et al. 1999). The message for the mNaDC-3 was abundantly expressed in the kidneys, and weakly in the brain; no message was detected in heart, stomach, intestine and skeletal muscle (Pajor et al. 2001). A similar distribution has been mentioned, but not explicitly shown, for hNaDC-3 (Wang et al. 2000). Taken together, NaDC-3 appears as a mainly kidney-specific transporter. With antibodies the localization of SDCT2/rNaDC-3 at the basolateral membrane of rat renal proximal tubule cells of the S3 segment has been shown (Chen et al. 1999). S1 and S2 segments were not labeled, suggesting that an Na^+-dicarboxylate cotransporter other than SDCT2/rNaDC-3 may be involved in the functional coupling to the

Table 3 Properties of the cloned NaDC-3 (SLC13A3)

Cloned from	Human, rat (NaDC-3, SDCT-2), mouse, flounder
Gene	Human: >80 kb; 13 exons; chromosome *20q12–13.1*
Expression	Human: placenta, brain, liver, kidneys, pancreas[a]
	Rat: kidneys >>liver >brain
	Mouse: kidneys >>brain (Northern blot); kidneys, liver, choroid plexus (RT-PCR)
Localization	Rat: proximal tubules, segment S3; liver: perivenous hepatocytes (in situ hybr.)
	Flounder kidneys: proximaltubules, segment II, basolateral membrane (antibodies)[b]
Protein	600–602 amino acids;11–12 transmembrane helices; 2–3 *N*-glycosylationsites; 3–8 protein kinase C sites; 7 casein kinase II sites(flounder)
Transport	Symport of three sodium ions withone divalent di- or tricarboxylate
Regulation	Down-regulation by protein kinaseC in *Xenopus laevis* oocytes

[a] Data not shown, but mentioned in the text Wang et al. 2000
[b] J. Steffgen, H. Hentschel et al. , manuscript submitted

organic anion transporter, OAT1. The fNaDC-3 was localized to the basolateral membrane of flounder renal proximal tubule segments II (J. Steffgen and H. Hentschel, unpublished data). In the rat liver, SDCT2/rNaDC-3 antibodies stained perivenous hepatocytes; in rat brain, the meningeal layers, some cells in the cerebellar granular layer, and scattered astrocytes in the cortex were positive (Chen et al. 1999; Huang et al. 2000)

NaDC-3 protein

The predicted NaDC-3 proteins contain 600 (rat, mouse), 601 (flounder) or 602 (human) amino acids (cf., Table 3). Setting the human NaDC-3 to 100%, rat, mouse, and flounder orthologs share 85%, 87%, and 61% identities, respectively. The identity between NaDC-3 and NaDC-1 is in the range of 45%. Secondary structure predictions revealed 11–12 putative transmembrane helices. It is likely that the topological model for NaDC-1 (11 transmembrane-spanning segments with intracellular N- and extracellular C-termini (Zhang and Pajor 2000) also holds true for NaDC-3; experimental data are, however, not yet available to support this hypothesis.

Posttranslational modifications of NaDC-3 may occur at two (all NaDC-3 s) to three (flounder) *N*-glycosylation sites, three (rat), five (flounder) or eight (human) protein kinase C sites, one site for cAMP/cGMP-dependent kinase (human), and seven casein kinase II sites (flounder). It is not clear at present which of the glycosylation and phosphorylation sites are actually used in vivo. A down-regulation by protein kinase C has so far been demonstrated only for NaDC-1 (Pajor and Sun 1999). In isolated rabbit renal proximal tubules, an activation of protein kinase A or protein kinase C had no influence on the activity of Na^+-dependent glutarate uptake (Röver et al. 1998), suggesting that the rabbit NaDC-3 is not affected by these protein kinases. We were recently able to show that fNaDC-3 is down-regulated by phorbol esters, when expressed in oocytes (Hagos et al. 2001). Whether species differences (rabbit vs flounder) or differences between the cells (intact tubules vs oocytes) are the reason for the discrepant results needs to be clarified.

Functional characterization of NaDC-3 substrates

Substrates of NaDC-3 are summarized in Table 4. Transport properties of NaDC-3 were studied after heterologous expression in *Xenopus laevis* oocytes and HRPE (human retinal

Table 4 Substrate specificity of the cloned NaDC-3

Compound	Species	Expression	Transport or K_m (μM)	Inhibition or IC_{50} (μM)	References
N-acetyl-aspartate	Human	HRPE cells		232	Huang et al. 2000
		Oocyte	300[a]		Huang et al. 2000
	Rat	HRPE cells		59	Huang et al. 2000
D- and L-aspartate	Human	HRPE cells		+[1]; 3,070[2]	[1] Wang et al. 2000; [2] Huang et al. 2000
	Rat	Oocyte	+[a]		Chen et al. 1999
		HRPE cells		+[1]; 1,490[2]	[1] Kekuda et al. 1999; [2] Huang et al. 2000
	Mouse	Oocyte	+[a]	–	Pajor et al. 2001
	Flounder	Oocyte		–	Steffgen et al. 1999
Butyrate	Human	HRPE cells		–	Wang et al. 2000
cis-Aconitate	Mouse	Oocyte		–	Pajor et al. 2001
	Flounder	Oocyte	+[a]	+	Steffgen et al. 1999
Citrate	Human	Oocyte	+[a]		Wang et al. 2000
		HRPE cells		+	Wang et al. 2000
	Rat	Oocyte	220[a]	+	Chen et al. 1999
		HRPE cells		+	Kekuda et al. 1999
	Mouse	Oocyte	+[a]	+[b]	Pajor et al. 2001
	Flounder	Oocyte		+	Steffgen et al. 1999
Dipicolinate	Rat	HRPE cells		–	Kekuda et al. 1999
Dimethylsuccinate[c]	Human	Oocyte	112		Wang et al. 2000
		HRPE cells		+	Wang et al. 2000
	Rat	HRPE cells		+	Kekuda et al. 1999
2,2-Dimethylsuccinate	Mouse	Oocyte		+	Pajor et al. 2001
	Flounder	Oocyte	7[a]		Burckhardt BC et al. 2002
2,3-Dimethylsuccinate	Rat	Oocyte	27[a]		Chen et al. 1999
	Mouse	Oocyte	+[a]	+	Pajor et al. 2001
	Flounder	Oocyte	+[a1]; 40[a2]	+[1]	[1]Steffgen et al. 1999; [2]Burckhardt BC et al. 2002
2,3-Dimercaptosuccinate (DMSA, succimer)	Rat	HRPE cells		+	Kekuda et al. 1999
	Flounder	Oocyte	93[a]		Burckhardt BC et al. 2002
2,3-Dimercapto-1-propane sulfonate (DMPS)	Flounder	Oocyte	+		Burckhardt BC et al. 2002
Fumarate	Human	Oocyte	+[a]		Wang et al. 2000
		HRPE cells		+	Wang et al. 2000
	Rat	Oocyte	+[a]		Chen et al. 1999
		HRPE cell		+	Kekuda et al. 1999
	Mouse	Oocyte	+[a]	+	Pajor et al. 2001
	Flounder	Oocyte		+	Steffgen et al. 1999
L-Glutamate	Human	HRPE cells		–	Huang et al. 2000
	Rat	Oocyte	+		Chen et al. 1999
		HRPE cells		–	Kekuda et al. 1999; Huang et al. 2000
	Mouse	Oocyte	–	+	Pajor et al. 2001
L-Glutamine	Mouse	Oocyte	–	–	Pajor et al. 2001
Glutarate	Human	HRPE cells		+	Wang et al. 2000
	Rat	HRPE cells		+	Kekuda et al. 1999
	Flounder	Oocyte		+	Steffgen et al. 1999

Table 4 (continued)

Compound	Species	Expression	Transport or K_m (μM)	Inhibition of IC_{50} (μM)	References
Isocitrate	Flounder	Oocyte		+	Steffgen et al. 1999
α-Ketoglutarate	Human	Oocyte	+[a]		Wang et al. 2000
		HRPE cells		+	Wang et al. 2000
	Rat	HRPE cells		+	Kekuda et al. 1999
		Oocyte	+[a]		Chen et al. 1999
	Mouse	Oocyte	+		Pajor et al. 2001
	Flounder	Oocyte		+	Steffgen et al. 1999
Lactate	Human	Oocyte	–[b]		Wang et al. 2000
		HRPE cells		–	Wang et al. 2000
	Rat	Oocyte	–[a]		Chen et al. 1999
		HRPE cells		–	Kekuda et al. 1999
	Flounder	Oocyte		–	Steffgen et al. 1999
Malate	Human	Oocyte	+[a]		Wang et al. 2000
		HRPE cells		+	Wang et al. 2000
	Rat	Oocyte	+[a]		Chen et al. 1999
		HRPE cells		+	Kekuda et al. 1999
	Mouse	Oocyte	+[a]	+	Pajor et al. 2001
	Flounder	Oocyte		–	Steffgen et al. 1999
Maleate	Human	Oocyte	–[a]		Wang et al. 2000
		HRPE cells		–	Wang et al. 2000
	Rat	HRPE cells		–	Kekuda et al. 1999
	Flounder	Oocyte		–	Steffgen et al. 1999
Malonate	Human	Oocyte	–[a]		Wang et al. 2000
		HRPE cells		–	Wang et al. 2000
	Rat	HRPE cells		–	Kekuda et al. 1999
Methylsuccinate	Rat	Oocyte	8		Chen et al. 1999
Oxalate	Human	Oocyte	–[a]		Wang et al. 2000
		HRPE cells		–	Wang et al. 2000
	Rat	Oocyte		–	Kekuda et al. 1999
Oxaloacetate	Rat	Oocyte	+		Chen et al. 1999
Pyruvate	Human	Oocyte	–[a]		Wang et al. 2000
		HRPE cells		–	Wang et al. 2000
	Rat	Oocyte	–		Chen et al. 1999
		HRPE cells		–	Kekuda et al. 1999
	Mouse	Oocyte		–	Pajor et al. 2001
	Flounder	Oocyte		–	Steffgen et al. 1999
Succinate	Human	HRPE cells	20.4	+	Wang et al. 2000
		Oocyte	102		Wang et al. 2000
	Rat	Oocyte	15	+	Chen et al. 1999
		HRPE cells	2		Kekuda et al. 1999
	Mouse	Oocyte	16[a], 142	+	Pajor et al. 2001
	Flounder	Oocyte	30[1]; 31[a2]; 60[a3]		[1]Steffgen et al. 1999; [2]Burckhardt BC et al. 2002; [3]Burckhardt et al. 2000a
Sulfate	Mouse	Oocyte	–	–	Pajor et al. 2001
	Flounder	Oocyte		–	Steffgen et al. 1999
Valerate	Human	HRPE cells		–	Wang et al. 2000

[a] Electrophysiological determination
[b] pH dependent
[c] Position of methyl groups not specified

pigmental epithelial) cells. In most studies, succinate served as a reference substrate. Thereby, either the uptake of radiolabeled succinate was measured or the current induced by unlabeled succinate was determined as a measure of sodium and succinate translocation. NaDC-3 is distinguished from sat-1 and from the OATs by its direct dependence on sodium. Succinate uptake increased in the presence of Na^+ and showed a sigmoidal dependence on extracellular Na^+ concentration, indicating the interaction of more than one sodium ion with NaDC-3 (Huang et al. 2000; Kekuda et al. 1999; Pajor et al. 2001; Wang et al. 2000). In voltage-clamped oocytes, an inward current was induced by succinate, which was again a sigmoidal function of the extracellular Na^+ concentration (Burckhardt BC et al. 2000a; Chen et al. 1999). The Hill coefficient was greater than 2 (Huang et al. 2000; Kekuda et al. 1999; Steffgen et al. 1999; Wang et al. 2000). These data collectively indicated a symport of three Na^+ ions with one molecule of succinate, which carries two negative charges. Thus, with each succinate molecule a surplus of one positive charge is taken up into the cell, leading to the observed, succinate-induced, inward current in voltage-clamped oocytes.

The half-maximal activation of succinate uptake/succinate-induced currents occurred at $K_{0.5}$ of 15–65 mM Na^+. In voltage-clamp experiments on human (Huang et al. 2000) and flounder (Burckhardt BC et al. 2000a) NaDC-3, the $K_{0.5}$ of sodium decreased with hyperpolarization, i.e., at increasing inside negativity. It appeared from these data that the binding sites for sodium are located within a deep pocket; Na^+ ions experience part of the transmembrane electrical potential difference when they travel to these sites. An inside more negative potential thus facilitates binding of Na^+ and decreases $K_{0.5}$ for Na^+.

At a constant Na^+ concentration, the translocation of succinate or the succinate-induced currents depended on the extracellular succinate concentration with simple Michaelis–Menten kinetics, as expected from the interaction of one succinate molecule with the transporter. Half-maximal transport or current was achieved at 2 or 15 μM (rNaDC-3/SDCT2; Chen et al. 1999; Kekuda et al. 1999), 16 and 142 μM (current and tracer uptake in mNaDC-3-expressing oocytes, respectively; Pajor et al. 2001), 20 μM (hNaDC-3; Wang et al. 2000), and 30 or 60 μM (tracer uptake or current on fNaDC3-expressing oocytes, respectively; Burckhardt BC et al. 2000a; Steffgen et al. 1999). The K_m for succinate was unaffected by membrane potential at inside negative clamp potentials (Burckhardt BC et al. 2000a) or increased, particularly at inside positive potentials (Pajor et al. 2001; Wang et al. 2000). If succinate also bound to a deep pocket within the transporter, an inside negative membrane potential should keep succinate out, i.e., increase the K_m. The opposite was observed, suggesting that succinate binding occurs close to the extracellular surface of NaDC-3.

In contrast to NaDC-1, which is pH-independent between pH 6 and 8, succinate transport by NaDC-3 increased with increasing pH (Chen et al. 1999; Kekuda et al. 1999; Pajor et al. 2001; Steffgen et al. 1999). Common to NaDC-1 and NaDC-3 was the inhibition of succinate transport by lithium. The IC_{50} ranged between 0.5 mM (fNaDC-3), <3 mM (rNaDC-3/SDCT2), and 2.6–5 mM (hNaDC-3); mNaDC-3 was not inhibited by 5 mM Li^+. Lithium is thought to bind to one of the Na^+ sites and to slow down the symport with succinate. In the complete absence of Na^+, Li^+ maintained a slow uptake of succinate.

Table 4 lists alphabetically the substrates which have been tested on NaDC-3 from various species. The citric acid cycle derivatives citrate, α-ketoglutarate, succinate, fumarate, malate, and oxaloacetate interacted with most of the NaDC-3s. Citrate uptake by SDCT2 (Chen et al. 1999) and rNaDC-3 (Kekuda et al. 1999) increased with decreasing pH, i.e., at increasing concentrations of the divalent form of citrate. Likewise, the inhibition by citrate of labeled succinate uptake into mNaDC-3-expressing oocytes was more marked at pH 5.5 than at pH 7.5 (Pajor et al. 2001). At pH 7.5, the currents induced by citrate were small (mNaDC-3; Pajor et al. 2001) or not measurable (hNaDC-3; Wang et al. 2000). Collectively, these data show that citrate interacts with NaDC-3 in its divalent form. α-Ketoglutarate, an α-substituted dicarboxylate with five carbons, was transported, and it inhibited succinate transport in all NaDC-3s. In voltage-clamp studies, α-ketoglutarate produced currents equal to or larger than those evoked by succinate (Chen et al. 1999; Pajor et al. 2001; Wang et al. 2000). These data indicate that α-ketoglutarate binds to NaDC-3 with high affinity and is translocated at high rates. The unsubstituted C5 dicarboxylate, glutarate, also inhibited NaDC-3-mediated succinate transport. Among the citric acid cycle intermediates with four carbons, succinate, fumarate, malate and oxaloacetate interacted with NaDC-3, as revealed by either uptake or inhibition experiments (see Table 4). Fumarate produced smaller (Chen et al. 1999; Pajor et al. 2001; Wang et al. 2000), and malate (Chen et al. 1999; Pajor et al. 2001; Wang et al. 2000) and oxaloacetate (Chen et al. 1999) larger currents than succinate. Maleate, a *cis*-dicarboxylate with four carbons did not interact with NaDC-3, as opposed to fumarate, a *trans*-dicarboxylate with four carbons. This finding indicates that NaDC-3 prefers dicarboxylates in the *trans*-configuration. Malonate (three carbons) and oxalate (two carbons) did not interact with NaDC-3 (Table 4), supporting the notion that dicarboxylates must have at least four carbons for interaction with NaDC-3. As already noted earlier in experiments on succinate uptake into renal proximal tubules in situ, the distance between the two negative charges is critical and must exceed 5 Å (Fritzsch et al. 1989; Ullrich et al. 1987b). The dicarboxylic amino acids D- and L-aspartate (four carbons) weakly interacted with NaDC-3, but L-glutamate (five carbons) did not. *N*-acetyl-aspartate inhibited succinate transport by hNaDC-3 half-maximally at 232 μM, i.e., at a roughly tenfold lower concentration than L-aspartate (IC_{50} 3,070 μM) (Huang et al. 2000). This finding shows that the positively charged nitro group (in aspartate and glutamate) strongly decreases the affinity of C4 and C5 dicarboxylic amino acids for NaDC-3. In *N*-acetyl-aspartate, this positive charge is not present. Since also 2,2-dimethylsuccinate was accepted by mouse and flounder NaDC-3, it appears that the transporter tolerates mono (=O or -N-CO-CH_3) and disubstitutions [-$(CH_3)_2$] at carbon #2.

A specific feature of NaDC-3s is their ability to interact with dicarboxylates which are substituted in the 2 and 3 position. 2,3-Dimethylsuccinate was transported by rat, mouse and flounder NaDC-3. Dimethylsuccinate transport was also shown for hNaDC-3; however, it is not clear whether the 2,2- or 2,3-dimethylsuccinate was used in that study (Wang et al. 2000). Dimercaptosuccinate inhibited succinate uptake by rNaDC-3 (Kekuda et al. 1999). *Meso*-2,3-Dimercaptosuccinate (DMSA) is a heavy metal chelator and is in clinical use for the treatment of lead intoxications in children (for a review see Zalups 2000). In laboratory animals, DMSA also increased the urinary excretion of mercury (Zalups 1993) and lead (Maiorino et al. 1993). We were recently able demonstrate that DMSA induces a Na^+-dependent and Li^+-inhibitable current in fNaDC-3-expressing oocytes (Burckhardt BC et al. 2002). Thus, it is conceivable that DMSA is taken up from the blood by NaDC-3, captures lead or mercury which has been deposited in proximal tubule cells, and leaves as a metal-DMSA complex the cells via as yet unidentified transport mechanism in the luminal membrane.

During the studies on DMSA, it became apparent that another heavy metal antidote, 2,3-dimercaptopropane-1-sulfonate (DMPS), produced currents in fNaDC-3-expressing oocytes. DMPS does not possess two carboxyl groups, but rather one sulfonate group. It seems that the second negative charge, which is necessary for interaction with NaDC-3, may be provided by a dissociated SH group (Burckhardt BC et al. 2002). That two charges are required for interaction can also be seen from Table 4: the monocarboxylates lactate, pyruvate, butyrate and valerate were not transported, nor did they inhibit dicarboxylate transport. As is also evident from Table 4, there are no drugs that have been tested as possible substrates for NaDC-3, with the exception of DMSA and DMPS. In the intact rat kidney, a considerable number of drugs was tested, but most of them did not inhibit succinate transport in vivo, with the exception of benzylpenicillin (app. K_i 15.8 mM) and furosemide (app. K_i 5.1 mM) (Ullrich 1999). Thus it seems that NaDC-3 is a transporter specialized for dicarboxylates. The relatively narrow substrate specificity may not allow for the interaction with a wider scope of xenobiotics. Nevertheless, by maintaining the intracellular-to-extracellular α-ketoglutarate concentration difference, NaDC-3 is instrumental in driving a large variety of endogenous and exogenous organic anions through OAT1, the organic anion/α-ketoglutarate exchanger.

The organic anion transporter 1

Transport of organic anions in the kidney

A considerable number of drugs and toxins entering the body carry a negative charge at physiological pH and, hence, are organic anions. In addition, metabolism of either endogenous or exogenous uncharged molecules, e.g., sulfation or glucuronidation, generates organic anions which have to be disposed of by either hepatic or renal excretion. For organic anion excretion, the kidneys are equipped with transport systems of great efficiency and broad substrate specificity (for reviews see Burckhardt G and Pritchard 2000; Pritchard and Miller 1993). These transport systems are located in the proximal tubules, where they participate in the secretion of organic anions into the tubule lumen. The model organic anion, *p*-aminohippurate (PAH) is nearly completely removed from the blood entering both kidneys, i.e., the PAH clearance approached renal plasma flow (Smith et al. 1945). PAH is secreted in the proximal tubules at high rates, particularly in the pars recta (Tanner and Isenberg 1970; Tune et al. 1969; Woodhall et al. 1978).

The secretion of organic anions (OA) can be divided into two processes: (1) uptake of OA across the basolateral membrane into proximal tubule cells and (2) release of OA from the cells into the tubule lumen via the luminal membrane (see Fig. 3). Here, we briefly consider the release mechanisms first and then turn to the transporters for OA in the basolateral membrane. At the luminal membrane of proximal tubule cells, the efflux mechanisms for organic anions differ with species. As summarized in (Burckhardt G and Pritchard 2000), the luminal membrane may be equipped with (1) organic anion uniporters ("channel"; flounder, rat, rabbit, pig kidneys), (2) one or more organic anion/Cl^- (OH^-, HCO_3^-) antiporters (dog, rat), (3) an organic anion/α-ketoglutarate antiporter (human, cow), and (4) ATP-driven organic anion transporters. Figure 3 shows all functionally described systems in a single cell disregarding individual combinations in different species. Three cloned transporters have been shown to translocate PAH: NPT1, mrp-2 and OAT4. The human sodium-phosphate cotransporter, NPT1, was expressed in HEK293 cells and performed uptake of radiolabeled PAH and urate (Uchino et al. 2000). The authors claimed that NPT1 is the equivalent for the organic anion uniporter in the luminal membrane. However, they did not demonstrate an influence of the membrane potential on PAH transport. The multidrug resistance protein 2, mrp-2, is mainly responsible for the ATP-

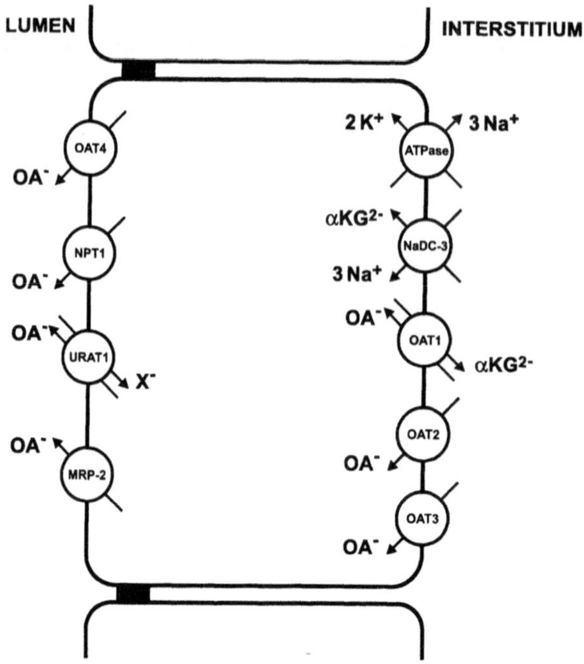

Fig. 3 Secretion of organic anions in proximal tubules. The first step of secretion is the uptake of organic anions (OA^-) from the blood/interstitium into proximal tubule cells. The classic secretory pathway for *p*-aminohippurate (PAH) involves OAT1, which exchanges an extracellular OA^- against an intracellular α-ketoglutarate (αKG^{2-}). The α-ketoglutarate released by OAT1 is pumped back into the cell by NaDC-3, and the three Na^+ ions cotransported with α-ketoglutarate are removed by the Na^+,K^+-ATPase. Additional organic anion transporters are OAT2 (only in human kidneys) and OAT3, which do not appear to operate as anion exchangers. In the luminal membrane, several transporters for organic anion secretion have been identified. URAT1 is a urate/anion exchanger, MRP-2 an ATP-driven export pump, NPT a phosphate transporter also accepting organic anions. OAT4 appears to be a uniporter, which has a low affinity for PAH. Depending on species, there are more transporters (electrogenic PAH-uniporter, PAH/anion exchanger), which have not yet been identified on a molecular level

driven translocation of sulfated and glucuronidated compounds (König et al. 1999). In human and rat kidneys, mrp-2 is located in the luminal membrane of proximal tubule cells (Schaub et al. 1997, 1999). When human or rabbit mrp-2 was expressed in sf9 or in HEK293 cells, an ATP-driven PAH uptake was found in inside-out vesicles prepared from these cells (Leier et al. 2000; Van Aubel et al. 2000). Typical substrates of mrp-2 such as leukotriene C4 or cyclosporin A inhibited PAH transport. Thus, mrp-2 contributes to the efflux of organic anions across the brush-border membrane. The extent to which mrp-2 is involved in total organic anion efflux, however, is unknown. OAT4 was cloned from a human kidney cDNA library (Cha et al. 2000) and shown to be localized in the luminal membrane (Babu et al. 2002a). OAT4 showed very little transport of PAH (Cha et al. 2000), and ochratoxin A transport by hOAT4 was not significantly inhibited by PAH (Babu et al. 2002a), suggesting that OAT4 may not be involved in the classic organic anion secretion pathway. Recently, URAT1 (SLC22A12) was cloned based on its sequence homologies to OAT4 (Enomoto et al. 2002a). URAT1 functions as an antiporter exchanging urate against chloride or against a series of monovalent organic anions. PAH was not accepted by this transporter. Thus, URAT1 is not involved in PAH secretion, but rather in urate absorption. A urate channel, UAT1 (Leal-Pinto et al. 1997, 2002), has so far not been tested for PAH as a possible substrate. Studies with bovine renal BBMV indicated that, at least in the cow, the PAH and urate conductance have pharmacological properties distinct from UAT1 (Krick et al. 2000).

PAH concentration in the tubule lumen exceeded that in the blood, suggesting that PAH secretion is an energy-requiring process (Grantham et al. 1974; Häberle 1981). Based on experiments with rat renal basolateral membrane vesicles, we (Shimada et al. 1987) and Pritchard (Pritchard 1988) proposed a model for energizing PAH uptake into proximal tubule cells. As shown in Fig. 3, three transporters cooperate in the basolateral membrane: the Na^+,K^+-ATPase creates and maintains an extracellular-to-intracellular concentration difference of Na^+ at the expense of ATP. The three Na^+ ions pumped out by the Na^+, K^+-ATPase are cotransported with one molecule α-ketoglutarate via NaDC-3. In turn, the intracellularly accumulated α-ketoglutarate exchanges with an extracellular organic anion (OA), e.g., PAH. Thus, PAH uptake is tertiary-active, with the Na^+, K^+-ATPase as the primary active (directly ATP-consuming) process, and the NaDC-3 as the secondary active transporter. Altogether, the uptake of one molecule PAH (or of any other organic anion) requires the hydrolysis of one molecule of ATP (Burckhardt G et al. 2001). The transporter performing organic anion/α-ketoglutarate antiport has been cloned and called organic anion transporter 1 (OAT1).

Figure 3 indicates the presence of two further organic anion transporters, OAT2 and OAT3 in the basolateral membrane. Since these OATs did not appear to exchange anions against α-ketoglutarate, it is not clear how uptake of organic anions into the cells is energized. Given the inside negative membrane potential, OAT2 and OAT3 should rather release anions from the cell into the interstitium, i.e., be involved in organic anion absorption. They may, however, work in the secretory direction, if the organic anions are pumped out across the luminal membrane in an energy-requiring transport step, e.g., by an ATP-driven anion transporter such as mrp-2.

Substrate specificity of the basolateral PAH transporter in the intact rat kidney

In microperfusion experiments, several hundred compounds were tested as possible inhibitors of the uptake of radiolabeled PAH from blood capillaries into proximal tubule cells in the intact rat kidney. The results of these studies were summarized in (Ullrich 1994, 1997, 1998; Ullrich and Rumrich 1988). The uptake of PAH itself was Na^+-independent and saturable with an apparent K_m of 80 µM (Ullrich et al. 1987a). The following classes of substrates were tested on PAH transport: monocarboxylates (Ullrich et al. 1987c), dicarboxylates (Ullrich et al. 1987b), substituted benzene analogs (Ullrich et al. 1988), diuretics (Ullrich et al. 1989a), β-lactam antibiotics (Ullrich et al. 1993a), glutathione- and cysteine conjugates, oligopeptides (Ullrich et al. 1989b), cyclic nucleotides, eicosanoids (Ullrich et al. 1991b), steroid hormones Ullrich et al. 1991c, 1993a), cyclophosphamides (Ullrich et al. 1993a), diphosphonates (Ullrich et al. 1997) and others. The interaction of an impressive list of drugs with the PAH transporter was summarized in (Ullrich 1999). From all these studies, the structural requirements for substrates interacting with the PAH transporter emerged:

1. Substrates must have a hydrophobic domain of 4–10 Å.
2. An increase in the strength of the negative charge, i.e., a decrease in pK_a, favors the interaction with the PAH transporter.

3. A full ionic charge, however, is not a must: uncharged molecules such as steroid hormones can interact with the PAH transporter, provided they are hydrophobic enough.
4. Electron-attracting side groups such as Cl, Br, NO_2 increased the affinity for the PAH transporter.

Taken together, binding of substrates to the PAH transporter most likely involves hydrophobic interactions and the formation of multiple H bonds. The negatively charged groups may bind to oppositely charged amino acid residues of the transporter protein or also form H bonds. Indeed, as shown for a series of amines, the logarithm of apparent K_i values was negatively correlated, i.e., the affinity increased, with the number of H-bond acceptors (Ullrich et al. 1993a). An interaction of a transporter with its substrates by H bonds has advantages. Since H bonds are relatively weak, release of substrates after translocation is easier. More important, the ability to form multiple H bonds enables a transporter to interact with molecules of different chemical classes. H bonds may, therefore, play the key role in polyspecific transporters.

Cloning, gene, and tissue distribution of OAT1

Table 5 summarizes the properties of the organic anion transporter 1, OAT1, or SLC22A6. In 1997, three groups independently cloned the orthologs from rat (ROAT, Sweet et al. 1997; OAT1, Sekine et al. 1997), mouse (NKT; Lopez-Nieto et al. 1997), and winter flounder (ROAT1; Wolff et al. 1997). The human ortholog was first cloned in 1998 (hROAT1; Reid et al. 1998), later also by other laboratories (hOAT1, Hosoyamada 1999; Race et al. 1999; Cihlar et al. 1999; PAHT, Lu et al. 1999). A more distantly related OAT1 was found in *Caenorhabditis elegans* (George et al. 1999). More recently, the OAT1 homologs from rabbit (Bahn et al. 2002) and pig (Hagos et al. 2002) kidneys have been described. Despite the different nomenclature initially used by the various laboratories, we shall use the term "OAT1" for all these orthologs and indicate the species by small letters preceding OAT1, e.g., hOAT1 for human OAT1.

The gene of the hOAT1 comprises 8.5 kb and is composed of 10 exons (Bahn et al. 2002). It is localized on chromosome *11q13.1–3* (Bahn et al. 2000; Hosoyamoda et al. 1999), in the neighborhood to the *OAT3* gene (Cha et al. 2001). As assessed by Northern blot studies, human (Cihlar et al. 1999; Hosoyamada et al. 1999; Lu et al. 1999; Race et al. 1999), rat (Buist et al. 2002; Kobayashi et al. 2002a; Sekine et al. 1997; Sweet et al. 1997) and mouse (Lopez-Nieto et al. 1997) OAT1 is highly expressed in the kidneys,

Table 5 Properties of the cloned OAT1 (SLC22A6)

Cloned from	Human, pig, rabbit, rat, mouse, flounder, *C. elegans*
Gene	7.9–8.6 kb; 10 exons; chromosome *11q13.1*
Expression	Kidneys >>brain
Localization	Human: proximal tubules, basolateral membrane Rat: proximal tubules, segments S1~S2>S3; basolateral membrane
Protein	526–563 aa; 12 transmembrane helices; 4–6 *N*-glycosylation sites; 0–2 protein kinase A sites; 4–5 protein kinase C sites; 3–7 casein kinase II sites; 0–1 tyrosine kinase site
Transport	Exchange of organic anions against α-ketoglutarate
Regulation	Down-regulation by protein kinase C; up-regulation by mitogen-activated kinase (MEK)/extracellular regulated kinase (ERK) 1/2

where it was found predominantly in the cortex. Some (Hosoyamada et al. 1999; Pavlova et al. 2000; Race et al. 1999; Sekine et al. 1997), but not all investigators (Cihlar et al. 1999; Kobayashi et al. 2002a; Lopez Nieto et al. 1997; Lu et al. 1999, Sweet et al. 1997), detected OAT1 transcripts in the brain. In two studies on human tissues, weak signals were found also in the liver (Cihlar et al. 1999), skeletal muscle, and placenta (Hosoyamada et al. 1999), whereas in most of the other studies these organs were negative, if tested. We have obtained data to suggest an expression of rabbit OAT1 in the retina (Bahn et al. 2001a).

In mice, the first renal OAT1 (NKT) transcripts were detected at fetal day 14 and the amount of OAT1 mRNA increased in parallel with the maturation of proximal tubules (Pavlova et al. 2000). In mouse central nervous system (choroid plexus, dura mater, root ganglions and spinal cord), OAT1 mRNA showed the highest levels at the embryonic stage, with decreasing signals toward adulthood, where OAT1 mRNA was detected only in the meninges (Pavlova et al. 2000). The transient expression during embryogenesis suggests a hitherto unknown role of OAT1 in the development of central nervous structures. In Sprague-Dawley rats, mRNA for OAT1 was detectable in the fetal kidney from day 18 on (Nakajima et al. 2000). Also in Sprague-Dawley rats, the amount of mRNA for rOAT1 steadily increased after birth and reached a maximum at around 30 days. Thereafter, the mRNA declined in females, whereas it stayed constant in males, leading to gender differences in rOAT1 mRNA amounts from day 36 on (Buist et al. 2002). In 35-day-old Wistar rats, similar amounts of OAT1 mRNA were detected in the kidneys from male and female animals (Kobayashi et al. 2002a). Unpublished immunohistochemical data by I. Sabolić suggest that in addition to mRNA, the amount of OAT1 protein is different in male and female rats, which may contribute to gender differences in the pharmacokinetics of anionic drugs.

As determined by immunohistochemistry, human (Hosoyamada et al. 1999; Motohashi et al. 2002) and rat (Kojima et al. 2002; Tojo et al. 1999) OAT1 protein is located in the basolateral membrane of proximal tubule cells. Labeling of the S1 and S3 segments in rat kidneys was present, but weak (I. Sabolić, unpublished data, and Kojima et al. 2002). In human kidneys, the S3 segment was either negative (Hosoyamada et al. 1999) or all segments were found to be labeled to comparable degrees (Motohashi et al. 2002). Besides proximal tubules, other nephron segments, glomeruli, and vessels did not react with the antibody, suggesting an exclusive location of OAT1 in proximal tubules. Immunohistochemical studies on other species (rabbit, mouse, pig, flounder, *C. elegans*) have not been published.

OAT1 protein

Sequencing of the various OAT1 clones revealed regions coding for 526 (*C. elegans*; George et al. 1999), 546 (mouse; Lopez-Nieto et al. 1997), 550 (hOAT1, short isoform; Hosoyamada et al. 1999; Race et al. 1999), 551 (rat, Sekine et al. 1997; Sweet et al. 1997; rabbit, Bahn et al. 2002), 562 (winter flounder; Wolff et al. 1997), or 563 amino acids (hOAT1, long isoform; Hosoyamada et al. 1999). Setting the hOAT1 (short isoform) to 100%, the homologies decreased in the order rabbit (91.3% homology to hOAT1) >rat (90.6%) >pig (89.6%) >mouse (87.9%) >>flounder (55.9%) >>*C. elegans* (28.1%). Thus it is clear that human, rabbit, rat, pig and mouse OAT1 represent orthologs. Whether the OAT1 from winter flounder and from *C. elegans* are orthologs or have rather to be considered as homologs is not clear. Interestingly, the human OAT1 appears in four splice variants in the intact kidney: OAT1–1 (563 amino acids), OAT1–2 (550 aa), OAT1–3 (506 aa), and OAT1–4 (519 aa) (Bahn et al. 2000). The deletion of 13 amino acids in OAT1–2 is the result of alternative splicing at the border between exons 9 and 10. OAT1–3 and OAT1–4 carry a further deletion within exon 9. Preliminary functional studies show no obvious differences between OAT1–1 and OAT1–2 with respect to transport, whereas OAT1–3 and OAT1–4 are nonfunctional (Buttler et al. 2001). Splice variants also occur in rabbit and pig kidneys (Hagos et al. 2002). The physiological significance of the presence of splice variants is not known at this time.

The topological organization of OAT1 proteins within the membrane is unknown. Secondary structure prediction programs revealed 8–12 transmembrane helices. Most authors favor the presence of 12 transmembrane helices (M1–M12), because OAT1 belongs to a superfamily of transporters (MFS), which are all believed to have 12 transmembrane segments (Pao et al. 1998). If this assumption is correct, then the N- and C-termini would be located intracellularly and the proteins would have six extracellular (E1–E6) and five intracellular loops (I1–I5) connecting the transmembrane helices. The large loop (E1) between M1 and M2 carries all putative *N*-glycosylation sites, and the large loop (I3) between M6 and M7 harbors multiple consensus sequences for phosphorylation by protein kinases (Burckhardt G et al. 2001; Burckhardt G and Wolff 2000; Sweet and Pritchard 1999). Additional phosphorylation consensus sites are located at the C-terminus, an argument for its intracellular location. In recent experiments using the insertion of the FLAG epitope into hOAT1, we obtained results in agreement with the extracellular location of E1, which is accessible to anti-FLAG antibodies in the intact cell, and the intracellular location of the C-terminus, which is accessible to antibodies only after solubilizing the cell (N.A. Wolff, unpublished data). At present it is not known which of the various sites for *N*-glycosylation and for phosphorylation are actually used for posttranslational modification of OAT1 proteins. For mOAT1, the glycosylation is necessary for functional expression in COS7 cells (Kuze et al. 1999). We have preliminary data to suggest that serine 271, which is located within a protein kinase C consensus sequence of hOAT1, is required for trafficking from the endoplasmic reticulum to the plasma membrane of *Xenopus laevis* oocytes (Wolff et al. 2002).

The repeats of motifs in the first six and the second six transmembrane helices and their connecting loops within the MFS molecules, including OAT1 (Burckhardt G and Wolff 2000), suggested that these transporters have evolved from gene duplication. The evolutionary conservation of these motifs indicated an important functional role, for example, in stabilizing the protein within the membrane for its proper transport function.

Functional characterization of OAT1 substrates

For functional studies, cloned OAT1 has been expressed in *Xenopus laevis* oocytes, COS7 cells, HeLa cells, LLC-PK1 cells, CHO cells, and in immortalized mouse proximal tubule cells (mouse PTC; cf. Table 6). Stable expression was achieved in CHO cells (Ho et al. 2000) and in mouse S2 proximal tubule cells isolated from transgenic mice harboring the virus 40 large T antigen and transfected with hOAT1 (Takeda et al. 2001). Table 6 shows more than 150 substances that have so far been tested with the cloned and heterologously expressed OAT1 from man, rat, mouse, flounder, and *C. elegans*. Among these substances are nonsteroidal anti-inflammatory drugs, β-lactam antibiotics, *N*-acetyl-L-cysteine conju-

Table 6 Substrate specificity of the cloned OAT1

Compound	Species	Expression system	Transport or K_m (μM)	Inhibition or IC_{50} (μM)	References
Acetazolamide	Rat	Oocyte	+	1,100	Uwai et al. 2000
	Flounder	Oocyte	–[a]		Burckhardt et al. 2000b
N-acetyl-L-cysteine	Rat	Oocyte		+	Pombrio et al. 2001
N-acctyl-L-norlcucinc	Rat	Oocytc		+	Pombrio et al. 2001
N-acetyl-*S*-allyl-L-cysteine	Rat	Oocyte		+	Pombrio et al. 2001
N-acetyl-*S*-benzyl-L-cysteine	Rat	Oocyte		400	Pombrio et al. 2001
N-acetyl-*S*-BrCl-diF-ethyl-L-cysteine	Rat	Oocyte		+	Pombrio et al. 2001
N-acetyl-*S*-butyl-L-cysteine	Rat	Oocyte		+	Pombrio et al. 2001
N-acetyl-*S*-Cl-triF-ethyl-L-cysteine	Rat	Oocyte		+	Pombrio et al. 2001
N-acetyl-*S*-diBr-diF-ethyl-L-cysteine	Rat	Oocyte		+	Pombrio et al. 2001
N-acetyl-*S*-dichlorovinyl-L-cysteine	Rat	Oocyte		+	Pombrio et al. 2001
N-acetyl-*S*-2,4-dinitrophenyl-L-cysteine	Rat	Oocyte	2.0	1.9	Pombrio et al. 2001
N-acetyl-*S*-dodecyl-L-cysteine	Rat	Oocyte		+	Pombrio et al. 2001
N-acetyl-*S*-ethyl-L-cysteine	Rat	Oocyte		1,100	Pombrio et al. 2001
N-acetyl-*S*-LTE_4	Rat	Oocyte		9	Pombrio et al. 2001
N-acetyl-*S*-pentaCl-butadiene-L-cysteine	Rat	Oocyte		+	Pombrio et al. 2001
N-acetyl-*S*-3-phenylpropyl-L-cysteine	Rat	Oocyte		+	Pombrio et al. 2001
N-acetyl-*S*-propanol-L-cysteine	Rat	Oocyte		+	Pombrio et al. 2001
N-acetyl-*S*-propyl-L-cysteine	Rat	Oocyte		+	Pombrio et al. 2001
Acetylsalicylate	Human	HeLa cells		–	Lu et al. 1999
	Rat	Oocyte	+	428	Apiwattanakul et al. 1999
Acyclovir (ACV)	Human	Mouse PTC	342.3	+	Takeda et al. 2002b
	Rat	Oocyte	242	+	Wada et al. 2000
Adefovir	Human	CHO cells	23.8	28	Cihlar and Ho 2000; Ho et al. 2000
		Oocyte	30		Cihlar et al. 1999
	Rat	Oocyte	270		Cihlar et al. 1999
Adipate	Rat	Oocyte		+	Uwai et al. 1998
Amantadine	Rat	Oocyte		+	Wada et al. 2000
Amiloride	Human	Oocyte		–	Race et al. 1999

Table 6 (continued)

Compound	Species	Expression system	Transport or K_m (μM)	Inhibition or IC_{50} (μM)	References
p-Aminohippurate (PAH)	Human	Oocyte	+[1]; 3.9[2]; 4.0[3]; 9.3[4]	+[2]	[1]Race et al. 1999; [2]Islinger et al. 2001; [3]Cihlar et al. 1999; [4]Hosoyamada et al. 1999
		CHO cells	15.4[1,2]	8.8[1]	[1]Cihlar and Ho 2000; [2]Ho et al. 2000
		HeLa cells	5.0		Lu et al. 1999
		Mouse PTC	15.8[1]; 20.1[2]	+[3]; 6.02[4]	[1]Takeda et al. 2000a; [2]Takeda et al. 2001; [3]Babu et al. 2002b; [4]Jung et al. 2001
		OK cells	22		Motojima et al. 2002
	Rat	Oocyte	11[1]; 14.3[2]; 31[3]; 70[4]	+[4,5]	[1]Pombrio et al. 2001; [2]Sekine et al. 1997; [3]Uwai et al. 2000; [4]Sweet et al. 1997; [5]Apiwattanakul et al. 1999; Cihlar et al. 1999; Tsuda et al. 1999; Uwai et al. 1998
		LLC-PK1	47[1]; 59.5[2]		[1]Nagata et al. 2002; [2]Hasegawa et al. 2002
		Mouse PTC	43.2[1]; 85.1[2]		[1]Takeda et al. 1999; [2]Sugiyama et al. 2001
	Mouse	COS cells	37	+	Kuze et al. 1999
		LLC-PK1	162		You et al. 2000
	Flounder	Oocyte	21[1]; 37[2]; 58[a3]	+[1]	[1]Wolff et al. 1997; [2]Wolff et al. 2001; [3]Burckhardt BC et al. 2000b
	C. elegans	HeLa cells	430	+	George et al. 1999
Aminopyrine	Rat	Oocyte		+	Apiwattanakul et al. 1999
Amoxicillin	Rat	Oocyte		+	Jariyawat et al. 1999
Ampicillin	Rat	Oocyte		–	Sekine et al. 1997
Anandamide	Human	HeLa cells		–	Lu et al. 1999
Antipyrine	Rat	Human		+	Apiwattanakul et al. 1999
Ascorbate	Mouse	COS cells		–	Kuze et al. 1999
	C. elegans	HeLa cells		–	George et al. 1999
Aspartame	Human	Mouse PTC		–	Jung et al. 2001
	Rat	Oocyte		–	Tsuda et al. 1999
Azelate	Rat	Oocyte		+	Uwai et al. 1998
3'-Azido-3'-deoxythymidine (AZT; zidovudine)	Human	Mouse PTC	45.9	+	Takeda et al. 2002b
	Rat	Oocyte	68	+	Wada et al. 2000
Benzydamine	Rat	Oocyte		+	Apiwattanakul et al. 1999

Table 6 (continued)

Compound	Species	Expression system	Transport or K_m (μM)	Inhibition or IC_{50} (μM)	References
Benzylpenicillin (penicillin G)	Human	Oocyte		+	Hosoyamada et al. 1999; Islinger et al. 2001
		HeLa cells		–	Lu et al. 1999
		Mouse PTC		+	Jung et al. 2001
	Rat	Oocyte	+[1]	+[2,3]; 1,680[1]	[1]Jariyawat et al. 1999; [2]Islinger et al. 2001; [3]Sekine et al. 1997
		LLC-PK1		418[1]; 800[2]	[1]Hasegawa et al. 2002; [2]Nagata et al. 2002
	C. elegans	HeLa cells		+	George et al. 1999
Betamipron	Human	CHO cells		6[1]; 16.2[2]	[1]Mulato et al. 2000; [2]Ho et al. 2000
		Mouse PTC		23.6	Takeda et al. 2001
Bromcresolgreen	Human	Oocyte		+	Cihlar et al. 1999
	Rat	Oocyte		+	Sweet et al. 1997
8-Bromo-cAMP	Human	HeLa cells		–	Lu et al. 1999
Bromosulfophthalein (BSP)	Human	Oocyte		+	Islinger et al. 2001
	Rat	LLC-PK1		2.74	Hasegawa et al. 2002
Bumetanide	Human	Oocyte		+	Race et al. 1999
	Rat	Oocyte		5.5	Uwai et al. 2000
	Flounder	Oocyte	+[a1]	+[2]	[1]Burckhardt BC et al. 2000b; [2]Wolff et al. 1997
cAMP	Rat	Oocyte	+	+	Sekine et al. 1997
Captopril	Mouse	COS cells		+	Kuze et al. 1999
Carbenicillin	Rat	Oocyte		500	Jariyawat et al. 1999
5-Carboxyfluorescein	Human	CHO cells	–		Cihlar and Ho 2000
6-Carboxyfluorescein	Human	CHO cells	3.93		Cihlar and Ho 2000
Carprofen	Mouse	COS cells		+	Kuze et al. 1999
Cefadroxil	Human	Mouse PTC		6,140[b]	Takeda et al. 2002a
	Rat	Oocyte		+	Jariyawat et al. 1999
		Mouse PTC		–	Takeda et al. 1999
Cefamandole	Human	Mouse PTC		30[b]	Takeda et al. 2002a
	Rat	Oocyte		+	Jariyawat et al. 1999
		Mouse PTC		+	Takeda et al. 1999
Cefazolin	Human	Mouse PTC		180[b]	Takeda et al. 2002a
	Rat	Oocyte		450	Jariyawat et al. 1999
		Mouse PTC		+	Takeda et al. 1999
Cefoperazone	Human	Mouse PTC		210[b]	Takeda et al. 2002a
	Rat	Oocyte		+	Jariyawat et al. 1999
		Mouse PTC		+	Takeda et al. 1999
Cefotaxime	Human	Mouse PTC		3,130[b]	Takeda et al. 2002a
	Rat	Oocyte		+	Jariyawat et al. 1999
		Mouse PTC		+	Takeda et al. 1999
Cefsulodin	Rat	Oocyte		+	Jariyawat et al. 1999

Table 6 (continued)

Compound	Species	Expression system	Transport or K_m (µM)	Inhibition or IC_{50} (µM)	References
Ceftazidime	Rat	Oocyte		+	Jariyawat et al. 1999
Ceftriaxone	Human	Mouse PTC		230[b]	Takeda et al. 2002a
	Rat	Oocyte		+	Jariyawat et al. 1999
		Mouse PTC		+	Takeda et al. 1999
Cephalexin	Rat	Oocyte		2,310	Jariyawat et al. 1999
		Mouse PTC		–	Takeda et al. 1999
Cephaloridine	Human	CHO cells		1,250	Cihlar and Ho 2000
		Mouse PTC		740[b]	Takeda et al. 2002a
	Rat	Oocyte	+[1]	+[2]; 2,330[1]	[1]Jariyawat et al. 1999; [2]Sekine et al. 1997
		Mouse PTC	+	+	Takeda et al. 1999
Cephalothin	Human	Mouse PTC		220[b]	Takeda et al. 2002a
	Rat	Oocyte		290	Jariyawat et al. 1999
		Mouse PTC		+	Takeda et al. 1999
Cephradine	Human	CHO cells		1,600	Cihlar and Ho 2000
cGMP	Rat	Oocyte	+	+	Sekine et al. 1997
Cidofovir	Human	Oocyte	46		Cihlar et al. 1999
		CHO cells	58[1]	60[2]	[1]Ho et al. 2000; [1]Cihlar and Ho 2000
	Rat	Oocyte	238		Cihlar et al. 1999
Cidofovir, prodrug	Human	CHO cells	309[1,2]	1,100[1]	[1]Cihlar and Ho 2000; [2]Ho et al. 2000
Cilastatin	Human	Mouse PTC		231	Takeda et al. 2001
Cimetidine	Human	Mouse PTC		+	Jung et al. 2001
	Rat	LLC-PK1		–	Nagata et al. 2002
	Flounder	Oocyte	+[a]		Burckhardt BC et al. 2003
Cinoxacin	Rat	Oocyte		+	Jariyawat et al. 1999
Cisplatin	Mouse	COS cells		–	Kuze et al. 1999
Citrate	Flounder	Oocyte		–	Burckhardt G et al. 1998
Citrinin	Human	Mouse PTC		3,080	Jung et al. 2001
	Rat	Oocyte		+	Tsuda et al. 1999
Chloramphenicol	Rat	Oocyte		+	Jariyawat et al. 1999
Chlorothiazide	Rat	Oocyte		+	Uwai et al. 2000
Cloxacillin	Rat	Oocyte		+	Jariyawat et al. 1999
Choline	Human	Oocyte		–	Race et al. 1999
Cyclosporin A	Human	Oocyte		–	Cihlar et al. 1999
	Rat	Oocyte		–	Sweet et al. 1997
Cyclothiazide	Rat	Oocyte		+	Uwai et al. 2000
2,4-Dichlorophenoxyacetate	Human	Oocyte		+	Cihlar et al. 1999
Diclofenac	Human	CHO cells		4	Mulato et al. 2000
	Rat	Oocyte		+	Apiwattanakul et al. 1999
	Mouse	CHO cells		+	Kuze et al. 1999
Didanosine	Rat	Oocyte	+	+	Wada et al. 2000

Table 6 (continued)

Compound	Species	Expression system	Transport or K_m (μM)	Inhibition or IC_{50} (μM)	References
Diflunisal	Human	CHO cells		0.85	Mulato et al. 2000
	Rat	Oocyte		+	Apiwattanakul et al. 1999
Digoxin	Rat	LLC-PK1		–	Sugiyama et al. 2001
2,3-Dimercapto-1-propane sulfonate (DMPS)	Human	Oocyte		22	Islinger et al. 2001
	Flounder	Oocyte	+[a]		Burckhardt BC et al. 2000a
Doxycycline	Human	Mouse PTC		+	Babu et al. 2002b
Enalapril	Mouse	COS cells		+	Kuze et al. 1999
Erythromycin	Rat	Oocyte		–	Jariyawat et al. 1999
17β-Estradiol-D-17β-glucuronide ($E_2 17\beta G$)	Rat	LLC-PK1		–	Nagata et al. 2002; Sugiyama et al. 2001
Ethacrynate	Rat	Oocyte		+	Sekine et al. 1997
		Oocyte		+	Uwai et al. 2000
	Flounder	Oocyte	+[a]		Burckhardt BC et al. 2000b
Ethoxzolamide	Rat	Oocyte		+	Uwai et al. 2000
Etodolac	Human	CHO cells		50	Mulato et al. 2000
Fluorescein	Human	Oocytes		+	Islinger et al. 2001
		CHO cells	+		Cihlar and Ho 2000
		HeLa cells		+	Lu et al. 1999
Flurbiprofen	Human	CHO cells		1.5	Mulato et al. 2000
	Rat	Oocyte		+	Apiwattanakul et al. 1999
Folate	Rat	Oocyte	+	+	Uwai et al. 1998
	Mouse	COS cells		+	Kuze et al. 1999
	C. elegans	HeLa cells		+	George et al. 1999
Foscarnet	Rat	Oocyte	–	–	Wada et al. 2000
Fumarate	Flounder	Oocyte		–	Burckhardt G et al. 1998; Wolff et al. 1997
Furosemide	Human	Oocyte		+	Hosoyamada et al. 1999; Race et al. 1999
		HeLa cells		+	Lu et al. 1999
	Rat	Oocyte	+[1]	+[2]; 9.5[1]	[1]Uwai et al. 2000; [2]Sekine et al. 1997
	Flounder	Oocyte	–[a]		Burckhardt BC et al. 2000b
	C. elegans	HeLa cells		+	George et al. 1999
Ganciclovir	Human	Mouse PTC	896	+	Takeda et al. 2002b
Gentamycin	Rat	Oocyte		–	Jariyawat et al. 1999
Glutarate	Human	CHO cells	10.7	4.9	Cihlar and Ho 2000
		HeLa cells		+	Lu et al. 1999
	Rat	Oocyte		+	Uwai et al. 1998
	Flounder	Oocyte		+	Burckhardt G et al. 1998; Wolff et al. 1997

Table 6 (continued)

Compound	Species	Expression system	Transport or K_m (μM)	Inhibition or IC_{50} (μM)	References
cGMP	Rat	Oocyte	+		Sekine et al. 1997
Grepafloxacin	Rat	Oocyte	–	+	Uwai et al. 1998
Hippurate	Human	HeLa cells		+	Lu et al. 1999
		OK cells		20	Motojima et al. 2002
Hydrochlorothiazide	Human	Oocyte		–	Race et al. 1999
	Rat	Oocyte		150	Uwai et al. 2000
o-Hydroxyhippurate	Human	OK cells		27	Motojima et al. 2002
p-Hydroxyhippurate	Human	OK cells		25	Motojima et al. 2002
p-Hydroxyphenyl acetate	Human	OK cells		+	Motojima et al. 2002
Ibuprofen	Human	CHO cells	–	8.0	Mulato et al. 2000
	Rat	Oocyte		3.5	Apiwattanakul et al. 1999
Indole acetate	Human	OK cells	+	83	Motojima et al. 2002
Indomethacin	Human	Oocyte		+	Hosoyamada et al. 1999
		CHO cells		3.0	Mulato et al. 2000
		HeLa cells		+	Lu et al. 1999
		Mouse PTC		+	Jung et al. 2001
	Rat	Oocyte	+[1]	+[2]; 10[1]	[1]Apiwattanakul et al. 1999; [2]Sekine et al. 1997
	Mouse	COS cells		+	Kuze et al. 1999
	C. elegans	HeLa cells		+	George et al. 1999
Indoxyl sulfate	Human	OK cells	+	83	Motojima et al. 2002
α-Ketoglutarate (2-oxoglutarate)	Human	Oocyte		+	Cihlar et al. 1999; Hosoyamada et al. 1999; Race et al. 1999
		HeLa cells	+	+	Lu et al. 1999
		OK cells		+	Motojima et al. 2002
	Rat	Oocyte	+[1]	+[1,2,3]	[1]Sekine et al. 1997; [2]Sweet et al. 1997; [3]Uwai et al. 1998
	Mouse	COS cells		+	Kuze et al. 1999
	Flounder	Oocyte		+	Burckhardt G et al. 1998; Wolff et al. 1997
Ketoprofen	Human	CHO cells	–[1]	1.3[1]; 1.4[2]	[1]Mulato et al. 2000; [2]Cihlar and Ho 2000
	Rat	Oocyte		+	Apiwattanakul et al. 1999
Lamivudine	Rat	Oocyte	+		Wada et al. 2000
Levofloxacin	Rat	Oocyte	–	+	Uwai et al. 1998
Lisinopril	Mouse	COS cells		–	Kuze et al. 1999
Losartan	Human	Oocyte		+	Race et al. 1999
Malonate	Rat	Oocyte		–	Uwai et al. 1998
	Flounder	Oocyte		–	Burckhardt et al. G 1998

Table 6 (continued)

Compound	Species	Expression system	Transport or K_m (μM)	Inhibition or IC_{50} (μM)	References
Meclofenamate	Rat	Oocyte		+	Apiwattanakul et al. 1999
Mercury-cysteine	Human	HeLa cells		–	Lu et al. 1999
Methazolamide	Rat	Oocyte		+	Uwai et al. 2000
Methotrexate	Human	HeLa cells	–	–	Lu et al. 1999
	Rat	Oocyte	+	+	Sekine et al. 1997; Uwai et al. 1998
	Mouse	COS cells		+	Kuze et al. 1999
N-methylnicotinamide	Human	CHO cells		–	Cihlar and Ho 2000
Minocycline	Human	Mouse PTC		+	Babu et al. 2002b
Nafcillin	Rat	Oocyte		+	Jariyawat et al. 1999
Nalidixate	Rat	Oocyte		+	Sekine et al. 1997
Naproxen	Human	CHO cells		5.8	Mulato et al. 2000
	Rat	Oocyte		2.0	Apiwattanakul et al. 1999
(Noradamantan-3-yl) 1,3-dipropylxanthine (KW-3902)	Human	Mouse PTC		7.82	Takeda et al. 2001
Norfloxacin	Rat	Oocyte		–	Jariyawat et al. 1999
Ochratoxin A	Human	Mouse PTC	0.42		Jung et al. 2001
	Rat	Oocyte	2.1		Tsuda et al. 1999
		Mouse PTC	0.57		Tsuda et al. 1999
	Flounder	Oocyte	+[a]		Burckhardt BC et al. 2000b
Octanoate	Human	Mouse PTC		5.41	Jung et al. 2001
	Rat	Oocyte		+	Tsuda et al. 1999
Ofloxacin	Rat	Oocyte		–	Jariyawat et al. 1999
Oxyphenbutazone	Rat	Oocyte		32	Apiwattanakul et al. 1999
Oxytetracycline	Human	Mouse PTC		+	Babu et al. 2002b
Pantothenate	Mouse	COS cells		+	Kuze et al. 1999
	C. elegans	HeLa cells		+	George et al. 1999
Paracetamol	Rat	Oocyte		2,099	Apiwattanakul et al. 1999
Phenacetin	Human	CHO cells		200	Mulato et al. 2000
	Rat	Oocyte		488	Apiwattanakul et al. 1999
Phenol red	Human	Oocyte		+	Hosoyamada et al. 1999
Phenylbutazone	Rat	Oocyte		+	Apiwattanakul et al. 1999
9-(2-Phosphonylmethoxyethyl) diaminopurine (PMEDAP)	Human	Oocyte	+		Cihlar et al. 1999
	Rat	Oocyte	+		Cihlar et al. 1999
9-(2-Phosphonylmethoxy-oxyethyl)guanidine (PMEG)	Human	Oocyte	+		Cihlar et al. 1999
	Rat	Oocyte	+		Cihlar et al. 1999
Pimelate	Rat	Oocyte		+	Uwai et al. 1998
Piperacillin	Rat	Oocyte		+	Jariyawat et al. 1999

Table 6 (continued)

Compound	Species	Expression system	Transport or K_m (μM)	Inhibition or IC_{50} (μM)	References
Piroxicam	Human	CHO cells		20.5	Mulato et al. 2000
		Mouse PTC		19.8	Jung et al. 2001
	Rat	Oocyte		+[1]; 52[2]	[1]Tsuda et al. 1999; [2]Apiwattanakul et al. 1999
Pravastatin	Rat	LLC-PK1	–	1,150	Hasegawa et al. 2002
Probenecid	Human	Oocyte		+	Cihlar et al. 1999; Hosoyamada et al. 1999; Islinger et al. 2001; Pombrio et al. 2001; Race et al. 1999
		CHO cells		6.3[1]; 6.5[2]; 7.4[3]	[1]Cihlar and Ho 2000; [2]Ho et al. 2000; [3]Mulato et al. 2000
		HeLa cells		+	Lu et al. 1999
		Mouse PTC		4.29[1]; 12.1[2]	[1]Jung et al. 2001; [2]Takeda et al. 2001
		OK cells		+	Motojima et al. 2002
	Rat	Oocyte	–[1]	+[1,2]	[1]Uwai et al. 1998; [2]Cihlar et al. 1999; Mulato et al. 2000; Pombrio et al. 2001; Sekine et al. 1997; Sweet et al. 1997; Tsuda et al. 1999; Uwai et al. 2000; Wada et al. 2000
		Mouse PTC		+	Takeda et al. 1999
		LLC-PK1		31	Sugiyama et al. 2001
	Mouse	COS cells		+	Kuze et al. 1999
	Flounder	Oocyte		+	Burckhardt G et al. 1998; Wolff et al. 1997
	C. elegans	HeLa cells		+	George et al. 1999
Prostaglandin E_2	Human	HeLa cells	–		Lu et al. 1999
		Mouse PTC	0.97	+	Kimura et al. 2002
	Rat	Oocyte	+	+	Sekine et al. 1997
Prostaglandin $F_{2\alpha}$	Human	HeLa cells		–	Lu et al. 1999
		Mouse PTC	0.58	+	Kimura et al. 2002
Pyranine	Human	CHO cells	–		Cihlar and Ho 2000
Riboflavin	Mouse	COS cells		+	Kuze et al. 1999
Salicylate	Human	Oocyte		+	Race et al. 1999
		CHO cells		280	Cihlar and Ho 2000
	Rat	Oocyte	+	341	Apiwattanakul et al. 1999
Salicylurate	Rat	Oocyte		11	Apiwattanakul et al. 1999
Sebacate	Rat	Oocyte		+	Uwai et al. 1998
Semustine	Mouse	COS cells		+	Kuze et al. 1999

Table 6 (continued)

Compound	Species	Expression system	Transport or K_m (μM)	Inhibition or IC_{50} (μM)	References
SITS	Flounder	Oocyte		+	Burckhardt G et al. 1998; Wolff et al. 1997
Stavudine	Rat	Oocyte	+	+	Wada et al. 2000
Streptomycin	rat	Oocyte		–	Jariyawat et al. 1999
Suberate	Rat	Oocyte		+	Uwai et al. 1998
	Flounder	Oocyte		+	Burckhardt G et al. 1998; Wolff et al. 1997
Succinate	Rat	Oocyte		–	Uwai et al. 1998
	Flounder	Oocyte		–	Burckhardt G et al. 1998
Sulfanilamide	Flounder	Oocyte	–[a]		Burckhardt BC et al. 2000b
Sulindac	Mouse	COS cells		+	Kuze et al. 1999
Taurocholate	Human	Oocyte		+	Islinger et al. 2001
	Rat	Oocyte	–	–	Sekine et al. 1997; Uwai et al. 1998
		LLC-PK1		2,770	Sugiyama et al. 2001
Tetracycline	Human	Mouse PTC	+	+	Babu et al. 2002b
Tetraethylammonium	Human	Oocyte		–	Cihlar et al. 1999; Islinger et al. 2001
		OK cells		–	Motojima et al. 2002
	Rat	Oocyte	–[1]	–[1,2]	[1]Uwai et al. 1998; [2]Sekine et al. 1997; Sweet et al. 1997; Tsuda et al. 1999
	Flounder	Oocyte		–	Wolff et al. 1997
Tienilate	Flounder	Oocyte	+[a]		Burckhardt BC et al. 2000b
Tolmetin	Rat	Oocyte		+	Apiwattanakul et al. 1999
Trifluridine	Rat	Oocyte	+	+	Wada et al. 2000
Urate	Human	Oocyte	–[1]	–[1]; +[2]	[1]Race et al. 1999; [2]Cihlar et al. 1999; Hosoyamada et al. 1999
		OK cells		+	Motojima et al. 2002
	Rat	Oocyte	+[1]	–[2]; +[1]	[1]Sekine et al. 1997; [2]Sweet et al. 1997
	Flounder	Oocyte	–	+	Wolff et al. 1997
	C. elegans	HeLa cells		+	George et al. 1999
Valacyclovir	Human	Mouse PTC	–		Takeda et al. 2002b
Valproate	Rat	Oocyte		+	Sekine et al. 1997
Vancomycin	Rat	Oocyte		–	Jariyawat et al. 1999
Vidarabine	Rat	Oocyte		+	Wada et al. 2000
Zalcitabine	Rat	Oocyte	+	+	Wada et al. 2000

[a] Substrate-induced current in two-electrode voltage-clamp studies
[b] Corrected K_i values (Takeda et al. 2002; Eur J Pharmacol 450:111)

gates, antiviral drugs, diuretics, as well as a number of hippurates, test agents, vitamins, hormones, 2nd messengers, non-anionic antibiotics, cytostatic agents, and toxins.

The basic characterization of expressed OATs was performed in uptake experiments using radiolabeled *p*-aminohippurate (PAH). The uptake of labeled PAH into OAT1-expressing cells was saturable with apparent K_m values in the range of 3.9 to 430 μM (Table 6). hOAT1 appeared to exhibit the highest affinity for PAH, with a mean K_m of 11.9 μM taken from eight different investigations (K_m range 3.9–22 μM). The rOAT1 showed a mean K_m of 45.1 μM (eight investigations; range 11–85.1 μM). Although it seems that the K_m values determined for hOAT1 or rOAT1 in oocytes were smaller than those found for the same transporters in other cells, it is premature to postulate a systematic influence of the expression system on the apparent K_m. The variations more likely reflect differences between laboratories, as exemplified by the nearly sevenfold range of apparent K_m values reported for rOAT1 using a single expression system, *Xenopus laevis* oocytes.

A hallmark of OAT1 is its capability to exchange an extracellular organic anion against an intracellular α-ketoglutarate molecule. The inside-to-outside concentration difference of α-ketoglutarate, which is maintained by metabolic production and by Na^+-coupled uptake of α-ketoglutarate into proximal tubule cells (see NaDC-3), provides the driving force for the uptake of organic anions against the inside negative membrane potential and possibly also against an inside-to-outside concentration difference of the organic anion itself. The preloading of cells with α-ketoglutarate or the nonmetabolizable analog of α-ketoglutarate, glutarate, stimulated the uptake of labeled PAH by the expressed human (Cihlar et al. 1999; Ho et al. 2000; Lu et al. 1999; Motojima et al. 2002), rat (Cihlar et al. 1999; Pombrio et al. 2001; Sekine et al. 1997; Sweet et al. 1997), and flounder OAT1 (Wolff et al. 1997). Likewise, the coexpression of a Na^+-dicarboxylate cotransporter with OAT1 caused an increased PAH uptake in the presence of external dicarboxylates, which were first accumulated in the cells by the cotransporter and then exchanged for PAH via the OAT1 (rOAT1, Sekine et al. 1997; ceOAT1, George et al. 1999). Of note, PAH/α-ketoglutarate exchange also occurred in the nonphysiological direction, i.e., PAH efflux from oocytes was stimulated by external α-ketoglutarate in oocytes expressing fOAT1 (Wolff et al. 2001). Thus, organic anions and α-ketoglutarate can interfere with OAT1 from both sides of the membrane, the net transport direction (organic anion uptake and α-ketoglutarate efflux) being determined by the inside-to-outside α-ketoglutarate concentration difference prevailing in renal proximal tubule cells (Pritchard 1995).

From earlier studies on the intact rat kidney it was known that dicarboxylates must have a backbone of at least five carbons, e.g. glutarate and α-ketoglutarate, to interact with the PAH transporter (Ullrich et al. 1987b). When a homologous series of dicarboxylates with increasing chain length was tested on the rOAT1 expressed in oocytes, the same finding was obtained with malonate (three carbons, C3) and succinate (C4), having no effect on PAH uptake, whereas glutarate (C5), adipate (C6), pimelate (C7), suberate (C8), azelate (C9), and sebacate (C10) in the extracellular buffer inhibited PAH uptake (Uwai et al. 1998). Likewise, malonate, succinate and fumarate did not inhibit PAH uptake into fOAT1-expressing oocytes, but α-ketoglutarate and suberate did (Burckhardt G et al. 1998). Thus, the PAH transporter in intact kidneys (Fritzsch et al. 1989) and OAT1 required a distance of 6–7 Å between the two negative charges. The evolutionary more distant ceOAT1 exhibited a different behavior, as it was *trans*-stimulated by intracellular fumarate (C4), but not by glutarate (C5) (George et al. 1999). α-Ketoglutarate did show a *trans*-stimulation in accordance with findings on OAT1s from other species. Why ceOAT1 interfered with the five-carbon dicarboxylate α-ketoglutarate, but not with glutarate, is surprising and awaits explanation.

Besides PAH, a number of radiolabeled substrates were translocated by expressed OAT1 or induced an inward current in fROAT-expressing oocytes (see Table 6 for details). These substrates included endogenous organic anions such as cAMP, cGMP, folate, indole acetate, indoxyl sulfate, α-ketoglutarate, prostaglandin E_2, as well as exogenous anions such as antiviral drugs (acyclovir, adefovir, AZT or zidovudine, cidofovir, ganciclovir, lamivudine, PMEDAP, PMEG, stavudine, trifluridine, zalcitabine), antibiotics (benzylpenicillin, cephaloridine, tetracycline), cytostatic agents (methotrexate), diuretics (acetazolamide, bumetanide, ethacrynate, furosemide, tienilate), heavy metal chelators (DMPS), nonsteroidal anti-inflammatory drugs (NSAIDs: acetylsalicylate, indomethacin, salicylate), and a toxin (ochratoxin A). OAT1 also transported the fluorescent organic anions, fluorescein (FL) and 6-carboxyfluorescein (6-CF), and the uptake of 6-CF into CHO cells expressing hOAT1 was increased by preloading of the cells with glutarate (Cihlar and Ho 2000). Likewise, the uptake of adefovir (hOAT1; Ho et al. 2000) was increased by intracellular glutarate. The uptake of salicylate, acetylsalicylate and indomethacin (Apiwattanakul et al. 1999) and of ochratoxin A (Tsuda et al. 1999) into rOAT1-expressing oocytes was stimulated by the coexpression of rNaDC-1. Thus, OAT1 exchanges a variety, probably all, of the extracellular organic anions against intracellular dicarboxylates. Of toxicological interest is the transport of ochratoxin A, for which the human and rat OAT1s reveal remarkable affinities in the range of 0.6 to 2.1 μM. Thus, besides other OATs, OAT1 contributes to the accumulation of this nephrotoxic cancerogenic agent in proximal tubule cells.

For methotrexate, transport was demonstrated for rOAT1, but not for hOAT1, suggesting species differences. Prostaglandin E_2 was translocated by hOAT1 in one study (Kimura et al. 2002), but not in another also conducted on hOAT1 (Lu et al. 1999). The reason for these different results is not known. Also for urate, conflicting results were obtained, showing either no transport (hOAT1, Race et al. 1999; fOAT1, Wolff et al. 1997) or uptake (rOAT1, Sekine et al. 1997]) When urate was added to the external medium to compete with labeled OAT1 substrates for uptake, inhibition was seen with flounder (Wolff et al. 1997) and *C. elegans* OAT1 (George et al. 1999). With hOAT1 and rOAT1, either no inhibition (Race et al. 1999; Sweet et al. 1997) or inhibition (Cihlar et al. 1999; Hosoyamada et al. 1999; Motojima et al. 2002; Sekine et al. 1997) by urate was seen. Because urate inhibited the PAH transporter in vivo with low affinity (app. K_i 1,300 μM; Ullrich and Rumrich 1988), the discrepant results may be due to that weak interaction with OAT1, which is only visible at high urate concentrations. The contribution of OAT1 to proximal tubular urate secretion remains to be determined.

The organic cation, tetraethylammonium (TEA), was not transported by rOAT1 nor did it show any inhibition of human, rat, and flounder OAT1 (see Table 6). TEA is a prototypical organic cation for the polyspecific organic cation transporters, OCT1 and OCT2, which share structural elements with the OATs and are also present in renal proximal tubules (Burckhardt G and Wolff 2000; Dresser et al. 2001; Inui et al. 2000; Koepsell et al. 1999; Sweet and Pritchard 1999). Choline also did not inhibit hOAT1 (Race et al. 1999). Thus, organic cations do not appear to interfere with OAT1 with the exception of cimetidine, which interacted with the flounder OAT1, most probably, however, in its uncharged form. Preliminary results indicate transport of cimetidine also by the hOAT1 (Burckhardt BC et al. 2003). In the intact rat kidney, cimetidine inhibited both PAH and tetraethylammonium transport, indicating an interaction of this compound with the systems for secretion of organic anions and organic cations (bisubstrate) (Ullrich et al. 1993a).

Probenecid, the classic inhibitor of the renal organic anion secretion system, inhibited human, rat, mouse, flounder and *C. elegans* OAT1. Where determined, the half-maximal concentration for inhibition, IC_{50}, ranged between 4.3 and 31 μM, suggesting that probenecid has a high affinity for OAT1. In the intact rat kidney, probenecid showed a K_i of 30 μM (Ullrich et al. 1989a), a value identical to the 31 μM found with rOAT1 expressed in LLC-PK1 cells (Sugiyama et al. 2001). Probenecid was not transported by rOAT1, suggesting that it binds to OAT1, but is not, or not measurably, translocated (Uwai et al. 1998). It should be noted, however, that probenecid is by no means a specific inhibitor of OAT1, as it also inhibited sat-1 (Lee et al. 1999; Markovich et al. 1994), OAT2, and OAT3 (see below).

The clinically relevant drugs which have been mostly tested as inhibitors of OAT1, are discussed in separate sections, which include the results of OAT1, OAT2 and OAT3 for better comparison. OAT1 is involved in secretion of anionic diuretics and thereby in drug targeting, since these compounds act at salt transporters located in the thick ascending limb of Henle's loop and in the distal convoluted tubules. Moreover, by transporting the heavy metal chelator, 2,3-dimercaptopropane sulfonate (Islinger et al. 2001; Bahn et al. 2002), OAT1 contributes to the detoxification of proximal tubules, the primary site of mercury deposition and cellular lesions (Zalups 2000). Furthermore, OAT1 plays an important role in the disposition of anionic drugs showing a mostly renal excretion, e.g., *β*-lactam antibiotics, antiviral drugs, and ACE inhibitors. OAT1 also interacts with NSAIDs, although these compounds are mostly disposed of by the liver. NSAIDs, however, can compete with other drugs for OAT1, leading to an impaired renal excretion and drug toxicity. Finally, by transporting nephrotoxic agents such as cephaloridine and antiviral drugs, OAT1 contributes to renal damage seen after application of these drugs. Drug interaction can be deliberately used to prevent toxicity, as has been demonstrated for rOAT1-expressing cells. Probenecid increased the rate of survival of these cells in the presence of cephaloridine (Takeda et al. 1999). Likewise, in hOAT1-expressing CHO cells, NSAIDs counteracted antiviral drugs in causing cytotoxicity (Mulato et al. 2000).

Summary of substrate specificity

A survey of Table 6 clearly reveals that OAT1 is a polyspecific transporter handling chemically unrelated compounds. Is OAT1 identical to the PAH transporter in the intact rat kidney? This question is not easy to answer, because besides OAT1, OAT3 is also present in the basolateral cell membrane of rat proximal tubules (Kojima et al. 2002; Motohashi et al. 2002) and, therefore, two PAH transporters operate in parallel. Furthermore, relatively few apparent K_i values have been determined with expressed OAT1 so that the available set of data does not allow for a quantitative comparison between rat OAT1 in vitro and rat tubules in vivo. There are, however, clear correspondences between both systems. The interaction of aliphatic dicarboxylates of different chain lengths with rOAT (Uwai et al. 1998) and with the PAH transporter (Ullrich et al. 1987b) followed exactly the same rule. Substrates with no or low affinity for OAT1 such as acetylsalicylate, ampicillin, cephaloridine, citrate, cyclosporin A, norfloxacin, ofloxacin, paracetamol, and urate (see Table 6 for references) exhibited apparent K_i values of greater than 1,000 μM in vivo (Ullrich and Rumrich 1988; Ullrich et al. 1987b, 1989a, 1989b, 1990, 1991a, 1993b). Conversely, substrates with high affinity for OAT1 or a strong inhibitory effect on OAT1-mediated transport such as PAH, α-ketoglutarate, indomethacin, ketoprofen, and probenecid had apparent K_i values of less than 100 μM in vivo (Ullrich and Rumrich 1990; Ullrich et al. 1987a, 1989a, 1993c). Finally, OAT3 interacted with high affinity with β-lactam antibiotics (cf. "Functional characterization of OAT3, substrates"), in contrast to PAH transport in vivo which showed low affinities (Ullrich et al. 1989a). Taken together, PAH transport across the basolateral membrane of rat renal proximal tubule cells most probably occurred through OAT1. Hence, the delineations for substrate interaction, i.e., hydrophobicity, charge strength, and ability to form hydrogen bonds, should also hold for OAT1.

Regulation of OAT1

As alluded to earlier, OAT1 proteins possess several potential phosphorylation sites for various protein kinases. Some of these sites are conserved (Sweet and Pritchard 1999), suggesting that they play a role in the regulation of OAT1. To date, however, the functional role of these sites has not been clarified. Studies on proximal tubules isolated from rabbit kidneys revealed either a stimulation (Hohage et al. 1994) or an inhibition (Gekle et al. 1999) of PAH uptake by phorbol esters, which activate protein kinase C. The inhibitor of protein kinase C, staurosporine, reversed the stimulation or inhibition by phorbol esters. In killifish tubules, phorbol esters inhibited the uptake of fluorescein (Miller 1998). In opossum kidney (OK) cells, an activation of protein kinase C by phorbol esters inhibited PAH uptake across the basolateral membrane and transcellular PAH secretion (Sauvant et al. 2001; Takano et al. 1996). Likewise, parathyroid hormone acting through protein kinase C, decreased PAH uptake into OK cells (Nagai et al. 1997). Taken together, the majority of results indicate an inhibition of PAH transport by activation of protein kinase C. In agreement with these studies, expressed human (Lu et al. 1998), rat (Uwai et al. 1998) and mouse OAT1 (You et al. 2000) were inhibited by phorbol esters. Staurosporine reversed the inhibition. In the study on mOAT1, activation of protein kinase C led to a decrease in V_{max} with no significant changes in K_m (You et al. 2000), suggesting an internalization of active transporters. Interestingly, the activation of protein kinase C was not associated with a phosphorylation of mOAT1. Accordingly, we could not prevent the down-regulation of hOAT1 by mutation of the protein kinase C consensus sequences (Wolff et al. 2002). Thus, protein kinase C may phosphorylate an OAT1 regulatory protein that is responsible for inhibition/internalization of OAT1. On the other hand, the phosphatase inhibitor, okadaic acid, increased the phosphorylation of mOAT1 by an as yet unknown protein kinase and caused an inhibition of PAH uptake (You et al. 2000). Thus, OAT1 may be directly and indirectly regulated by protein kinases; we are, however, far from understanding the molecular mechanism of this regulation.

The organic anion transporter 2

Cloning, gene, and tissue distribution of OAT2

A summary of molecular properties is shown in Table 7. Actually, OAT2 was the first OAT family member to be cloned. Simonson et al. (Simonson et al. 1994) isolated from a rat liver cDNA library a clone that coded for a protein containing motifs of bacterial H^+-

sugar symporters, H^+-citrate symporters, and mammalian sugar uniporters (GLUTs). Hence, the clone was named novel liver transporter (NLT), although initially no functional data were presented. The NLT gene, called *TI-LTP* (type-I liver transporting protein) was shown to be composed of 10 exons and 9 introns (Simonson and Iwanij 1995). A functional expression of NLT in *Xenopus laevis* oocytes was found later and did reveal that the protein translocated a multitude of organic anions, and NLT was renamed OAT2 (Sekine et al. 1998b) or, alternatively, SLC22A7 (Kok et al. 2000). The sequence of hOAT2 was deposited in the gene bank (accession no. L27651), but details on the protein have not been published in a journal. The occurrence of two splice variants, hOAT2A and hOAT2B, has been reported, but sequences have not been shown (Sun et al. 2001). In preliminary experiments, we found liver- and kidney-specific splice variants of hOAT2, differing in their exon 10-encoded C-terminal amino acids (Bahn et al. 2001b). The *SLC22A7* gene spans 6.8 kb and is located on chromosome *6p21*.1 (Kok et al. 2000) (Table 7). The mouse OAT2 has also been cloned and localized to chromosome *17C* (Kobayashi et al. 2002b).

In Northern blot studies on humans, mRNA for OAT2 was found more in liver than in kidneys (Sun et al. 2001). In male rats, the OAT2 transcript was also found in liver with little expression in kidneys (Sekine et al. 1998b; Simonson et al. 1994; Sun et al. 2001). Very low levels were detected in adipose tissue (Simonson et al. 1994) as well as in the gastrointestinal tract (Buist et al. 2002). In rat liver, transcripts appeared at gestation days 19–21 and increased markedly after birth, reaching a maximum at 2 weeks post-partum (Simonson et al. 1994). During fetal life, OAT2 transcripts transiently appeared in a multitude of other organs, including the gastrointestinal tract, bones and cartilage (Pavlova et al. 2000). This expression of OAT2 suggested an impact on the development of organs; the exact role of OAT2 in organogenesis, however, is unknown at present.

In adult rats, there is a striking difference between male and female animals. As opposed to males, female rats show a high expression of OAT2 in the kidneys in addition to OAT2 in the liver (Buist et al. 2002; Kobayashi et al. 2002a). This gender difference appeared at an age of 35 days, the onset of estrogen production in female rats, suggesting that female sex hormones are involved in the regulation of renal, but not in hepatic expression of OAT2 (Buist et al. 2002). If present also in man, the difference of renal OAT2 lev-

Table 7 Properties of the cloned OAT2 (SLC22A7)

Cloned from	Human, rat ("NLT"), mouse
Gene	Human: 6.0 kb; 10 exons; chromosome *6p21,1*; mouse: chromosome *17C*
Expression	Human: liver >kidneys Male rat: liver >kidneys; female rat[a]: kidneys >>liver Male mouse: kidneys >>liver; female mouse: liver ≈ kidneys
Localization	Human kidneys: basolateral membrane of proximal tubules Rat kidneys: luminal membrane of collecting ducts; cortical thick ascending limb of Henle's loop Rat liver: sinusoidal membrane
Protein	Human: 538–548 aa splice variants) Rat: 535 aa; 12 transmembrane helices; 2 *N*-glycosylation sites; 2 protein kinase A sites; 2 protein kinase C sites Mouse: 540 aa; 12 transmembrane helices; 3 *N*-glycosylation sites; 6 protein kinase C sites
Transport	Uniport of organic anions (*trans*-stimulation by anions not demonstrated)
Regulation	Not determined

[a] From an age of 35 days on (Buist et al. 2002)

els could explain gender-specific differences in salicylate clearance, which is higher in women than in men (Kobayashi et al. 2002a). Also in mice, gender-specific expression of mOAT1 RNA was observed. As opposed to rats, both sexes had a high renal mOAT1 expression; female mice also showed transcripts in the liver, whereas in male mice no signal was detected (Kobayashi et al. 2002b). Thus, expression of OAT1 mRNA shows both gender and species differences.

Immunohistochemistry yielded results that differ between rats and humans. In rats, the OAT2 protein was found to colocalize with Tamm-Horsefall protein in the medullary thick ascending limb of Henle's loop and with aquaporin 3 in the cortical and medullary collecting ducts (Kojima et al. 2002). In the collecting ducts, OAT2 was localized to the luminal membrane. In human kidneys, however, OAT2 antibodies stained the basolateral membrane of proximal tubules (Enomoto et al. 2002b). The authors did not discuss their own discrepant findings concerning the luminal vs basolateral localization, nor did they investigate whether in human kidneys OAT2 is also present in collecting ducts. Nevertheless, we shall include OAT2 in our consideration of proximal tubular organic anion transporters of the basolateral membrane, being aware that this may be true only for human kidneys. As already alluded to earlier, OAT2 protein in the liver is located in the sinusoidal (basolateral) membrane of hepatocytes (Simonson et al. 1994).

OAT2 protein

Rat NLT or OAT2 is a protein of 535 amino acids (Simonson et al. 1994), whereas the mOAT2 has 540 amino acids (Kobayashi et al. 2002b). Both sequences are 88% identical and, hence, closely related. The two splice variants of hOAT2, hOAT2A and hOAT2B, had 546 and 538 amino acids (Sun et al. 2001). In our studies, the liver-specific splice variant had 548 amino acids, and the kidney-specific form 538 amino acids (Bahn et al. 2001b). The Kyte-Doolittle algorithm predicted 12 transmembrane domains. The large first extracellular loop, E1, between the first and second transmembrane helix (M1 and M2) of rOAT2 and mOAT2 contains two potential sites for *N*-glycosylation; in mOAT2, there is a third potential *N*-glycosylation site at Asn356, which is located in the 4th extracellular loop E4 between M7 and M8. Furthermore, two potential sites for phosphorylation by both protein kinase A and protein kinase C were reported for rOAT2, and six potential protein kinase C sites for mOAT1. It is not known whether OAT2 is regulated by protein kinases and, hence, the importance of the putative phosphorylation sites remains to be determined.

Functional characterization of OAT2 substrates

For functional characterization, rat and human OAT2 were expressed in *Xenopus laevis* oocytes, LLC-PK1 cells, EcR293 cells, and in mouse proximal tubule (S2) cells (see Table 8). The uptake of radiolabeled acetylsalicylate, PAH, α-ketoglutarate, methotrexate, prostaglandin E_2, and salicylate was approximately two to eight times higher in rOAT2-expressing oocytes than in water-injected control oocytes (Sekine et al. 1998b). The uptake of labeled salicylate was Na^+-independent and saturable with an apparent K_m of 89 µM. The K_m for α-ketoglutarate was reported to be 17.8 µM. Surprisingly however, salicylate uptake was not *trans*-stimulated by intracellular glutarate (Sekine et al. 1998b). In a follow-up paper, no transport of labeled α-ketoglutarate was found when rOAT2 was expressed in LLC-PK1 cells (Morita et al. 2001). This latter finding fits better with the fact that OAT2 apparently does not exchange organic anions against intracellular (α-keto)glutarate. In our own experiments on rOAT2, we did not find an inhibition of PAH transport by dicarboxylates, which supports the notion that rat OAT2 does not interact with dicarboxylates (N.A. Wolff, E. Petzinger, unpublished data). Oocytes expressing the mouse OAT1 transported radiolabeled α-ketoglutarate and glutarate two to three times faster than nonexpressing control oocytes (Kobayashi et al. 2002b). It has not been tested whether mOAT2 is capable of exchanging C5 dicarboxylates against an organic anion as does OAT1.

Again conflicting results were reported for methotrexate, which was transported by rOAT2 expressed in oocytes, but not by the rOAT2 expressed in LLC-PK1 cells (Morita et al. 2001; Sekine et al. 1998b). Further labeled substrates transported by rOAT2 included some antiviral drugs (AZT, 2',3'-dideoxycytidine or ddC) and an NSAID (indomethacin). Mouse OAT1 translocated labeled allopurinol, *p*-aminohippurate, methotrexate, ochratoxin A, prostaglandin E_2, and valproate, but – unlike rOAT2 – not salicylate. Transport of ochratoxin A, methotrexate and valproate into mOAT2-expressing oocytes was only slightly (less than twofold) higher than in nonexpressing oocytes (Kobayashi et al. 2002b), raising the question of whether the observed transport was due to the expressed OAT2 or rather to the up-regulation of an endogenous system in oocytes. The human OAT2 transported labeled PAH, α-ketoglutarate, AZT, cAMP, methotrexate, prostaglandins E_2 and $F_{2\alpha}$, and tetracycline, but not the antiviral drugs acyclovir, ganciclovir, and valacyclovir, which do not carry a negative charge. Transport of radiolabeled tetracycline was hardly greater than uptake into nonexpressing mouse S2 cells (Babu et al. 2002b), raising the same concerns as with mOAT2. It is probably safe to conclude that OAT2 does not contribute much to proximal tubular transport of ochratoxin A, methotrexate, valproate, and tetracyclines.

A splice variant, hOAT2A, was expressed in EcR293 cells under the control of an ecdysone promoter. In ponasterone-induced cells, PAH uptake was 13-fold higher than in noninduced cells (for comparison, 100-fold higher PAH uptake for OAT1 expressed under identical conditions) (Sun et al. 2001). Induced cells transported also cAMP, methotrexate, and α-ketoglutarate, but not dehydroepiandrosterone sulfate and salicylate. Again, the data with α-ketoglutarate are puzzling, because unlabeled α-ketoglutarate in the medium did not inhibit hOAT2A-mediated PAH uptake in the same study.

Inhibition studies revealed that NSAIDs (diclofenac, ibuprofen, ketoprofen), antibiotics (benzylpenicillin, cefoperazone, minocycline, oxytetracycline, tetracycline, rifampicin), a diuretic (bumetanide), a cardiotonic (digoxin), and a cytostatic drug (methotrexate) interacted with rat and/or human OAT2 (see Table 8). In a few cases, K_i values were determined; indocyanine green and ketoprofen showed K_i values below 2 μM, suggesting a very high affinity of rat OAT2 to these two compounds (Morita et al. 2001). The K_i values for benzoate and diclofenac were below 100 μM and those for ibuprofen, tolbutamide and verapamil below 200 μM. Probenecid inhibited rat and human OAT2, the latter with a low affinity (K_i 766 μM) (Enomoto et al. 2002b; Morita et al. 2001). Mouse OAT2 was inhibited by bromosulfophthalein, cimetidine, indocyanine green, cefoperazone, enalapril, oxaloacetate and, most potently, by erythromycin (Kobayashi et al. 2002b). Enalapril and erythromycin did not inhibit rOAT2 (Sekine et al. 1998b), suggesting the presence of species differences.

Table 8 Substrate specificity of the cloned OAT2

Compound	Species	Expression system	Transport or K_m (μM)	Inhibition or IC_{50} (μM)	References
Acetylsalicylate	Rat	Oocyte	+		Sekine et al. 1998b
Acyclovir (ACV)	Human	Mouse PTC	–	–	Takeda et al. 2002b
Allopurinol	Mouse	Oocyte	+		Kobayashi et al. 2002b
p-Aminohippurate	Human	EcR293	+		Sun et al. 2001
		LLC-PK1	+	–	Morita et al. 2001
	Rat	Oocyte	+		Sekine et al. 1998b
	Mouse	Oocyte	+		Kobayashi et al. 2002b
3'-Azido-3'-deoxythymidine	Human	Mouse PTC	26.8	+	Takeda et al. 2002b
(AZT; zidovudine)	Rat	LLC-PK1	26		Morita et al. 2001
Benzoate	Rat	LLC-PK1		86.9	Morita et al. 2001
Benzylpenicillin	Rat	LLC-PK1		+	Morita et al. 2001
Betamipron	Human	Mouse PTC		–	Enomoto et al. 2002b
Bromosulfophthalein	Human	EcR293		+	Sun et al. 2001
	Rat	Oocyte		+	Sekine et al. 1998b
	Mouse	Oocyte		+	Kobayashi et al. 2002b
Bumetanide	Rat	Oocyte		+	Sekine et al. 1998b
cAMP	Human	EcR293	+		Sun et al. 2001
Cefoperazone	Rat	Oocyte		+	Sekine et al. 1998b
	Mouse	Oocyte		+	Kobayashi et al. 2002b
Cilastatin	Human	Mouse PTC		–	Enomoto et al. 2002b
Cimetidine	Rat	LLC-PK1		+	Morita et al. 2001
	Mouse	Oocyte		+	Kobayashi et al. 2002b
Cholate	Rat	Oocyte		+	Sekine et al. 1998b
Dehydroepiandrosterone sulfate (DHEA-S)	Human	EcR293	–		Sun et al. 2001
Diclofenac	Rat	LLC-PK1		49.3	Morita et al. 2001
2',3'-Dideoxycytidine	Rat	LLC-PK1	3,080		Morita et al. 2001
Digoxin	Rat	LLC-PK1		+	Morita et al. 2001
Doxycycline	Human	Mouse PTC		–	Babu et al. 2002b
Enalapril	Rat	Oocyte		–	Sekine et al. 1998b
	Mouse	Oocyte		+	Kobayashi et al. 2002b
Erythromycin	Rat	Oocyte		–	Sekine et al. 1998b
	Mouse	Oocyte		+	Kobayashi et al. 2002b
Ganciclovir =GCV	Human	Mouse PTC	–	+	Takeda et al. 2002b
Glibenclamide	Rat	LLC-PK1	–	12.3	Morita et al. 2001
Glutarate	Mouse	Oocyte	15.8		Kobayashi et al. 2002b
Ibuprofen	Rat	LLC-PK1		155	Morita et al. 2001
Indocyanine green	Rat	Oocyte		+	Sekine et al. 1998b
		LLC-PK1		1.15	Morita et al. 2001
	Mouse	Oocyte		+	Kobayashi et al. 2002b
Indomethacin	Rat	LLC-PK1	0.37		Morita et al. 2001
α-Ketoglutarate	Human	EcR293	+	–	Sun et al. 2001
	Rat	Oocyte	17.8		Sekine et al. 1998b
		LLC-PK1	–	–	Morita et al. 2001
	Mouse	Oocyte	+		Kobayashi et al. 2002b

Table 8 (continued)

Compound	Species	Expression system	Transport or K_m (μM)	Inhibition or IC_{50} (μM)	References
Ketoprofen	Rat	Oocyte		+	Sekine et al. 1998b
		LLC-PK1		1.84	Morita et al. 2001
KW-3902	Human	Mouse PTC		–	Enomoto et al. 2002b
Methotrexate	Human	EcR293	+		Sun et al. 2001
	Rat	Oocyte	+	+	Sekine et al. 1998b
		LLC-PK1	–	+	Morita et al. 2001
	Mouse	Oocyte	+		Kobayashi et al. 2002b
Minocycline	Human	Mouse PTC		+	Babu et al. 2002b
Ochratoxin A	Mouse	Oocyte	+		Kobayashi et al. 2002b
Oxaloacetate	Mouse	Oocyte		+	Kobayashi et al. 2002b
Oxytetracycline	Human	Mouse PTC		+	Babu et al. 2002b
Phenytoin	Rat	LLC-PK1		–	Morita et al. 2001
Probenecid	Human	Mouse PTC		766	Enomoto et al. 2002b
	Rat	LLC-PK1		+	Morita et al. 2001
Propionate	Rat	LLC-PK1		+	Morita et al. 2001
Prostaglandin E_2	Human	Mouse PTC	0.72	+	Kimura et al. 2002
	Rat	Oocyte	+		Sekine et al. 1998b
		LLC-PK1	38.5		Morita et al. 2001
	Mouse	Oocyte	0.005		Kobayashi et al. 2002b
Prostaglandin $F_{2\alpha}$	Human	Mouse PTC	+[1]; 0.43[2]	+[3]	[1]Babu et al. 2002b; [2]Enomoto et al. 2002b; [3]Kimura et al. 2002
Rifampicin	Rat	Oocyte		+	Sekine et al. 1998b
		LLC-PK1		+	Morita et al. 2001
Salicylate	Human	EcR293	–		Sun et al. 2001
	Rat	Oocyte	88.9		Sekine et al. 1998b
		LLC-PK1	81.2		Morita et al. 2001
	Mouse	Oocyte	–		Kobayashi et al. 2002b
Taurocholate	Rat	Oocyte	–		Sekine et al. 1998b
		LLC-PK1		+	Morita et al. 2001
Tetracycline	Human	Mouse PTC	439.9	+	Babu et al. 2002b
Tolbutamide	Rat	LLC-PK1		176	Morita et al. 2001
Valacyclovir	Human	Mouse PTC	–		Takeda et al. 2002b
Valproate	Mouse	Oocyte	+		Kobayashi et al. 2002b
Verapamil	Rat	LLC-PK1		142	Morita et al. 2001

Since relatively few potential substrates were tested with OAT2, it is premature to consider its substrate specificity in comparison to OAT1 and OAT3. It appears, however, that OAT2 has a low affinity for the inhibitor of organic anion secretion, probenecid. Our own data indicate a low affinity also for PAH (N.A. Wolff, unpublished results). Thus, if indeed present in the basolateral membrane of proximal tubule cells, OAT2 is unlikely to contribute much to the classic organic anion secretion in proximal tubules. The lack of inhibition by estrone sulfate and dehydroepiandrosterone sulfate further suggests that OAT2 is not involved in proximal tubular excretion of sulfated compounds. More work must be done to investigate the driving force for organic anion transport, to clarify whether dicarboxylates are accepted or not, and to find out whether gender differences with respect to hepatic and renal expression also exist in humans.

The organic anion transporter 3

Cloning, gene, and tissue distribution of OAT3

The molecular properties of OAT3 are summarized in Table 9. This OAT isoform was first cloned from a rat kidney cDNA library using degenerate primers based on conserved sequences in OAT1, OAT2 and OCT1 (Kusuhara et al. 1999). The clone shared a high identity (92%) with Roct (reduced in *oc* transporter), a gene detected by differential display techniques in the kidneys of mice suffering from juvenile cystic kidney (Brady et al. 1999). Roct is most probably the murine homolog of OAT3, mOAT3. The localization of the *Roct* gene on mouse chromosome *19* at the locus for osteopetrosis/osteosclerosis prompted the authors to investigate abnormalities associated with this bone malformation. In osteosclerosis mice, the Roct protein was missing in kidneys and developing bone. Mutations leading to the defect have not yet been found. The human homolog, hOAT3, was first cloned by Race et al. (Race et al. 1999). The protein, however, was not functional after expression in oocytes. A functional hOAT3 was then cloned from a human kidney cDNA library based on sequences derived from rOAT3 (Cha et al. 2001). rOAT3 shares 79%, and mOAT3 78% identity with hOAT3.

Northern blot studies revealed a strong band in rat liver, weaker bands in kidneys and brain and a faint band in the eye (Kusuhara et al. 1999). In contrast, a strong band in kidneys and weak bands in the brain and in skeletal muscle were found with human tissues (Cha et al. 2001). When OAT3 mRNA was quantified using branched-DNA signal amplification, rat kidneys showed the highest signal, followed in decreasing order by liver (male rats; approximately one-fifth of the kidney signal), lungs, cerebellum and cerebral cortex (Buist et al. 2002). Renal OAT3 message in rats increased sharply after birth and reached a maximum at the end of the 2nd week, with similar levels in male and female animals (Buist et al. 2002). In human kidneys, mRNA of hOAT3 was nearly three times higher than that for hOAT1, suggesting a high expression, as was indeed verified by Western blot

Table 9 Properties of the cloned OAT3 (SLC22A8)

Cloned from	Human, rat, mouse
Gene	16–19 kb; 10 exons; *11q12–13.3*
Expression	Human: kidneys >>skeletal muscle, brain Rat: liver >>kidneys >brain[a]; or kidneys >>liver >lungs >brain[b]
Localization	Rat kidneys: proximal tubules, segments S1>S2 = S3, basolateral membrane; thick ascending limb of Henle's loop, connecting tubule, collecting duct
Protein	536–543 aa; 12 transmembrane helices; 4 *N*-glycosylation sites; 3–8 protein kinase C sites
Transport	Uniport of OA or selected OC (antiport not determined)
Regulation	Down-regulation by protein kinase C

[a] Northern blot in initial study (Kusuhara et al. 1999)
[b] Quantitative PCR (Buist et al. 2002)

studies (Motohashi et al. 2002). The human OAT3 or *SLC22A8* gene was localized to chromosome *11q12–13.3* (Cha et al. 2001; Race et al. 1999).

Studies with antibodies localized rat (Hasegawa et al. 2002) and human OAT3 (Cha et al. 2001) to the basolateral membrane of renal proximal tubules. In a more recent study on rat kidney (Kojima et al. 2002), OAT3 was not only found in all three segments of proximal tubules, but also in the thick ascending limbs, the connecting tubules and the collecting ducts, where it was localized at the basolateral membrane. The physiological role of OAT3 at locations outside of proximal tubules is unknown at present. On the other hand, OAT3 in human kidneys was confined to the proximal tubules and was colocalized with hOAT1 at the basolateral membrane (Motohashi et al. 2002). Some proximal tubules were positive for OAT1, but not for OAT3, showing that, in human kidneys, OAT1 has a somewhat wider distribution in proximal tubules than OAT3. In the brain, rOAT3 is located in the luminal membrane of choroid plexus cells (Nagata et al. 2002), where it is most probably involved in the vectorial transport of organic anions from liquor into the blood.

OAT3 protein

The clones coded for proteins of 536 (rOAT3), 537 (Roct/mOAT3), and either 543 or 568 amino acids (hOAT3). Rat OAT3 was 49% and 39% identical to rOAT1 and rOAT2, respectively. The respective numbers for hOAT3 are 51% and 36% identity to hOAT1 and hOAT2, respectively. Human OAT3 (543 amino acids) was 85% identical to hOAT3 (568 amino acids). Since only the 543 amino acid hOAT3 proved to be functional, it more likely represents the transporter expressed in vivo. The Kyte-Doolittle algorithm predicted 12 transmembrane domains with a large, putatively extracellular loop located between the first (M1) and second transmembrane domain (M2), carrying four potential sites for *N*-glycosylation. Rat OAT3 contains three and hOAT3 eight potential protein kinase C phosphorylation sites. As with OAT1 and OAT2, it is not known which of these sites is actually used for regulation of OAT3. There are also no experimental data available on the topological organization of OAT3 prevailing within the membrane.

Functional characterization of OAT3 substrates

Table 10 gives an overview of the substrates of OAT3. OAT3 was expressed in *Xenopus laevis* oocytes, LLC-PK1 cells, and mouse proximal tubule cells. The initial characterization of rat and human OAT3 was performed on oocytes. The uptake of radiolabeled PAH exceeded that in nonexpressing control oocytes (Cha et al. 2001; Kusuhara et al. 1999). In rOAT3-expressing oocytes, half-maximal uptake rates were reached at 65 μM PAH, a value at the higher end of the K_m values reported for rOAT1 (11–85.1 μM). In another study (Feng et al. 2001), a K_m of 278 μM was found. rOAT3 expressed in LLC-PK1 cells rather than in oocytes showed no PAH uptake at all, and half-maximal inhibition of 17β-estradiol-D-17β-glucuronide ($E_2 17\beta G$) uptake occurred at as much as 1,350 μM PAH. Others reported lower K_i values for PAH, 301 (Sugiyama et al. 2001) or 398 μM (Nagata et al. 2002). Collectively, these studies suggest that the affinity of rOAT3 for PAH is lower than that of rOAT1. This is also true for the hOAT3, which exhibited a K_m for PAH of 87 μM (Cha et al. 2001), whereas hOAT1 showed K_m values between 3.9 and 22 μM.

Table 10 Substrate specificity of cloned OAT3

Compound	Species	Expression system	Transport or K_m (μM)	Inhibition or IC_{50} (μM)	References
Acyclovir (ACV)	Human	Oocyte	–		Cha et al. 2001
		Mouse PTC	–	–	Takeda et al. 2002b
p-Aminohippurate	Human	Oocyte	87.2		Cha et al. 2001
		Mouse PTC		19.6	Jung et al. 2001
	Rat	Oocyte	64.7[1]; 278[2]	+[3]	[1]Kusuhara et al. 1999; [2]Feng et al. 2001; [3]Deguchi et al. 2002
		LLC-PK1	–[1]; +[2]	301[3]; 398[2]; 1,350[1]	[1]Hasegawa et al. 2002; [2]Nagata et al. 2002; [3]Sugiyama et al. 2001
Aspartame	Human	Mouse PTC		–	Jung et al. 2001
3'-Azido-3'-deoxythymidine (AZT; zidovudine)	Human	Oocyte		–	Cha et al. 2001
		Mouse PTC	145.1	+	Takeda et al. 2002a
	Rat	Oocyte		+	Kusuhara et al. 1999
Benzylpenicillin (penicillin G)	Human	Oocyte		+	Cha et al. 2001
		Mouse PTC		+	Jung et al. 2001
	Rat	Oocyte		+	Deguchi et al. 2002; Kusuhara et al. 1999
		LLC-PK1	82.6[1]	52.8[2]; 82.6[1]	[1]Nagata et al. 2002; [2]Hasegawa et al. 2002
Betamipron	Human	Mouse PTC		48.3	Takeda et al. 2001
Bromosulfophthalein	Human	Oocyte		+	Cha et al. 2001
	Rat	Oocyte		+	Kusuhara et al. 1999
		LLC-PK1		3.09	Hasegawa et al. 2002
Bumetanide	Human	Oocyte		+	Cha et al. 2001
	Rat	Oocyte		+	Kusuhara et al. 1999
cAMP	Human	Oocyte	+		Cha et al. 2001
3-Carboxy-4-methyl-5-propyl-2-furanpropanoate (CMPF)	Rat	Oocyte		4.01	Deguchi et al. 2002
Cefadroxil	Human	Mouse PTC		8,620[a]	Takeda et al. 2002a
Cefamandol	Human	Mouse PTC		50[a]	Takeda et al. 2002a
Cefazolin	Human	Mouse PTC		550[a]	Takeda et al. 2002a
Cefoperazone	Human	Mouse PTC		1,890[a]	Takeda et al. 2002a
	Rat	Oocyte		+	Kusuhara et al. 1999
Cefotaxime	Human	Mouse PTC		290[a]	Takeda et al. 2002a
Ceftriaxone	Human	Mouse PTC		4,390[a]	Takeda et al. 2002a
Cephaloridine	Human	Mouse PTC		2,460[a]	Takeda et al. 2002a
Cephalothin	Human	Mouse PTC		40[a]	Takeda et al. 2002a
Cholate	Human	Oocyte		+	Cha et al. 2001
	Rat	Oocyte		+	Kusuhara et al. 1999
Cilastatin	Human	Mouse PTC		1,470	Takeda et al. 2001
Cimetidine	Human	Oocyte	57.4	+	Cha et al. 2001
		Mouse PTC		+	Jung et al. 2001
	Rat	Oocyte	+[1]; 40[2]	+[1–3]	[1]Kusuhara et al. 1999; [2]Feng et al. 2001; [3]Deguchi et al. 2002
		LLC-PK1		46.8	Nagata et al. 2002

Table 10 (continued)

Compound	Species	Expression system	Transport or K_m (μM)	Inhibition or IC_{50} (μM)	References
Citrinin	Human	Mouse PTC		15.4	Jung et al. 2001
Corticosteron	Human	Oocyte		+	Cha et al. 2001
Dehydroepiandrosterone sulfate (DHEA-S)	Human	Oocyte	+	+	Cha et al. 2001
Diclofenac	Human	Oocyte		+	Cha et al. 2001
DIDS (4,4'-diisothiocyanatostilbene-2,2'-disulfonate)	Rat	Oocyte		+	Kusuhara et al. 1999
Digoxin	Rat	Oocyte		–	Kusuhara et al. 1999
		LLC-PK1		–	Sugiyama et al. 2001
3,4-Dihydroxymandelate	Rat	Oocyte		+	Kusuhara et al. 1999
3,4-Dihydroxyphenylacetate	Rat	Oocyte		+	Kusuhara et al. 1999
Doxycycline	Human	Mouse PTC		–	Babu et al. 2002b
Estradiol glucuronide	Human	Oocyte	+		Cha et al. 2001
17β-Estradiol-D-17β-glucuronide ($E_2 17\beta G$)	Rat	LLC-PK1	8.43[1]	35.6[2]	[1]Sugiyama et al. 2001; [2]Nagata et al. 2002
Estrone sulfate	Human	Oocyte	3.1		Cha et al. 2001
		Mouse	2.2[1]; 7.5[2]	+[3]	[1]Takeda et al. 2001; [2]Takeda et al. 2000a; [3]Babu et al. 2002b
	Rat	Oocyte	2.34[1]; 33[2]	+[2,3]	[1]Kusuhara et al. 1999; [2]Feng et al. 2001; [3]Deguchi et al. 2002
		Mouse PTC	6.18		Takeda et al. 2000b
		LLC-PK1		9.1	Nagata et al. 2002
Furosemide	Human	Oocyte		+	Cha et al. 2001
	Rat	Oocyte		+	Feng et al. 2001; Kusuhara et al. 1999
Ganciclovir (GCV)	Human	Mouse PTC		+	Takeda et al. 2002b
Glucuronate	Human	Oocyte	–		Cha et al. 2001
Glutarate	Human	Oocyte	+	+	Cha et al. 2001
Guanidine	Human	Oocyte		–	Cha et al. 2001
	Rat	Oocyte		–	Kusuhara et al. 1999
Hippurate	Rat	Oocyte		+	Deguchi et al. 2002
4-Hydroxybenzoate	Rat	Oocyte		+	Deguchi et al. 2002
2-Hydroxycinnamate	Rat	Oocyte		+	Deguchi et al. 2002
5-Hydroxyindole-3-acetate	Rat	Oocyte		+	Kusuhara et al. 1999
4-Hydroxy-3-methoxymandelate	Rat	Oocyte		+	Deguchi et al. 2002; Kusuhara et al. 1999
4-Hydroxy-3-methoxyphenylacetate	Rat	Oocyte		+	Deguchi et al. 2002; Kusuhara et al. 1999
4-Hydroxyphenylacetate	Rat	Oocyte		+	Deguchi et al. 2002
Hippurate	Rat	Oocyte		11.9	Deguchi et al. 2002
Ibuprofen	Human	Oocyte	–	+	Cha et al. 2001
Indocyanine green	Rat	Oocyte		+	Kusuhara et al. 1999
Indole acetate	Rat	Oocyte		+	Deguchi et al. 2002
Indomethacin	Human	Oocyte		+	Cha et al. 2001

Table 10 (continued)

Compound	Species	Expression system	Transport or K_m (μM)	Inhibition or IC_{50} (μM)	References
	Rat	Oocyte		–[1]; +[2]	[1]Deguchi et al. 2002; Kusuhara et al. 1999; [2]Feng et al. 2001; Jung et al. 2001
Indoxyl sulfate	Rat	Oocyte	158		Deguchi et al. 2002
KW-3902	Human	Mouse PTC		3.7	Takeda et al. 2001
Melatonin	Rat	Oocyte		+	Kusuhara et al. 1999
Methotrexate	Human	Oocyte	10.9		Cha et al. 2001
	Rat	Oocyte		+	Kusuhara et al. 1999
5-Methoxyindole-3-acetate	Rat	Oocyte		+	Kusuhara et al. 1999
5-Methoxytryptophol	Rat	Oocyte		+	Kusuhara et al. 1999
1-Methyl-4-phenylpyridinium (MPP)	Rat	Oocyte	+		Feng et al. 2001
Minocycline	Human	Mouse PTC		–	Babu et al. 2002b
Ochratoxin A	Human	Oocyte	+		Cha et al. 2001
		Mouse PTC	0.75		Jung et al. 2001
	Rat	Oocyte	0.74	+	Kusuhara et al. 1999; Feng et al. 2001
Octanoate	Human	Mouse PTC		8.6	Jung et al. 2001
Ouabain	Human	Oocyte		–	Cha et al. 2001
	Rat	Oocyte		–	Kusuhara et al. 1999
Oxytetracycline	Human	Mouse PTC		–	Babu et al. 2002b
Piroxicam	Human	Mouse PTC		4.88	Jung et al. 2001
	Rat	Oocyte		+	Kusuhara et al. 1999
Pravastatin	Rat	LLC-PK1	13.4		Hasegawa et al. 2002
Probenecid	Human	Oocyte		+	Cha et al. 2001
		Mouse PTC		4.4[1]; 9[2]	[1]Jung et al. 2001; [2]Takeda et al. 2001
	Rat	Oocyte		+[1]; 1.68[2]	[1]Feng et al. 2001; Kusuhara et al. 1999; [2]Deguchi et al. 2002
		LLC-PK1		4.43[1]; 20[2]	[1]Nagata et al. 2002; [2]Sugiyama et al. 2001
Prostaglandin E_2	Human	Oocyte	+		Cha et al. 2001
		Mouse PTC	0.35	+	Kimura et al. 2002
Prostaglandin $F_{2\alpha}$	Human	Mouse PTC	1.09	+	Kimura et al. 2002
Quinidine	Human	Oocyte		+	Cha et al. 2001
	Rat	Oocyte		+	Feng et al. 2001; Kusuhara et al. 1999
Quinolinate	Rat	Oocyte		–	Deguchi et al. 2002
Salicylate	Human	Oocyte	+	+	Cha et al. 2001
	Rat	Oocyte		+	Deguchi et al. 2002
Succinate	Human	Oocyte	–		Cha et al. 2001
Sulfobromophthalein	Rat	Oocyte		+	Kusuhara et al. 1999
Taurocholate	Human	Oocyte	+	+	Cha et al. 2001
	Rat	Oocyte		+	Kusuhara et al. 1999
		LLC-PK1		790	Sugiyama et al. 2001

Table 10 (continued)

Compound	Species	Expression system	Transport or K_m (μM)	Inhibition or IC_{50} (μM)	References
Tetracycline	Human	Mouse PTC	566.2	–	Babu et al. 2002b
Tetraethylammonium	Human	Oocyte	–	+	Cha et al. 2001
	Rat	Oocyte		–	Feng et al. 2001; Kusuhara et al. 1999
3,5,3'-Triiodo-L-thyronine	Rat	Oocyte		–	Deguchi et al. 2002
Urate	Human	Oocyte	+		Cha et al. 2001
Valacyclovir	Human	Mouse PTC	+		Takeda et al. 2002b
Valproate	Human	Oocyte	–		Cha et al. 2001
Verapamil	Rat	Oocyte		–	Kusuhara et al. 1999

[a] Corrected K_i values (Takeda et al. 2002; Eur J Pharmacol 450:111)

Besides PAH, rOAT3 transported radiolabeled benzylpenicillin, cimetidine, $E_2 17\beta G$, estrone sulfate, indoxyl sulfate, ochratoxin A, and pravastatin. With human OAT3, transport of labeled AZT, cAMP, cimetidine, dehydroepiandrosterone sulfate (DHEA-S), estradiol glucuronide, estrone sulfate, methotrexate, ochratoxin A, prostaglandins E_2 and $F_{2\alpha}$, salicylate, taurocholate, urate, and valacyclovir was demonstrated (see Table 10). Glutarate uptake was only twofold above background and succinate uptake was not detectable (Cha et al. 2001). The ability to translocate sulfate or glucuronide conjugates of steroid hormones distinguished OAT3 from OAT1 and labeled estrogen conjugates proved a suitable test substrate for studying the function of rat or human OAT3. Another seemingly specific feature of OAT3 was the interaction with the organic cation cimetidine. The apparent affinity for cimetidine was 40 or 57.4 μM for rat and human OAT3 (Cha et al. 2001; Feng et al. 2001). Inhibition of OAT3-mediated estrone-sulfate transport (Cha et al. 2001; Kusuhara et al. 1999) or cimetidine transport (Feng et al. 2001) by quinidine and by tetraethylammonium was small, but significant, suggesting that OAT3 does not discriminate between organic anions and organic cations as sharply as does OAT1. However, we have recently seen that human and flounder OAT1 transport labeled cimetidine, most probably not as a cation, but in its uncharged form (Burckhardt BC et al. 2003). Thus, cimetidine is not a substrate specific for OAT1.

Site-directed mutagenesis revealed that transport of PAH and of cimetidine require the interaction with different amino acids within the rOAT3 protein. Single replacement of the conserved, cationic residue arginine 454 (R454) in the 11th transmembrane helix (M11) by aspartate (R454D) or by glutamine (R454 N) impaired PAH transport, but left cimetidine uptake unaffected (Feng et al. 2001). Cimetidine transport was no longer inhibited by unlabeled PAH, suggesting that R454 is important for PAH binding to rOAT3. The single replacement of another conserved residue, lysine 370 (K370) in M8 by alanine, also diminished PAH transport without affecting cimetidine translocation. Unlabeled PAH, however, inhibited cimetidine transport in the K370A mutant. Thus it appeared that, if R454 is intact but K370 is mutated, PAH can bind to rOAT3, but is not translocated. The double mutant, K370A;R454D still transported cimetidine, but also 1-methyl-4-phenylpyridinium (MPP^+), an organic cation, which is hardly accepted by the wild type. Interestingly, hydrophobic organic anions such as estrone sulfate and ochratoxin A, but not PAH, inhibited organic cation transport by the double mutant (Feng et al. 2001). Taken together, these findings indicate that rOAT3 has probably more than one substrate binding site, e.g., one for small hydrophilic organic anions such as PAH, one for more hydrophobic organic anions shared with cimetidine. The cationic amino acids which are critical to substrate recognition and transport in rOAT3 correspond to those which are critical for the interaction of dicarboxylates in fOAT1 (Wolff et al. 2001). Hence, positively charged amino acids in M8 and M11 are probably part of the substrate binding sites.

Interestingly, rOAT3 has been shown to interact with melatonin and a variety of acidic neurotransmitter metabolites such as 3,4-dihydroxymandelate, 4-hydroxy-3-methoxymandelate (metabolites of norepinephrine/epinephrine), 3,4-dihydroxyphenylacetate (metabolite of dopamine), 5-hydroxyindole acetate and 5-methoxyindole acetate (metabolites of 5-hydroxytryptamine) (Kusuhara et al. 1999). This finding is of considerable interest, because these metabolites are removed from the cerebrospinal fluid by a process which is inhibited by PAH and probenecid (Guldberg et al. 1966; Cserr and van Dyke 1971). The location of rOAT3 in the luminal membrane of choroid plexus cells is in accordance with such a role. In addition, OAT3 is most probably involved in the elimination of β-lactam antibiotics from the cerebrospinal fluid, since it translocates benzylpenicillin with a comparably high affinity (K_m 82.6 μM; Nagata et al. 2002). A series of cephalosporins (cefadroxil, cefamandole, cefazolin, cefoperazone, cefotaxime, ceftriaxone, cephaloridine and cephalothin) have also been shown to interact with the hOAT3 (Takeda et al. 2002a). The reported K_i values, from 0.04 up to 8.62, are not in micromoles per liter as published, but in millimoles per liter (Takeda et al. 2002; Eur J Pharmacol 450:111). Most of the tested cephalosporins are shared by OAT1 and OAT3; however, OAT1 is most probably not present in the rat choroid plexus (Nagata et al. 2002) so that OAT3 is alone responsible for the clearance of β-lactam antibiotics from the cerebrospinal fluid.

Rat OAT3 expressed in oocytes transported labeled indoxyl sulfate, a compound that promotes the progression of renal sclerosis. Indoxyl sulfate uptake was not only inhibited by PAH, benzylpenicillin, and cimetidine, but also by the uremic toxins hippurate, 3-carboxy-4-methyl-5-propyl-2-furanpropanoate, 2-hydroxycinnamate, 4-hydroxyphenylacetate, and 4-hydroxybenzoate (Deguchi et al. 2002). Since indoxyl sulfate, hippurate and hydroxy-hippurates interacted with high affinity with human OAT1 (Motojima et al. 2002), both OAT1 and OAT3 may be involved in the transport of uremic toxins.

A fundamental difference between OAT1 and OAT3 is possibly the mode of organic anion transport. OAT1 acts as an anion exchanger, whereas estrone sulfate transport by rOAT3 could not be *trans*-stimulated by unlabeled estrone sulfate, PAH or ochratoxin (Kusuhara et al. 1999). Likewise, estrone sulfate efflux mediated by hOAT3 was not *trans*-stimulated by unlabeled estrone sulfate in the medium (Cha et al. 2001). These findings have led to the notion that OAT3, unlike OAT1, does not act as an anion exchanger. Surprisingly, there are no published data on a possible *trans*-stimulation of OAT3-mediated transport by intracellular α-ketoglutarate or glutarate. Human OAT3 does transport labeled glutarate (Cha et al. 2001) and, hence, is obviously capable of interacting with dicarboxylates. If OAT3 indeed does not perform anion exchange as commonly believed, the driving force for organic anion uptake against an inside negative membrane potential in proximal tubule cells remains unclear.

Regulation of OAT3

When the rOAT3 was expressed in mouse proximal tubule cells, phorbol 12-myristate 13-acetate (PMA) dose- and time-dependently inhibited the uptake of labeled estrone sulfate (Takeda et al. 2000b). Chelerytrine chloride, a specific inhibitor of protein kinase C, reversed the effect of PMA. The inhibition of rOAT3 was due to a decrease in V_{max} at unchanged K_m, suggesting either an internalization of transporters or an inhibition of delivery of preformed OATs to the cell membrane. It is not known which of the potential protein kinase C phosphorylation sites is involved in the observed inhibition.

Interaction of selected drug classes with the transporters

Antibiotics

β-Lactam antibiotics, i.e., penicillins and cephalosporins, possess a negatively charged carboxyl group. Renal excretion accounts for 60–95% of their total elimination from the body (Knauf and Mutschler 2000). The inhibition of penicillin excretion by probenecid (Beyer et al. 1951; Cunningham et al. 1981) and by other organic anions (Nierenberg 1987) suggested early on the involvement of organic anion transporters in proximal tubules. Some cephalosporins, particularly cephaloridine, exert cytotoxic effects on proximal tubule cells, most likely because they are avidly taken up into proximal tubule cells across the basolateral membrane and can not leave quickly enough the cell across the luminal membrane (Tune 1997; Walker 2000). Most tetracyclines are also eliminated by the kidneys (Knauf and Mutschler 2000). Unlike β-lactam antibiotics, these compounds do not carry a negative charge. Quinolone gyrase inhibitors possess a carboxyl group and are renally excreted to various extents (Knauf and Mutschler 2000). Due to their negative charge they can be substrates of renal organic anion transporters.

OAT1

Uptake of labeled PAH into oocytes expressing human OAT1 (Hosoyamada et al. 1999) was inhibited by 2 mM benzylpenicillin (penicillin G), whereas hOAT1 expressed in HeLa cells (Lu et al. 1999) was unaffected by 0.1 mM benzylpenicillin. Since the affinity of OATs for β-lactam antibiotics is small (see below), the latter finding is almost certainly due to the low penicillin concentration applied and does not rule out an interaction of human OAT1 with benzylpenicillin. The uptake of the fluorescent anion 6-carboxyfluorescein into hOAT1-expressing CHO cells was inhibited by cephaloridine (IC_{50} 1.25 mM) as well as by cephradine (IC_{50} 1.6 mM), indicating that these antibiotics interact with human OAT1 with low affinity (Cihlar and Ho 2000). Doxycycline, minocycline, oxytetracycline, and tetracycline inhibited PAH uptake into mouse S2 proximal tubule cells stably expressing hOAT1 (Babu et al. 2002b). The uptake of [^{3}H]tetracycline itself was only slightly enhanced in OAT1-expressing S2-cells as compared to mock-transfected cells, indicating that translocation of tetracycline by OAT1 is very slow.

Among the penicillins, amoxillin, benzylpenicillin, carbenicillin, cloxacillin, nafcillin and piperacillin inhibited PAH uptake into oocytes expressing the rat OAT1. At 2 mM concentration, benzylpenicillin had the smallest inhibitory effect, whereas piperacillin, cloxacillin and nafcillin afforded a greater than 90% inhibition. Ampicillin did not inhibit PAH uptake, which may be due to the presence of a positively charged amino group (Ullrich et al. 1989a). The inhibition by benzylpenicillin and carbenicillin was competitive with K_i values of 1.68 and 0.5 mM, respectively, and radiolabeled benzylpenicillin was taken into rOAT1-expressing oocytes up only two to three times faster than into nonexpressing control oocytes (Jariyawat et al. 1999). In another study, benzylpenicillin inhibited PAH uptake into rOAT1-expressing LLC-PK1 cells with a K_i of 0.42 mM (Hasegawa et al. 2002). When also added at 2 mM to a [^{14}C]PAH-containing medium, cefadroxil, cefotaxime, ceftazidime, cephaloridine, cefoperazone, ceftriaxone, cephalexin, cefsulodin, cefazolin, cephalothin and cefamandole inhibited PAH uptake into rOAT1-expressing oocytes in this sequence with increasing potency (Jariyawat et al. 1999). The inhibition for selected compounds was competitive and the K_i values (0.29 mM for cephalothin; 0.45 mM for cefazolin; 2.31 mM for cephalexin; and 2.33 mM for cephaloridine) indicated a low affinity for rOAT1. Uptake of labeled cephaloridine was demonstrated for rOAT1 expressed in oocytes or mouse proximal tubule cells (Jariyawat et al. 1999; Takeda et al. 1999). Cephaloridine decreased the viability of the rOAT1-expressing cells and probenecid afforded protection against the cytotoxic effect, suggesting a crucial role of OAT1 in the nephrotoxic affect of this antibiotic also in vivo (Takeda et al. 1999).

At a 2-mM concentration, the gyrase inhibitors ofloxacin and norfloxacin did not significantly inhibit PAH uptake into rOAT1-expressing oocytes, whereas cinoxacin showed a strong inhibitory effect. The non-anionic antibiotics chloramphenicol, erythromycin, gentamycin, streptomycin and vancomycin showed no or a very small inhibition (Jariyawat et al. 1999). In the intact rat kidney, the affinities for ofloxacin (app. K_i 2,500 μM) and norfloxacin (1,900 μM; Ullrich et al. 1993b) were low.

Taken together, a number of β-lactam antibiotics interfere with human and rat OAT1, albeit with relatively low affinity. In addition, the transport rate of benzylpenicillin is low, resulting, possibly, in only a small contribution of OAT1 to secretion of β-lactam antibiotics. Likewise, the transport rate of tetracycline into hOAT1-expressing cells was too low to suggest an important role of OAT1 in tubular tetracycline secretion.

OAT2

Tetracyclines were tested with human OAT2 stably expressed in mouse S2 proximal tubule cells (Babu et al. 2002b). The uptake of radiolabeled tetracycline into hOAT2-expressing cells was at all time points larger than in nonexpressing control cells, was saturable, and showed an apparent K_m of 440 μM. The uptake of 5 nM labeled prostaglandin $F_{2\alpha}$ by hOAT2 was inhibited significantly, but no more than 20%, by a 1000-fold excess (5 μM) of tetracycline, oxytetracycline and minocycline; doxycycline did not show an inhibition (Babu et al. 2002b).

Among the β-lactam antibiotics, benzylpenicillin and cefoperazone inhibited the uptake of labeled salicylate by rat OAT2 expressed in LLC-PK1 cells (Morita et al. 2001) or *Xenopus laevis* oocytes (Sekine et al. 1998b), respectively. Rifampicin also interacted with rOAT2 in both studies, whereas erythromycin did not inhibit salicylate uptake (Sekine et al. 1998b). On the other hand, mouse OAT2 expressed in oocytes showed a strong inhibition of glutarate transport by erythromycin (Kobayashi et al. 2002b).

OAT3

Benzylpenicillin moderately inhibited estrone-3-sulfate uptake by human and rat OAT3-expressing *Xenopus laevis* oocytes (Cha et al. 2001; Kusuhara et al. 1999). Half-maximal inhibition of pravastatin uptake by benzylpenicillin was observed at 52.8 μM in LLC-PK1 cells expressing rOAT3 (Hasegawa et al. 2002). Radiolabeled tetracycline was taken up with an apparent K_m of 566 μM into mouse proximal tubule cells transfected with hOAT3 slightly faster than into mock-transfected cells (Babu et al. 2002b). For unknown reasons, unlabeled tetracycline did not inhibit the uptake of labeled estrone sulfate by hOAT3. Likewise, oxytetracycline, minocycline and doxycycline were without an effect on hOAT3. Thus, hOAT3 most probably has too low an affinity for tetracyclines to be of any importance for their renal handling.

Antiviral drugs

Purine derivatives such as acyclovir (ACV), adefovir, ganciclovir, valacyclovir, and pyrimidine derivatives, e.g., cidofovir and AZT (zidovudine), are clinically used to treat a number of viral diseases including AIDS (Yarchoan et al. 1989). Adefovir (Cundy et al. 1995a) and cidofovir (Cundy et al. 1995b) carry a negatively charged phosphonate group and are excreted by the kidneys in their unchanged form. Renal excretion of the negatively charged cidofovir was inhibited by probenecid, suggesting the involvement of transporters for organic anions (Cundy et al. 1995b). Probenecid also inhibited the renal excretion of acyclovir (Laskin et al. 1982) and AZT (Chatton et al. 1990; Mays et al. 1991), although these two compounds do not carry a negative charge. One limitation of antiviral therapy is nephrotoxicity, which is most prominent in proximal tubule cells, the site of organic anion secretion. Therefore, the role of transporters for organic anions in translocation of antiviral drugs needs to be thoroughly understood to prevent nephrotoxicity and other side effects.

OAT1

The influence of antiviral drugs has been studied with human (Cihlar and Ho 2000; Cihlar et al. 1999; Ho et al. 2000; Takeda et al. 2002b) and rat OAT1 (Cihlar et al. 1999; Wada et al. 2000). The uptake of labeled PAH into oocytes (Cihlar et al. 1999) or CHO cells (Cihlar and Ho 2000; Ho et al. 2000) expressing human OAT1 was inhibited by acyclovir, adefovir, AZT, cidofovir and its cyclic prodrug, and by ganciclovir. Half-maximal inhibition of 6-carboxyfluorescein uptake into hOAT1-expressing CHO cells was achieved at 28 μM adefovir and 60 μM cidofovir (Cihlar and Ho 2000), suggesting an interaction of these drugs with OAT1 with a reasonably high affinity. That antiviral drugs were actually translocated by human OAT1 has been shown in experiments with radiolabeled compounds (Cihlar and Ho 2000; Cihlar et al. 1999; Ho et al. 2000; Takeda et al. 2002b). The uptake of adefovir, AZT and cidofovir by hOAT1 was saturable and exhibited apparent K_m values in the range of 24–58 μM. Therefore, the concentrations needed for half-maximal inhibition of 6-carboxyfluorescein uptake and for half-maximal uptake of the drug itself matched nicely: the IC_{50} for adefovir was 28 μM and the K_m for adefovir uptake was 24 μM. The analogous numbers for cidofovir were 60 and 58 μM (Cihlar and Ho 2000). Such a perfect match provides strong evidence for transport of these drugs by hOAT1.

The cidofovir prodrug, which lacks a negative charge, showed a considerably weaker affinity for hOAT1 (K_m 309 μM).

Incubation of CHO or V79 cells for 5 days with increasing concentrations of adefovir or cidofovir caused an increasing cytotoxic effect (Ho et al. 2000). When human OAT1 was expressed in these cells, approximately 100-fold smaller concentrations of adefovir or cidofovir sufficed to induce cytotoxicity. Probenecid or betamipron, both of which inhibited adefovir and cidofovir uptake, protected OAT1-expressing cells. The NSAIDs, diclofenac, diflunisal, flurbiprofen, ketoprofen, and naproxen, were equally or even more potent in inhibiting [^{3}H]adefovir uptake into hOAT1-expressing CHO cells and in their protection against adefovir-induced cytotoxicity (Mulato et al. 2000). These data provide a rationale for the use of probenecid, betamipron or NSAIDs for prevention of nephrotoxicity by antiviral drugs in vivo.

Antiviral drugs were also tested with rat OAT1 expressed in oocytes (Wada et al. 2000). A weak inhibition of rOAT1-mediated transport was observed with acyclovir, amantadine, AZT, didanosine, ganciclovir, stavudine, trifluridine, vidarabine and zalcitabine, with AZT being the most potent inhibitor. IC_{50} values were not reported. However, apparent K_m values for the uptake of radiolabeled acyclovir (242 μM), adefovir (270 μM), AZT (68 μM), cidofovir (238 μM) were given (Wada et al. 2000). The uptake of labeled zalcitabine, didanosine, and trifluridine was markedly, and that of lamivudine and stavudine slightly higher in OAT1-expressing oocytes than in water-injected controls. Foscarnet was not taken up and did not inhibit PAH transport (Wada et al. 2000).

OAT2

Unlike hOAT1, human OAT2 did not show any uptake of labeled acyclovir, ganciclovir and valacyclovir beyond mock-transfected control cells. However, AZT (zidovudine) was taken up by hOAT2 with an apparent K_m of 26.8 μM (Takeda et al. 2002b). AZT was also transported by rat OAT2 with the very same K_m (26 μM), whereas 2',3'-dideoxycytidine (ddC) exhibited a weak affinity (K_m 3,080 μM) (Morita et al. 2001)). Since the maximal transport rate for ddC (4,020 pmol/minmg protein) was considerably greater than for AZT (66 pmol/minmg protein), rOAT2 may transport ddC quite efficiently.

OAT3

Acyclovir and ganciclovir were not transported by human OAT3 (Cha et al. 2001; Takeda et al. 2002b), but the valyl ester of acyclovir, valacyclovir, and AZT showed uptake into hOAT3-expressing mouse proximal tubule cells (Takeda et al. 2002b). Half-maximal rate was achieved at 145 μM AZT. At a 1-mM concentration, acyclovir, ganciclovir and AZT inhibited uptake of 50 nM radiolabeled estrone sulfate by only 25–35%, although these antiviral drugs were present in the medium at a 20,000-fold higher concentration than the label. In another study, 5 μM AZT did not significantly inhibit estrone sulfate uptake into hOAT3-expressing oocytes (Cha et al. 2001). On the other hand, the rat OAT3 was strongly inhibited by 1 mM AZT (Kusuhara et al. 1999). A systematic survey of possible species differences does not exist.

Conclusion

The above-mentioned experiments suggest that OAT1 is involved in uptake of antiviral drugs from the blood into the cell. Therefore, OAT1 plays a role in renal excretion of the drugs, but unfortunately also in the nephrotoxicity caused by uptake of these drugs into proximal tubule cells. Drugs competing with antiviral drugs at OAT1 will help to prevent nephrotoxicity, but they will also prolong the duration of drug action within the body due to an impaired renal excretion. OAT2 and OAT3 transport valacyclovir and zidovudine, but not acyclovir and ganciclovir. Hence, these transporters may contribute to renal excretion of a subset of antiviral drugs.

Diuretics

Except for the Na^+ channel blocker amiloride, diuretics/saluretics are organic anions, which interact either with the carbonic anhydrase of proximal tubules or with cation-chloride cotransporters of the distal tubule (Okusa and Ellison 2000). Acetazolamide, ethoxzolamide and methazolamide are examples of carbonic anhydrase blockers. Bumetanide and furosemide inactivate the Na^+,K^+,2Cl-symporter (BST or NKCC2 or SLC12A1; Russell 2000) located in the luminal membrane of the cells of the thick ascending limb of Henle's loop, whereas thiazides such as (hydro)chlorothiazide and cyclothiazide block a NaCl symporter (TSC or NCC or SLC12A13; Okusa and Ellison 2000) in the luminal membrane of the distal convoluted tubule. To gain access to the cotransporters, the anionic diuretics must be present in the tubule lumen. Due to the high degree of binding to plasma proteins, glomerular filtration of diuretics is very limited. Proximal tubular secretion is hence the main route to deliver diuretics to the tubule lumen and to direct them to their place of action (drug targeting). Furthermore, water absorption along proximal tubule and thin limbs of Henle's loop increases the concentration of diuretics at their target salt transporter. This local concentrating effect allows for low systemic concentrations of diuretics, which prevents side effects.

OAT1

Uptake of radiolabeled PAH into oocytes (Hosoyamada et al. 1999; Race et al. 1999) or HeLa cells (Lu et al. 1999) expressing the human OAT1 was inhibited by addition of unlabeled furosemide to the medium, indicating an interaction of this loop diuretic with the organic anion transporter. In addition, bumetanide, but not hydrochlorothiazide, inhibited hOAT1 expressed in oocytes (Race et al. 1999). Since kinetic studies have not been performed, the apparent affinity of furosemide and bumetanide for hOAT1 is unknown. Transport of radiolabeled furosemide or bumetanide in hOAT1-expressing cells was not shown. Thus, it remains unclear whether the hOAT1 is only inhibited by furosemide and bumetanide, or whether hOAT1 indeed translocates these potent diuretics. Amiloride did not inhibit PAH uptake most probably because this diuretic is an organic cation and thus does not interfere with hOAT1 (Race et al. 1999).

Acetazolamide, ethoxzolamide, methazolamide, bumetanide, ethacrynate, furosemide, chlorothiazide, cyclothiazide and hydrochlorothiazide were tested as possible substrates of the rat OAT1 expressed in *Xenopus laevis* oocytes (Uwai et al. 2000). When present in the medium, all compounds *cis*-inhibited uptake of labeled PAH into oocytes. The inhibition by acetazolamide, bumetanide, and hydrochlorothiazide was competitive, suggesting an interaction of these diuretics with the PAH transport site. Although acetazolamide had the lowest inhibitory potency (K_i 1.1 mM), it *trans*-stimulated PAH efflux from [^{14}C]PAH-loaded oocytes and was itself taken up into rOAT1-expressing oocytes in a probenecid-sensitive way. Thus, acetazolamide was indeed translocated by rOAT1. Hydrochlorothiazide (K_i 150 μM) had an intermediate affinity for rOAT1, and hydrochlorothiazide and chlorothiazide also *trans*-stimulated [^{14}C]PAH efflux, suggesting that thiazides are substrates for OAT1. The loop diuretics furosemide and bumetanide strongly inhibited PAH uptake, when present together with [^{14}C]PAH in the medium. The K_i values, 9.5 μM for furosemide and 5.5 μM for bumetanide, demonstrated a high affinity of these diuretics for rOAT1. Although furosemide and bumetanide *cis*-inhibited [^{14}C]PAH uptake, they did not *trans*-stimulate [^{14}C]PAH efflux from oocytes. This does not rule out that furosemide and bumetanide are translocated by rOAT1, because the turnover of rOAT1 in the PAH/furosemide or PAH/bumetanide exchange mode may be smaller than in the PAH/PAH exchange mode. Furosemide uptake was only slightly faster in rOAT1-expressing oocytes, suggesting its slow translocation by rOAT1. It remained open from these studies (Uwai et al. 2000), whether there is a second transport system for loop diuretics explaining their efficient proximal tubular secretion.

Translocation of diuretics was also studied on the flounder renal OAT, fROAT, using the two-electrode voltage-clamp technique (Burckhardt BC et al. 2000b). At holding potentials more negative than -20 mV, 0.1 mM of ethacrynate, tienilate, and bumetanide induced inward currents of the same magnitude as did 0.1 mM PAH, indicating their translocation by fROAT. [^{3}H]Bumetanide was taken up into fROAT expressing oocytes faster than in nonexpressing control oocytes, supporting the assumption that bumetanide is translocated by the flounder organic anion transporter. In contrast, the carbonic anhydrase inhibitors acetazolamide and sulfanilamide did not evoke currents, suggesting that these compounds are not measurably transported by fROAT. Furosemide, at 0.1 mM, even reversed the inward current into an outward current, a behavior similar to the classic inhibitor of OATs, probenecid. Thus, furosemide appears to inhibit fROAT without being translocated (Burckhardt BC et al. 2000b).

OAT2

Unpublished data (N. A. Wolff, E. Petzinger) suggest that rat OAT2 expressed in oocytes transports radiolabeled bumetanide. Thus OAT2 may represent the hepatic transporter for sodium-independent bumetanide uptake.

OAT3

Five micromolar of either furosemide or bumetanide strongly inhibited the uptake of labeled estrone sulfate into oocytes expressing the human OAT3 (Cha et al. 2001). Likewise, estrone sulfate transport mediated by rat OAT3 in oocytes was markedly decreased by 1 mM bumetanide or furosemide (Kusuhara et al. 1999). Although these data clearly indicate an interaction of loop diuretics with OAT3, it is not known whether these diuretics only inhibit or are also translocated by OAT3.

Conclusion

Many anionic diuretics interfere with OAT1, OAT2, and OAT3 and thereby inhibit their activities. Translocation, however, has been demonstrated directly for only a subset of diuretics in rOAT1 (acetazolamide, furosemide) and fROAT (ethacrynate, tienilate, bumetanide). The importance of OATs for the translocation of furosemide remains to be determined. For human OATs, direct evidence for the translocation of diuretics is missing, as are data on the affinities of these compounds for the transporter.

Nonsteroidal anti-inflammatory drugs

Renal excretion is not the major route of elimination of NSAIDs in their unchanged form (Knauf and Mutschler 2000). These drugs are rather metabolized by the liver and secreted with the bile or, as water-soluble sulfated or glucuronidated compounds, also with the urine. Nevertheless, salicylate was actively secreted by isolated proximal tubules and secretion was inhibited by probenecid (Schild and Roch-Ramel 1988). NSAIDs also interfered with the excretion of penicillin (Nierenberg 1987). These findings suggested that NSAIDs interfere with renal transporters for organic anions in proximal tubules.

OAT1

In a number of studies, the heterologously expressed human OAT1 was inhibited by diclofenac, diflunisal, etodolac, flurbiprofen, ibuprofen, indomethacin, ketoprofen, naproxen, phenacetin, piroxicam and salicylate, when these NSAIDs were added to the incubation medium (Hosoyamada et al. 1999; Lu et al. 1999; Race et al. 1999; Cihlar and Ho 2000; Mulato et al. 2000; Jung et al. 2001). Acetylsalicylate exerted no inhibition (Lu et al. 1999). The uptake of the radiolabeled mycotoxin ochratoxin A by mouse proximal tubule cells expressing the long isoform of human OAT1 (OAT1–1) was competitively inhibited by piroxicam (Jung et al. 2001). In experiments with CHO cells expressing hOAT1, many NSAIDs inhibited [^{3}H]adefovir uptake with a very high affinity (IC_{50} values in brackets): diflunisal (0.85 μM), ketoprofen (1.3 μM), flurbiprofen (1.5 μM), indomethacin (3 μM), diclofenac (4 μM), naproxen (5.8 μM), ibuprofen (8 μM). Other drugs such as piroxicam (21 μM) oxyphenbutazone (32 μM), etodolac (50 μM) had a still high affinity (Mulato et al. 2000). Ketoprofen and ibuprofen, however, were not measurably transported by hOAT1-expressing CHO cells. These compounds appear to bind to OAT1 and inhibit it, but their rate of translocation is slow, probably due to a tight binding to the transporter, i.e., with high association and low dissociation rates. As compared to the above-mentioned NSAIDs, phenacetin (IC_{50} 200 μM) and salicylate (280 μM) showed a relatively low affinity to hOAT1 (Cihlar and Ho 2000; Mulato et al. 2000). Whether these drugs are transported was not reported.

NSAIDs were also systematically tested on rat OAT1 expressed in oocytes (Apiwattanakul et al. 1999). [^{14}C]PAH uptake was inhibited (IC_{50} values in parentheses) by acetylsalicylate (428 μM), aminopyrine, antipyrine, benzydamine, diclofenac, diflunisal, flurbiprofen, ibuprofen (3.5 μM), indomethacin (10 μM), ketoprofen, meclofenamate, naproxen (2 μM), oxyphenbutazone (32 μM), paracetamol (2,099 μM), phenacetin (488 μM), phenylbutazone, piroxicam (52 μM), salicylate (341 μM), salicylurate (11 μM), and tolmetin. The inhibition by ibuprofen and salicylurate was competitive, suggesting a displacement of labeled PAH from the transport site at rOAT1. The uptake of labeled salicylate, acetylsalicylate and indomethacin into rOAT1-expressing oocytes was slightly, but significantly higher than uptake into noninjected control oocytes. Coexpression of NaDC-1 to increase the intracellular content of α-ketoglutarate stimulated salicylate, acetylsalicylate and indomethacin uptake, providing evidence for an exchange of NSAIDs with α-ketoglutarate by the rat OAT1. Acetylsalicylate, salicylate and salicylurate *trans*-stimulated PAH efflux from [^{14}C]PAH preloaded oocytes, which is a further indication that these compounds are indeed translocated by OAT1. Indomethacin *trans*-inhibited PAH efflux, possibly by locking the transporter. Finally, diclofenac, indomethacin, and sulindac also inhibited the mouse OAT1 expressed in COS7 cells.

OAT2

[^{14}C]salicylate was transported by rat OAT2 expressed in *Xenopus laevis* oocytes (Sekine et al. 1998b) and in LLC-PK1 cells (Morita et al. 2001) with closely matching apparent K_m values, 88.9 and 81.2 μM, respectively. The uptake of [^{14}C]acetylsalicylate by rOAT2 in oocytes was less than twice the background. rOAT2 in LLC-PK1 cells had an exquisite affinity for indomethacin: uptake of labeled indomethacin exhibited a K_m of 0.37 μM. Other NSAIDs inhibited salicylate uptake, when added to the incubation medium (K_i values): diclofenac (49.3 μM), ibuprofen (155 μM), and ketoprofen (1.84 μM), indicating that OAT2 is able to interact with a series of these drugs with higher affinity than does the rat OAT1.

OAT3

At a 5-μM concentration, diclofenac, ibuprofen, and indomethacin inhibited uptake of 50-nM labeled estrone sulfate into human OAT3-expressing oocytes by more than 80%; salicylate was less effective and inhibited by about 40% (Cha et al. 2001). The inhibition of hOAT3-mediated ochratoxin A uptake by piroxicam was competitive and exhibited a K_i of 4.9 μM, indicating a high affinity for hOAT3 for this NSAID (Jung et al. 2001). Since, however, the therapeutic plasma concentration of piroxicam (4.8 nM) is much lower than the K_i, this drug cannot be used clinically to prevent ochratoxin A uptake by OAT3 and nephrotoxicity. The uptake of labeled salicylate was slightly enhanced in hOAT3-expressing oocytes, whereas that of ibuprofen was not significantly different from nonexpressing control oocytes (Cha et al. 2001). Thus, translocation of NSAIDs by OAT3 may be slow, if it takes place at all.

Piroxicam (1 mM) was also a strong inhibitor of rat OAT3-mediated estrone sulfate uptake into oocytes (Kusuhara et al. 1999). For indomethacin, conflicting data have been reported. At 1 mM concentration, indomethacin did not affect estrone sulfate uptake (Kusuhara et al. 1999), but inhibited cimetidine uptake by rOAT3 expressing oocytes (Feng et al. 2001). The reason for this discrepancy is unknown.

Conclusion

Collectively, the results indicate that OATs interact with many NSAIDs. Therefore, NSAIDs inhibit with high efficiency the transport of other organic anions, i.e., PAH, 6-carboxyfluorescein, and, of note, nephrotoxic agents such as antiviral drugs and ochratoxin A. They may thus be used to protect patients from organic anion-induced nephrotoxicity, if their therapeutic plasma level is high enough to effectively compete with the toxic compounds. The human OAT1 interacts with high affinity with NSAIDs containing a negatively charged carboxyl group and two hydrophobic aromatic rings connected by a small spacer as, for example, in diflunisal, ketoprofen or flurbiprofen; compounds lacking the carboxyl group (phenacetin, piroxicam) had a smaller affinity (Mulato et al. 2000). The rat OAT1 preferred hydrophobic NSAIDs over hydrophilic ones. Again, the presence of a carboxyl group appeared to increase the affinity to rOAT1 (Apiwattanakul et al. 1999). Some compounds (acetylsalicylate, indomethacin, salicylate, salicylurate) showed a slow transport by OAT1, others such as ibuprofen and ketoprofen were not measurably translocated. Thus it remains open whether OAT1 does contribute in vivo much to the renal excretion of NSAIDs. Data on OAT2 and OAT3 are relatively scarce. Since in most cases, the affinities for the NSAIDs have not been determined, it is unclear whether all OATs require similar structural features of these drugs for high-affinity interaction. hOAT3 appeared to slowly translocate salicylate, but not ibuprofen. As for OAT1, it remains unclear whether OAT3 does contribute to the renal secretion of NSAIDs.

General conclusions

We have described the anion transporters sat-1, NaDC-3, OAT1, and OAT3, which are present in the basolateral membrane of proximal tubule cells. OAT2 is located in the basolateral membrane in humans, but has not been found at this location in rats. Given the age and gender differences of *OAT2* gene expression, more experiments are needed to study OAT2 in the kidneys of male and female animals and humans. The substrate specificities of sat-1 and NaDC-3 were investigated, but so far no drugs and metabolites were tested as possible substrates of these systems. The broadest spectrum of substrates was tested with OAT1. Although in most cases a systematic survey on a series of homologous compounds was not performed, the available data are compatible with the structural requirements defined for the PAH transport system in the intact rat kidney proximal tubule: OAT1 prefers amphiphilic substrates with one or two negative charges at proper distances. Structure-affinity and structure-velocity relationships can not yet be deduced from the available data, because OAT1s were expressed in various cells, leading to experimental differences. In addition, only a few studies have been conducted on the V_{max} of substrates in a single expression system.

The substrate specificities of the OATs overlap. This raises the question of how these transporters can function in parallel. If, for example, PAH is taken up into the cell by exchange against α-ketoglutarate through OAT1, it could flow back out through OAT3, which is not driven into the inward direction by α-ketoglutarate. This futile cycle most probably does not occur in the proximal tubule cell. It is conceivable that OAT3 has a very low affinity for PAH (and other OAT1 substrates) at its intracellular side, making a backleak unlikely. Alternatively, OAT3, and perhaps also OAT2, do operate in vivo as anion exchangers, the counter ion being unknown at this time. Clearly, more experiments are needed to study driving forces for organic anion transport through OAT2 and OAT3.

The characterization of isolated, heterologously expressed carriers provided ample insights into their substrate specificities. As outlined in this review, a considerable number of anionic drugs is transported by one or more of the OATs. However, the relative contributions of the organic anion transporters to overall renal secretion of organic anions can not be answered with this type of experiment. In the future, the results should be reintegrated into the intact tubule, and finally into the intact kidney. We urgently need to know the impact of organic anion transporters in the luminal membrane of proximal tubule cells, before we can understand the process of organic anion secretion. If all involved transporters and their interplay become known, we can fully appreciate renal drug excretion and drug interaction during secretion.

Very few experiments have been undertaken to find out which amino acid residues of the transporter proteins are involved in binding and translocation of the organic anions. Two positively charged residues in transmembrane helices 8 and 11 are part of the substrate binding site in OAT1 and OAT3. It is, however, not clear how these and other polyspecific transporters accommodate a great number of chemically distinct substrates. At present it remains a dream to crystallize OAT1 or OAT3 in the presence of different substrates. More realistic are experiments using site-directed mutagenesis and generation of chimera to narrow down the regions involved in substrate recognition. These experiments, however, are tricky and time-consuming, and are not guaranteed of success and funding. The detection of inborn mutations of the organic anion transporters with altered transport properties, however, will undoubtedly increase the interest in the molecular structure of the organic anion binding sites.

References

Apiwattanakul N, Sekine T, Chairoungdua A, Kanai Y, Nakajima N, Sophasan S, Endou H (1999) Transport properties of nonsteroidal anti-inflammatory drugs by organic anion transporter 1 expressed in Xenopus laevis oocytes. Mol Pharmacol 55:847–854

Babu E, Takeda M, Narikawa S, Kobayashi Y, Enomoto A, Tojo A, Cha SH, Sekine T, Sakthisekaran D, Endou H (2002a) Role of human organic anion transporter 4 in the transport of ochratoxin A. Biochim Biophys Acta 1590:64–75

Babu E, Takeda M, Narikawa S, Kobayashi Y, Yamamoto T, Cha SH, Sekine T, Sakthisekaran D, Endou H (2002b) Human organic anion transporters mediate the transport of tetracycline. Jpn J Pharmacol 88:69–76

Bahn A, Prawitt D, Butler D, Reid G, Enklaar T, Wolff NA, Ebbinghaus C, Hillemann A, Schulten H-J, Gunawan B, Füzesi L, Zabel B, Burckhardt G (2000) Genomic structure and in vivo expression of the human organic anion transporter 1 (hOAT1) gene. Biochem Biophys Res Commun 275:623–630

Bahn A, Quondamatteo F, Knabe M, Godehardt S, Hillemann R, Herken G, Burckhardt G (2001a) Evidences for an important role of the organic anion transporter 1 (OAT1) in handling of neurotransmitter metabolites in the eye. Pflügers Arch Eur J Physiol 441:R126

Bahn A, Reid G, Schulten H-J, Ebbinghaus C, Hillemann A, Füzesi L, Burckhardt G, Wolff NA (2001b) Cloning and characterization of the human organic anion transporter 2 (hOAT2). FASEB J 15:A434

Bahn A, Knabe M, Hagos Y, Rödiger M, Godehardt S, Grober-Neufeld Ds, Evans KK, Burckhardt G, Wright SH (2002) Interaction of the metal chelator 2,3-dimercapto-1-propanesulfonate with the rabbit multispecific organic anion transporter 1 (rbOAT1). Mol Pharmacol 62:1128–1136

Baruch SB, Burich RL, Eun CK, King VF (1975) Renal metabolism of citrate. Med Clin North Am 59:569–582

Bästlein C, Burckhardt G (1986) Sensitivity of rat renal luminal and contraluminal sulfate transport systems to DIDS. Am J Physiol Renal Physiol 250:F226–F234

Beck L, Markovich D (2000) The mouse Na+-sulfate cotransporter gene Nas1. Cloning, tissue distribution, gene structure, chromosomal assignment, and transcriptional regulation by vitamin D. J Biol Chem 275:11880–11890

Beck L, Silve C (2001) Molecular aspects of renal tubular handling and regulation of inorganic sulfate. Kidney Int 59:835–845

Beyer KH, Russo HF, Tillson EK, Miller AK, Verwey WF, Gass SR (1951) 'Benemid', p-(di-n-propylsulfamyl)-benzoic acid: its renal affinity and its elimination. Am J Physiol 166:625–640

Bissig M, Hagenbuch B, Stieger B, Koller T, Meier PJ (1994) Functional expression cloning of the canalicular sulfate transport system of rat hepatocytes. J Biol Chem 269:3017–3021

Brady KP, Dushkin H, Förnzler D, Koike T, Magner F, Her H, Gullans S, Segre GV, Green RM, Beier DR (1999) A novel putative transporter maps top the osteosclerosis (oc) mutation and is not expressed in the oc mutant mouse. Genomics 56:254–261

Brazy PC, Dennis VW (1981) Sulfate transport in rabbit proximal convoluted tubules: presence of an anion exchange. Am J Physiol Renal Physiol 241:F300–F307

Buist SCN, Cherrington NJ, Choudhuri S, Hartley DP, Klaasen CD (2002) Gender-specific and developmental influences on the expression of rat organic anion transporters. J Pharmacol Exp Ther 301:145–151

Burckhardt BC, Sato K, Frömter E (1984) Electrophysiological analysis of bicarbonate permeation across the peritubular cell membrane of rat kidney proximal tubule. I. Basic observations. Pflügers Arch 401:34–42

Burckhardt BC, Steffgen J, Langheit D, Müller GA, Burckhardt G (2000a) Potential-dependent steady-state kinetics of a dicarboxylate transporter cloned from winter flounder kidney. Pflügers Arch 441:323–330

Burckhardt BC, Wolff NA, Burckhardt G (2000b) Electrophysiologic characterization of an anion transporter cloned from Winter flounder kidney (fROAT). J Am Soc Nephrol 11:9–17

Burckhardt BC, Drinkuth B, Menzel C, König A, Steffgen J, Wright SH, Burckhardt G (2002) The renal Na+-dependent dicarboxylate transporter; NaDC-3, translocates dimethyl- and disulfhydryl compounds and contributes to renal heavy metal detoxification. J Am Soc Nephrol 13:2628–2638

Burckhardt BC, Brai S, Wallis S, Krick W, Wolff NA, Burckhardt G (2003) Transport of cimetidine by flounder and human renal organic anion transporter 1. Am J Physiol Renal Physiol 284 (in press)

Burckhardt G (1984) Sodium-dependent dicarboxylate transport in rat renal basolateral membrane vesicles. Pflügers Arch 401:254–261

Burckhardt G, Pritchard JB (2000) Organic anion and cation antiporters. In: Seldin DW, Giebisch G (eds) The kidney. Physiology and pathophysiology. Lippincott Williams & Wilkins, Philadelphia, pp 193–222

Burckhardt G, Wolff NA (2000) Structure of renal organic anion and cation transporters. Am J Physiol Renal Physiol 278: F853–F866

Burckhardt G, Porth J, Wolff NA (1998) Functional and molecular characterization of renal transporters for p-aminohippurate (PAH). Nova Acta Leopoldina 78:35–40

Burckhardt G, Bahn A, Wolff NA (2001) Molecular physiology of renal p-aminohippurate secretion. News Physiol Sci 16:114–118

Buttler D, Ebbinghaus C, Hillemann A, Wolff NA, Füzesi L, Burckhardt G, Bahn A (2001) In vivo studies, cloning and functional characterization of the isoforms of the human organic anion transporter 1 (hOAT1). Pflügers Arch Eur J Physiol 441:R170

Cha SH, Sekine T, Kusuhara H, Yu E, Kim JY, Kim DK, Sugiyama Y, Kanai Y, Endou H (2000) Molecular cloning and characterization of multispecific organic anion transporter 4 expressed in the placenta. J Biol Chem 275:4507–4512

Cha SH, Sekine T, Fukushima J-I, Kanai Y, Kobayashi Y, Goya T, Endou H (2001) Identification and characterization of human organic anion transporter 3 expressing predominantly in the kidney. Mol Pharmacol 59:1277–1286

Chatton J-Y, Odone M, Besseghir K, Roch-Ramel F (1990) Renal secretion of 3'-azido-3'-deoxythymidine by the rat. J Pharmacol Exp Ther 255:140–145

Chen XM, Tsukaguchi H, Chen XZ, Berger UV, Hediger MA (1999) Molecular and functional analysis of SDCT2, a novel rat sodium-dependent dicarboxylate transporter. J Clin Invest 103:1159–1168

Chen X-Z, Shayakul C, Berger UV, Tian W, Hediger MA (1998) Characterization of a rat Na+-dicarboxylate cotransporter. J Biol Chem 273:20972–20981

Cihlar T, Ho ES (2000) Fluorescence-based assay for the interaction of small molecules with the human renal organic anion transporter 1. Anal Biochem 283:49–55

Cihlar T, Lin DC, Pritchard JB, Fuller MD, Mendel DB, Sweet DH (1999) The antiviral nucleotide analogs cidofovir and adefovir are novel substrates for human and rat renal organic anion transporter 1. Mol Pharmacol 56:570–580

Clarke LA, Nasir J, McDonald H, Applegarth DA, Hayden MR (1994) Murine α-L-iduronidase: cDNA isolation and expression. Genomics 15:311–316
Cole DEC, Scriver CR (1980) Age-dependent serum sulfate levels in children and adolescents. Clin Chim Acta 107:135–139
Cserr HF, van Dyke DH (1971) 5-Hydroxyindoleacetic acid accumulation by isolated choroid plexus. Am J Physiol 220:718–723
Cundy KC, Barditch-Crovo P, Walker RE, Collier AC, Ebeling D, Toole J, Jaffe HS (1995a) Clinical pharmacokinetics of adefovir in human immunodeficiency virus type 1-infected patients. Antimicrob Agents Chemother 39:2401–2405
Cundy KC, Petty BG, Flaherty J, Fisher PE, Polis MA, Wachsman M, Lietman PS, Lalezari JP, Hitchcock PS, Jaffe HS (1995b) Clinical pharmacokinetics of cidofovir in human immunodeficiency virus-infected patients. Antimicrob Agents Chemother 39:1247–1252
Cunningham RF, Israili ZH, Dayton PG (1981) Clinical pharmacokinetics of probenecid. Clinical Pharmacokinetics 6:135–151
David C, Ullrich KJ (1992) Substrate specificity of the luminal Na+-dependent sulphate transport system in the proximal renal tubule as compared to the contraluminal sulphate exchanger. Pflügers Arch 421:455–465
Deguchi T, Ohtsuki S, Otagiri M, Takanaga H, Asaba H, Mori S, Terasaki T (2002) Major role of organic anion transporter 3 in the transport of indoxyl sulfate in the kidney. Kidney Int 61:1760–1768
Dresser MJ, Leabman MK, Giacomini KM (2001) Transporters involved in the elimination of drugs in the kidney: organic anion transporters and organic cation transporters. J Pharmaceut Sci 90:397–421
Enomoto A, Kimura H, Chairoungdua A, Shigeta Y, Jutabha P, Cha SH, Hosoyamada M, Takeda M, Sekine T, Igarashi T, Matsuo H, Kikuchi Y, Oda T, Ichida K, Hosoya T, Shimotaka K, Niwa T, Kanai Y, Endou H (2002a) Molecular identification of a renal urate-anion exchanger that regulates blood urate levels. Nature 417:447–452
Enomoto A, Takeda M, Shimoda M, Narikawa S, Kobayashi Y, Kobayashi Y, Yamamoto T, Sekine T, Cha SH, Niwa T, Endou H (2002b) Interaction of human organic anion transporters 2 and 4 with organic anion transport inhibitors. J Pharmacol Exp Ther 301:797–802
Everett LA, Glaser B, Beck JC, Idol JR, Buchs A, Heyman M, Adawi F, Hazani E, Nassir E, Baxevanis AD, Sheffield VC, Green ED (1997) Pendred syndrome is caused by mutations in a putative sulphate transporter gene (PDS). Nature Genet 17:411–422
Feng B, Dresser MJ, Shu Y, Johns SJ, Giacomini KM (2001) Arginine 454 and lysine 370 are essential for the anion specificity of the organic anion transporter, rOAT3. Biochemistry 40:5511–5520
Fritzsch G, Rumrich G, Ullrich KJ (1989) Anion transport through the contraluminal cell membrane of renal proximal tubule. The influence of hydrophobicity and molecular charge distribution on the inhibitory activity of organic anions. Biochim Biophys Acta 978:249–256
Gekle M, Mildenberger S, Sauvant C, Bednarczyk D, Wright SH, Dantzler WH (1999) Inhibition of initial transport rate of basolateral organic anion carrier in renal PT by BK and phenylephrine. Am J Physiol Renal Physiol 277:F251–F256
George RL, Wu X, Fei Y-J, Leibach FH, Ganapathy V (1999) Molecular cloning and functional characterization of a polyspecific organic anion transporter from Caenorhabditis elegans. J Pharmacol Exp Ther 291:596–603
Grantham JJ, Qualizza PB, Irwin RL (1974) Net fluid secretion in proximal straight renal tubules in vitro: role of PAH. Am J Physiol 226:191–197
Grinstein S, Turner RJ, Silverman M, Rothstein A (1980) Inorganic anion transport in kidney and intestinal brush-border and basolateral membranes. Am J Physiol 238:F452–F460
Guder W, Wiesner W, Stukowski B, Wieland O (1971) Metabolism of isolated kidney tubules. Hoppe-Seyler's Z Physiol Chem 352:1319–1328
Guldberg HC, Ashcroft GW, Crawford TBB (1966) Concentration of 5-hydroxyindolylacetic acid and homovanillic acid in the cerebrospinal fluid of the dog before and during treatment with probenecid. Life Sci 5:1571–1575
Häberle DA (1981) Characteristics of p-aminohippurate transport in the mammalian kidney. In: Greger R, Lang F, Silbernagl S (eds) Renal transport of organic substances. Springer Verlag, Heidelberg, pp 189–209
Hagenbuch B, Stange G, Murer H (1985) Transport of sulphate in rat jejunal and rat proximal tubular basolateral membrane vesicles. Pflügers Arch 405:202–208
Hagos Y, Larsen A, Schindelmann S, Burckhardt G, Steffgen J (2001) Protein kinase C mediated down regulation of sodium dicarboxylate cotransporter fNaDC-3. Kidney Blood Press Res 24:351
Hagos Y, Bahn A, Asif AR, Krick W, Sendler M, Burckhardt G (2002) Cloning of the pig renal organic anion transporter 1 (pOAT1). Biochimie (in press)

Hasegawa M, Kusuhara H, Sugiyama D, Ito K, Ueda S, Endou H, Sugiyama Y (2002) Functional involvement of rat organic anion transporter 3 (rOAT3; Slc22a8) in the uptake of organic anions. J Pharmacol Exp Ther 300:746–753

Hästbacka J, de la Chapelle A, Mahtani MM, Clines G, Reeve-Daly MP, Daly M, Hamilton BA, Kusumi K, Trivedi B, Weaver A, Coloma A, Lovett M, Buckler A, Kaitila I, Lander ES (1994) The diastrophic dysplasia gene encodes a novel sulfate transporter: positional cloning by fine-structure linkage disequilibrium mapping. Cell 78:1073–1078

Hierholzer K, Cade R, Gurd R, Kessler R, Pitts R (1960) Stop-flow analysis of renal reabsorption and excretion of sulfate in the dog. Am J Physiol 198:833–837

Ho ES, Lin DC, Mendel DB, Cihlar T (2000) Cytotoxicity of antiviral nucleotides adefovir and cidofovir is induced by the expression of human renal organic anion transporter 1. J Am Soc Nephrol 11:383–393

Hohage H, Löhr M, Querl U, Greven J (1994) The renal basolateral transport system for organic anions: properties of the regulation mechanism. J Pharmacol Exp Ther 269:659–664

Hosoyamada M, Sekine T, Kanai Y, Endou H (1999) Molecular cloning and functional expression of a multispecific organic anion transporter from human kidney. Am J Physiol Renal Physiol 276:F122–F128

Huang W, Wang H, Kekuda R, Fei Y-J, Friedrich A, Wang J, Conway SJ, Cameron RS, Leibach FH, Ganapathy V (2000) Transport of N-Acetylaspartate by the Na+-dependent high-affinity dicarboxylate transporter NaDC3 and its relevance to the expression of the transporter in the brain. J Pharmacol Exp Ther 295:392–403

Inui K-I, Masuda S, Saito H (2000) Cellular and molecular aspects of drug transport in the kidney. Kidney Int 58:944–958

Islinger F, Gekle M, Wright SH (2001) Interaction of 2,3-dimercapto-1-propane sulfonate with the human organic anion transporter hOAT1. J Pharmacol Exp Ther 299:741–747

Jariyawat S, Sekine T, Takeda M, Apiwattanakul N, Kanai Y, Sophasan S, Endou H (1999) The interaction and transport of β-lactam antibiotics with the cloned rat renal organic anion transporter 1. J Pharmacol Exp Ther 290:672–677

Jørgensen KE, Kragh-Hansen U, Roigaard-Petersen H, Sheikh MI (1983) Citrate uptake by basolateral and luminal membrane vesicles from rabbit kidney cortex. Am J Physiol 244:F686–F695

Jung KY, Takeda M, Kim DK, Tojo A, Narikawa S, Yoo BS, Hosoyamada M, Cha SH, Sekine T (2001) Characterization of ochratoxin A transport by human organic anion transporters. Life Sci 69:2123–2135

Karniski LP, Lötscher M, Fucentese M, Hilfiker H, Biber J, Murer H (1998) Immunolocalization of sat-1 sulfate/oxalate/bicarbonate anion exchanger in the rat kidney. Am J Physiol Renal Physiol 275:F79–F87

Kekuda R, Wang H, Huang W, Pajor AM, Leibach FH, Devoe LD, Prasad PD, Ganapathy V (1999) Primary structure and functional characteristics of a mammalian sodium-coupled high affinity dicarboxylate transporter. J Biol Chem 274:3422–3429

Kimura H, Takeda M, Narikawa S, Enomoto A, Ichida K, Endou H (2002) Human organic anion transporters and human organic cation transporters mediate renal transport of prostaglandins. J Pharmacol Exp Ther 301:293–298

Knauf H, Mutschler E (2000) Renal elimination of drugs. In: Seldin DW, Giebisch G (eds) The kidney. Physiology and pathophysiology. Lippincott Williams & Wilkins, Philadelphia, pp 2861–2875

Kobayashi A, Hirokawa N, Ohshiro N, Sekine T, Sasaki T, Tokuyama S, Endou H, Yamamoto T (2002a) Differential gene expression of organic anion transporters in male and female rats. Biochem Biophys Res Commun 290:482–487

Kobayashi Y, Ohshiro N, Shibusawa A, Sasaki T, Tokuyama S, Sekine T, Endou H, Yamamoto T (2002b) Isolation, characterization and differential gene expression of multispecific organic anion transporter 2 in mice. Mol Pharmacol 62:7–14

Koepsell H, Gorboulev V, Arndt P (1999) Molecular pharmacology of organic cation transporters in kidney. J Membr Biol 167:103–117

Kojima R, Sekine T, Kawachi M, Cha SH, Suzuki Y, Endou H (2002) Immunolocalization of multispecific organic anion transporters, OAT1, OAT2, and OAT3, in rat kidney. J Am Soc Nephrol 13:848–857

Kok LDS, Siu SS, Fung KP, Tsui SKW, Lee CY, Waye MMY (2000) Assignment of liver-specific organic anion transporter (SLC22A7) to human chromosome 6 bands p21.2 $\rightarrow$ p21.1 using radiation hybrids. Cytogenet Cell Genet 88:76–77

König J, Nies AT, Cui Y, Leier I, Keppler D (1999) Conjugate export pumps of the multidrug resistance protein (MRP) family: localization, substrate specificity, and MRP2-mediated drug resistance. Biochim Biophys Acta 1461:377–394

Krick W, Wolff NA, Burckhardt G (2000) Voltage-driven p-aminohippurate, chloride, and urate transport in porcine renal brush-border membrane vesicles. Pflügers Arch Eur J Physiol 441:125–132

Kuo S-M, Aronson PS (1988) Oxalate transport via the sulfate/HCO3– exchanger in rabbit renal basolateral membrane vesicles. J Biol Chem 263:9710–9717

Kusuhara H, Sekine T, Utsunomiya-Tate N, Tsuda M, Kojima R, Cha SH, Sugiyama Y, Kanai Y, Endou H (1999) Molecular cloning and characterization of a new multispecific organic anion transporter from rat brain. J Biol Chem 274:13675–13680

Kuze K, Graves P, Leahy A, Wilson P, Stuhlmann H, You G (1999) Heterologous expression and functional characterization of a mouse renal organic anion transporter in mammalian cells. J Biol Chem 274:1519–1524

Laskin OL, De Miranda P, King DH, Page DA, Longstreth JA, Rocco L, Lietman PS (1982) Effects of probenecid on the pharmacokinetics and elimination of acyclovir in humans. Antimicrob Agents Chemother 21:804–807

Leal-Pinto E, Tao WJ, Rappaport J, Richardson M, Knorr BA, Abramson RG (1997) Molecular cloning and functional reconstitution of a urate transporter/channel. J Biol Chem 272:617–625

Leal-Pinto E, Cohen BE, Lipkowitz MS, Abramson RG (2002) Functional analysis and molecular model of the human urate transporter/channel, hUAT. Am J Physiol Renal Physiol 283:F150–F163

Lee A, Beck L, Brown RJ, Markovich D (1999) Identification of a mammalian brain sulfate transporter. Biochem Biophys Res Commun 263:123–129

Lee A, Beck L, Markovich D (2000) The human renal Na+-sulfate cotransporter (SLC3A1; hNaSi-1) cDNA and gene: organization, chromosomal localization and functional characterization. Genomics 70:354–363

Leier I, Hummel-Eisenbeiss J, Cui Y, Keppler D (2000) ATP-dependent para-aminohippurate transport by apical multidrug resistance protein MRP2. Kidney Int 57:1636–1642

Lopez-Nieto CE, You GF, Bush KT, Barros EJG, Beier DR, Nigam SK (1997) Molecular cloning and characterization of NKT, a gene product related to the organic cation transporter family that is almost exclusively expressed in the kidney. J Biol Chem 272:6471–6478

Löw I, Friedrich T, Burckhardt G (1984) Properties of an anion exchanger in rat renal basolateral membrane vesicles. Am J Physiol 246:F334–F342

Lu R, Chan BS, Schuster VL (1999) Cloning of the human kidney PAH transporter: narrow substrate specificity and regulation by protein kinase C. Am J Physiol Renal Physiol 276:F295–F303

Lücke H, Stange G, Murer H (1980) Sulphate-ion/sodium-ion co-transport by brush-border membrane vesicles isolated from rat kidney cortex. Biochem J 182:223–229

Maiorino RM, Aposhian MM, Xu Z-F, Li Y, Polt RL, Aposhian HV (1993) Determination and metabolism of dithiol chelating agents. XV. The meso-2,3-dimercaptosuccinic acid-cysteine (1:2) mixed disulfide, a major urinary metabolity of DMSA in the human, increases the urinary excretion of lead in the rat. J Pharmacol Exp Ther 267:1221–1226

Markovich D (2001) Physiological roles and regulation of mammalian sulfate transporters. Physiol Rev 81:1499–1533

Markovich D, Regeer RR (1999) Expression of membrane transporters in cane toad bufo marinus oocytes. J Exp Biol 202:2217–2223

Markovich D, Forgo J, Stange G, Biber J, Murer H (1993) Expression cloning of rat renal Na+/SO42– cotransport. Proc Natl Acad Sci U S A 90:8073–8077

Markovich D, Bissig M, Sorribas V, Hagenbuch B, Meier PJ, Murer H (1994) Expression of rat renal sulfate transport systems in Xenopus laevis oocytes: functional characterization and molecular identification. J Biol Chem 269:3022–3026

Martin M, Ferrier B, Baverel G (1989) Transport and utilization of α-ketoglutarate by the rat kidney in vivo. Pflügers Arch 413:217–224

Mays DC, Dixon KF, Balboa A, Pawluk LJ, Bauer MR, Nawoot S, Gerber N (1991) A nonprimate animal model applicable to zidovudine pharmacokinetics in humans: inhibition of glucuronidation and renal excretion of zidovudine by probenecid in rats. J Pharmacol Exp Ther 259:1261–1270

Miller DS (1998) Protein kinase C regulation of organic anion transport in renal proximal tubule. Am J Physiol Renal Physiol 274:F156–F164

Morita N, Kusuhara H, Sekine T, Endou H, Sugiyama Y (2001) Functional characterization of rat organic anion transporter 2 in LLC-PK1 cells. J Pharmacol Exp Ther 298:1179–1184

Morris ME, Murer H (2001) Molecular mechanisms in renal and intestinal sulfate (re)absorption. J Membr Biol 181:1–9

Moseley RH, Höglund P, Wu GD, Silberg DG, Haila S, de la Chapelle A, Holmberg C, Kere J (1999) Downregulated in adenoma gene encodes a chloride transporter defective in congenital chloride diarrhea. Am J Physiol Gastrointest Liver Physiol 276:G185–G192

Motohashi H, Sakurai Y, Saito H, Masuda S, Urakami Y, Goto M, Fukatsu A, Ogawa O, Inui K-I (2002) Gene expression levels and immunolocalization of organic ion transporters in human kidney. J Am Soc Nephrol 13:866–874

Motojima K, Hosokawa A, Yamato H, Muraki T, Yoshioka T (2002) Uraemic toxins induce proximal tubular injury via organic anion transporter 1-mediated uptake. Br J Pharmac 135:555–563

Mulato AS, Ho ES, Cihlar T (2000) Nonsteroidal anti-inflammatory drugs efficiently reduce the transport and cytotoxicity of adefovir mediated by the human renal organic anion transporter 1. J Pharmacol Exp Ther 295:10–15

Nagai J, Yano I, Hashimoto Y, Takano M, Inui K-I (1997) Inhibition of PAH transport by parathyroid hormone in OK cells: involvement of protein kinase C pathway. Am J Physiol Renal Physiol 273:F674–F679

Nagata Y, Kusuhara H, Endou H, Sugiyama Y (2002) Expression and functional characterization of rat organic anion transporter 3 (rOAT3) in the choroid plexus. Mol Pharmacol 61:982–988

Nakajima N, Sekine T, Cha SH, Tojo A, Hosoyamada M, Kanai Y, Yan K, Awa S, Endou H (2000) Developmental changes in multispecific organic anion transporter 1 expression in rat kidney. Kidney Int 57:1608–1616

Nierenberg DW (1987) Drug inhibition of penicillin tubular secretion: concordance between in vitro and clinical findings. J Pharmacol Exp Ther 240:712–716

Nieth H, Schollmeyer P (1966) Substrate-utilization of the human kidney. Nature 209:1244–1245

Okusa MD, Ellison DH (2000) Physiology and pathophysiology of diuretic action. In: Seldin DW, Giebisch G (eds) The kidney. Physiology and pathophysiology. Lippincott Williams & Wilkins, Philadelphia, pp 2877–2922

Pajor AM (1995) Sequence and functional characterization of a renal sodium/dicarboxylate cotransporter. J Biol Chem 270:5779–5785

Pajor AM (1996) Molecular cloning and functional expression of a sodium-dicarboxylate cotransporter from human kidney. Am J Physiol Renal Physiol 270:F642–F648

Pajor AM (1999) Sodium-coupled transporters for Krebs cycle intermediates. Annu Rev Physiol 61:663–682

Pajor AM (2000) Molecular properties of sodium/dicarboxylate cotransporters. J Membr Biol 175:1–8

Pajor AM, Sun N (1999) Protein kinase C-mediated regulation of the renal Na+/dicarboxylate cotransporter, NaDC-1. Biochim Biophys Acta 1420:223–230

Pajor AM, Sun NN (2000) Molecular cloning, chromosomal organization, and functional characterization of a sodium-dicarboxylate cotransporter from mouse kidney. Am J Physiol Renal Physiol 279:F482–F490

Pajor AM, Gangula R, Yao X (2001) Cloning and functional characterization of a high-affinity Na+/dicarboxylate cotransporter from mouse brain. Am J Physiol Cell Physiol 280:C1215–C1223

Pao SS, Paulsen IT, Saier MHJr (1998) Major facilitator superfamily. Microbiol Mol Biol Rev 62:1–34

Pavlova A, Sakurai H, Leclercq B, Beier DR, Yu ASL, Nigam SK (2000) Developmentally regulated expression of organic ion transporters NKT (OAT1), OCT1, NLT (OAT2), and Roct. Am J Physiol Renal Physiol 278:F635–F643

Pombrio JM, Giangreco A, Li L, Wempe MF, Anders MW, Sweet DH, Pritchard JB, Ballatori N (2001) Mercapturic acids (N-acetylcysteine S-conjugates) as endogenous substrates for the renal organic anion transporter-1. Mol Pharmacol 60:1091–1099

Pritchard JB (1988) Coupled transport of p-aminohippurate by rat kidney basolateral membrane vesicles. Am J Physiol Renal Physiol 255:F597–F604

Pritchard JB (1995) Intracellular α-ketoglutarate controls the efficacy of renal organic anion transport. J Pharmacol Exp Ther 274:1278–1284

Pritchard JB, Miller DS (1993) Mechanisms mediating renal secretion of organic anions and cations. Physiol Rev 73:765–796

Pritchard JB, Renfro JL (1983) Renal sulfate transport at the basolateral membrane is mediated by anion exchange. Proc Natl Acad Sci U S A 80:2603–2607

Race JE, Grassl SM, Williams WJ, Holtzman EJ (1999) Molecular cloning and characterization of two novel human renal organic anion transporters (hOAT1 and hOAT3). Biochem Biophys Res Commun 255:508–514

Reid G, Wolff NA, Dautzenberg FM, Burckhardt G (1998) Cloning of a human renal p-aminohippurate transporter, hROAT1. Kidney Blood Press Res 21:233–237

Renfro JL, Pritchard JB (1982) H+-dependent sulfate secretion in the marine teleost renal tubule. Am J Physiol 243:F150–F159

Renfro JL, Clark NB, Metts RE, Lynch MA (1987) Sulfate transport by chick renal tubule brush-border and basolateral membranes. Am J Physiol Regul Integr Comp Physiol 252:R85–R93

Rocchiccioli F, Leroux JP, Cartier PH (1984) Microdetermination of 2-ketoglutaric acid in plasma and cerebrospinal fluid by capillary gas chromatography mass spectrometry: application to pediatrics. Biomed Mass Spectrom 11:24–28

Röver N, Krämer C, Stärk U, Gabriëls G, Greven J (1998) Basolateral transport of glutarate in proximal S2 segments of rabbit kidney: kinetics of the uptake process and effect of activators of protein kinase A and C. Pflügers Arch 436:423–428

Russell JM (2000) Sodium-potassium-chloride cotransport. Physiol Rev 80:211–276

Sauvant C, Holzinger H, Gekle M (2001) Modulation of the basolateral and apical step of transepithelial organic anion secretion in proximal tubular opossum kidney cells. J Biol Chem 276:14695–14703

Schaub TP, Kartenbeck J, König J, Vogel O, Witzgall R, Kriz W, Keppler D (1997) Expression of the conjugate export pump encoded by the mrp2 gene in the apical membrane of kidney proximal tubules. J Am Soc Nephrol 8:1213–1221

Schaub TP, Kartenbeck J, König J, Spring H, Dörsam J, Staehler G, Störkel S, Thon WF, Keppler D (1999) Expression of the MRP2 gene-encoded conjugate export pump in human kidney proximal tubules and in renal cell carcinoma. J Am Soc Nephrol 10:1159–1169

Schild L, Roch-Ramel F (1988) Transport of salicylate in proximal tubule (S2 segment) isolated from rabbit kidney. Am J Physiol 254:F554–F561

Schneider EG, Durham JC, Sacktor B (1984) Sodium-dependent transport of inorganic sulfate by rabbit renal brush-border membrane vesicles. J Biol Chem 259:14591–14599

Sekine T, Watanabe N, Hosoyamada M, Kanai Y, Endou H (1997) Expression cloning and characterization of a novel multispecific organic anion transporter. J Biol Chem 272:18526–18529

Sekine T, Cha SH, Hosoyamada M, Kanai Y, Watanabe N, Furuta Y, Fukuda K, Igarashi T, Endou H (1998a) Cloning, functional characterization, and localization of a rat renal Na+-dicarboxylate transporter. Am J Physiol Renal Physiol 275:F289–F305

Sekine T, Cha SH, Tsuda M, Apiwattanakul N, Nakajima N, Kanai Y, Endou H (1998b) Identification of multispecific organic anion transporter 2 expressed predominantly in the liver. FEBS Lett 429:179–182

Sheikh MI, Kragh-Hansen U, Jørgensen KE, Roigaard-Petersen H (1982) An efficient method for the isolation and separation of basolateral-membrane and luminal-membrane vesicles from rabbit kidney cortex. Biochem J 208:377–382

Sheridan E, Rumrich G, Ullrich KJ (1983) Reabsorption of dicarboxylic acids from the proximal convolution of rat kidney. Pflügers Arch 399:18–28

Shimada H, Burckhardt G (1986) Kinetic studies of sulfate transport in basolateral membrane vesicles from rat renal cortex. Pflügers Arch 407:S160–S167

Shimada H, Moewes B, Burckhardt G (1987) Indirect coupling to Na+ of p-aminohippuric acid uptake into rat renal basolateral membrane vesicles. Am J Physiol Renal Physiol 253:F795–F801

Simonson GD, Iwanij V (1995) Genomic organization and promoter sequence of a gene encoding a rat liver-specific type-I transport protein. Gene 154:243–247

Simonson GD, Vincent AC, Roberg KJ, Huang Y, Iwanij V (1994) Molecular cloning and characterization of a novel liver- specific transport protein. J Cell Sci 107:1065–1072

Simpson DP (1983) Citrate excretion: a window on renal metabolism. Am J Physiol 244:F223–F234

Smith HW, Finkelstein H, Aliminosa L, Crawford B, Graber M (1945) The renal clearance of substituted hippuric acid derivatives and other aromatic acid in the dog and man. J Clin Invest 24:388–404

Steffgen J, Burckhardt BC, Langenberg C, Kühne L, Müller GA, Burckhardt G, Wolff NA (1999) Expression cloning and characterization of a novel sodium-dicarboxylate cotransporter from winter flounder kidney. J Biol Chem 274:20191–20196

Sugiyama D, Kusuhara H, Shitara Y, Abe T, Meier PJ, Sekine T, Endou H, Sugiyama Y (2001) Characterization of the efflux transport of 17β-estradiol-D-17β-glucuronide from the brain across the blood-brain barrier. J Pharmacol Exp Ther 298:316–322

Sun W, Wu RR, van Poelje PD, Erion MD (2001) Isolation of a family of organic anion transporters from human liver and kidney. Biochem Biophys Res Commun 283:417–422

Sweet DH, Pritchard JB (1999) The molecular biology of renal organic anion and organic cation transporters. Cell Biochem Biophys 31:89–118

Sweet DH, Wolff NA, Pritchard JB (1997) Expression cloning and characterization of ROAT1. J Biol Chem 272:30088–30095

Takano M, Nagai J, Yasuhara M, Inui K-I (1996) Regulation of p-aminohippurate transport by protein kinase C in OK kidney epithelial cells. Am J Physiol 271:F469–F475

Takeda M, Tojo A, Sekine T, Hosoyamada M, Kanai Y, Endou H (1999) Role of organic anion transporter 1 (OAT1) in cephaloridine (CER)-induced nephrotoxicity. Kidney Int 56:2128–2136

Takeda M, Hosoyamada M, Cha SH, Sekine T, Endou H (2000a) Hydrogen peroxide downregulates human organic anion transporters in the basolateral membrane of the proximal tubule. Life Sci 68:679–687

Takeda M, Sekine T, Endou H (2000b) Regulation by protein kinase C of organic anion transport driven by rat organic anion transporter 3 (rOAT3). Life Sci 67:1087–1093

Takeda M, Narikawa S, Hosoyamada M, Cha SH, Sekine T, Endou H (2001) Characterization of organic anion transport inhibitors using cells stably expressing human organic anion transporters. Eur J Pharmacol 419:113–120

Takeda M, Babu E, Narikawa S, Endou H (2002a) Interaction of human organic anion transporters with various cephalosporin antibiotics. Eur J Pharmacol 438:137–142

Takeda M, Khamdang S, Narikawa S, Kimura H, Kobayashi Y, Yamamoto T, Cha SH, Sekine T, Endou H (2002b) Human organic anion transporters and human organic cation transporters mediate renal antiviral transport. J Pharmacol Exp Ther 300:918–924

Talor Z, Gold RM, Yang W-C, Arruda JAL (1987) Anion exchanger is present in both luminal and basolateral renal membranes. Eur J Biochem 164:695–702

Tanner GA, Isenberg MT (1970) Secretion of p-aminohippurate by rat kidney proximal tubules. Am J Physiol 219:889–892

Tojo A, Sekine T, Nakajima N, Hosoyamada M, Kanai Y, Kimura K, Endou H (1999) Immunohistochemical localization of multispecific renal organic anion transporter 1 in rat kidney. J Am Soc Nephrol 10:464–471

Tsuda M, Sekine T, Takeda M, Cha SH, Kanai Y, Kimura M, Endou H (1999) Transport of ochratoxin A by renal multispecific organic anion transporter 1. J Pharmacol Exp Ther 289:1301–1305

Tune BM (1997) Nephrotoxicity of beta-lactam antibiotics: mechanisms and strategies for prevention. Pediatr Nephrol 11:768–772

Tune BM, Burg MB, Patlak CS (1969) Characteristics of p-aminohippurate transport in proximal renal tubules. Am J Physiol 217:1057–1063

Turner RJ (1984) Sodium-dependent sulfate transport in renal outer cortical brush border membrane vesicles. Am J Physiol 247:F793–F798

Uchino H, Tamai I, Yamashita K, Minemoto Y, Sai Y, Yabuuchi H, Miyamoto K, Takeda E, Tsuji A (2000) p-Aminohippuric acid transport at renal apical membrane mediated by human inorganic phosphate transporter NPT1. Biochem Biophys Res Commun 270:254–259

Ullrich KJ (1994) Specificity of transporters for 'organic anions' and 'organic cations' in the kidney. Biochim Biophys Acta 1197:45–62

Ullrich KJ (1997) Renal transporters for organic anions and organic cations. Structural requirements for substrates. J Membr Biol 158:95–107

Ullrich KJ (1998) Features of substrates for interaction with renal transporters of organic anions and cations. Nova Acta Leopoldina 78:23–34

Ullrich KJ (1999) Affinity of drugs to the different renal transporters for organic anions and organic cations. In: Amidon GL, Sadée W (eds) Membrane transporters as drug targets. Kluwer Academic/Plenum Publishers, New York, pp 159–179

Ullrich KJ, Rumrich G (1988) Contraluminal transport systems in the proximal renal tubule involved in secretion of organic anions. Am J Physiol Renal Physiol 254:F453–F462

Ullrich KJ, Rumrich G (1990) Contraluminal renal anion and cation transport systems: interaction with fatty acids, eicosanoids, Krebs cycle intermediates, amino acids and analogues, cyclic nucleotides and steroid hormones. In: Berliner RW (ed) The frontiers of nephrology. Elsevier Science Publishers B.V., Amsterdam, pp 55–65

Ullrich KJ, Rumrich G, Klöss S (1980) Active sulfate reabsorption in the proximal convolution of the rat kidney: specificity, Na+ and HCO3– dependence. Pflügers Arch 383:159–163

Ullrich KJ, Fasold H, Rumrich G, Klöss S (1984a) Secretion and contraluminal uptake of dicarboxylic acids in the proximal convolution of rat kidney. Pflügers Arch 400:241–249

Ullrich KJ, Rumrich G, Klöss S (1984b) Contraluminal sulfate transport in the proximal tubule of the rat kidney. 1. Kinetics, effects of K+, Na+, Ca2+, H+, and anions. Pflügers Arch 402:264–271

Ullrich KJ, Rumrich G, Klöss S (1985a) Contraluminal sulfate transport in the proximal tubule of the rat kidney. II. Specificity: sulfate-esters, sulfonates and amino sulfonates. Pflügers Arch 404:293–299

Ullrich KJ, Rumrich G, Klöss S (1985b) Contraluminal sulfate transport in the proximal tubule of the rat kidney. III. Specificity: disulfonates, di- and tri-carboxylates and sulfocarboxylates. Pflügers Arch 404:300–306

Ullrich KJ, Rumrich G, Klöss S (1985c) Contraluminal sulfate transport in the proximal tubule of the rat kidney. IV. Specificity: salicylate analogs. Pflügers Arch 404:307–310

Ullrich KJ, Rumrich G, Klöss S, Lang H-J (1985d) Contraluminal sulfate transport in the proximal tubule of the rat kidney. V. Specificity: phenolphthaleins, sulfonphthaleins, and other sulfo dyes, sulfamoyl-compounds and diphenylamine 2 carboxylates. Pflügers Arch 404:311–318

Ullrich KJ, Rumrich G, Fritzsch G, Klöss S (1987a) Contraluminal para-aminohippurate (PAH) transport in the proximal tubule of the rat kidney. I. Kinetics, influence of cations, anions, and capillary preperfusion. Pflügers Arch 409:229–235

Ullrich KJ, Rumrich G, Fritzsch G, Klöss S (1987b) Contraluminal para-aminohippurate (PAH) transport in the proximal tubule of the rat kidney. II. Specificity: aliphatic dicarboxylic acids. Pflügers Arch 408:38–45

Ullrich KJ, Rumrich G, Klöss S (1987c) Contraluminal para-aminohippurate transport in the proximal tubule of the rat kidney. III. Specificity: monocarboxylic acids. Pflügers Arch 409:547–554

Ullrich KJ, Rumrich G, Klöss S (1988) Contraluminal para-aminohippurate (PAH) transport in the proximal tubule of the rat kidney. IV. Specificity: mono- and polysubstituted benzene analogs. Pflügers Arch 413:134–146

Ullrich KJ, Rumrich G, Klöss S (1989a) Contraluminal organic anion and cation transport in the proximal renal tubule: V. Interaction with sulfamoyl- and phenoxy diuretics, and with β-lactam antibiotics. Kidney Int 36:78–88

Ullrich KJ, Rumrich G, Wieland T, Dekant W (1989b) Contraluminal para-aminohippurate (PAH) transport in the proximal tubule of the rat kidney. VI. Specificity: amino acids, their N-methyl-, N-acetyl- and N-benzoyl derivatives, glutathione- and cysteine conjugates, di- and oligopeptides. Pflügers Arch 415:342–350

Ullrich KJ, Rumrich G, Gemborys M, Dekant W (1990) Transformation and transport: how does metabolic transformation change the affinity of substrates for the renal contraluminal anion and cation transporters? Toxicol Lett 53:19–27

Ullrich KJ, Rumrich G, Gemborys MW, Dekant W (1991a) Renal transport and nephrotoxicity. In: Bach PH, Gregg NJ, Wilks MF, Delacruz L (eds) Nephrotoxicity. Marcel Dekker, New York, pp 1–8

Ullrich KJ, Rumrich G, Papavassiliou F, Klöss S, Fritzsch G (1991b) Contraluminal p-aminohippurate transport in the proximal tubule of the rat kidney. VII. Specificity: cyclic nucleotides, eicosanoids. Pflügers Arch 418:360–370

Ullrich KJ, Rumrich G, Papavassiliou F, Hierholzer K (1991c) Contraluminal p-aminohippurate transport in the proximal tubule of the rat kidney. VIII. Transport of corticosteroids. Pflügers Arch 418:371–382

Ullrich KJ, Rumrich G, David C, Fritzsch G (1993a) Bisubstrates: substances that interact with renal contraluminal organic anion and organic cation transport systems. I. Amines, piperidines, piperazines, azepines, pyridines, quinolines, imidazoles, thiazoles, guanidines and hydrazines. Pflügers Arch 425:280–299

Ullrich KJ, Rumrich G, David C, Fritzsch G (1993b) Bisubstrates: substances that interact with both renal contraluminal organic anion and organic cation transport systems. II. Zwitterionic substrates: dipeptides, cephalosporins, quinolone-carboxylate gyrase inhibitors and phosphamide thiazine carboxylates. Nonionizable substrates: steroid hormones and cyclophosphamides. Pflügers Arch 425:300–312

Ullrich KJ, Rumrich G, David C, Fritzsch G (1993c) Interaction of xenobiotics with organic anion and cation transport systems in renal proximal tubule cells. In: Anders MW, Dekant W, Henschler D, Oberleithner H, Silbernagl S (eds) Renal disposition and nephrotoxicity of xenobiotics. Academic Press, San Diego, pp 97–115

Ullrich KJ, Fritzsch G, Rumrich G, David C (1994) Polysubstrates: substances that interact with renal contraluminal PAH, sulfate, and NMeN transport: sulfamoyl-, sulfonylurea-, thiazide- and benzenaminocarboxylate (nicotinate) compounds. J Pharmacol Exp Ther 269:684–692

Ullrich KJ, Rumrich G, Burke TR, Shirazi-Beechey SP, Lang HJ (1997) Interaction of alkyl/arylphosphonates, phosphonocarboxylates and diphosphonates with different anion transport systems in the proximal renal tubule. J Pharmacol Exp Ther 283:1223–1229

Uwai Y, Okuda M, Takami K, Hashimoto Y, Inui K-I (1998) Functional characterization of the rat multispecific organic anion transporter OAT1 mediating basolateral uptake of anionic drugs in the kidney. FEBS Lett 438:321–324

Uwai Y, Saito H, Hashimoto Y, Inui K-I (2000) Interaction and transport of thiazide diuretics, loop diuretics, and acetazolamide via rat renal organic anion transporter rOAT1. J Pharmacol Exp Ther 295:261–265

Van Aubel RAMH, Peters JGP, Masereeuw R, van Os CH, Russel FGM (2000) Multidrug resistance protein Mrp2 mediates ATP-dependent transport of classical renal organic anion p-aminohippurate. Am J Physiol Renal Physiol 279:F713–F717

Wada S, Tsuda M, Sekine T, Cha SH, Kimura M, Kanai Y, Endou H (2000) Rat multispecific organic anion transporter 1 (rOAT1) transports zidovudine, acyclovir, and other antiviral nucleoside analogs. J Pharmacol Exp Ther 294:844–849

Walker RJ (2000) Cellular mechanisms of drug nephrotoxicity. In: Seldin DW, Giebisch G (eds) The kidney. Physiology and pathophysiology. Lippincott Williams & Wilkins, Philadelphia, pp 2835–2860

Wang H, Fei Y-J, Kekuda R, Yang-Feng TL, Devoe LD, Leibach FH, Prasad PD, Ganapathy V (2000) Structure, function, and genomic organization of human Na+-dependent high-affinity dicarboxylate transporter. Am J Physiol Cell Physiol 278:C1019–C1030

Wolff NA, Werner A, Burkhardt S, Burckhardt G (1997) Expression cloning and characterization of a renal organic anion transporter from winter flounder. FEBS Lett 417:287–291

Wolff NA, Grünwald B, Friedrich B, Lang F, Godehardt S, Burckhardt G (2001) Mutational analysis of fROAT, the flounder renal organic anion transporter. J Am Soc Nephrol 12:2012–2018

Wolff NA, Thies K, Friedrich B, Lang F, Reid G, Burckhardt G (2002) PKC-regulation of hOAT1 leads to internalization of the carrier and is independent of several conserved OAT1 PKC consensus sites. Pflügers Arch 443:S357

Woodhall PB, Tisher CC, Simonton CA (1978) Relationship between para-aminohippurate secretion and cellular morphology in rabbit proximal tubules. J Clin Invest 61:1320–1329

Wright EM, Wright SH, Hirayama B, Kippen I (1982) Interactions between lithium and renal transport of Krebs cycle intermediates. Proc Natl Acad Sci U S A 79:7514–7517

Wright SH, Kippen I, Klinenberg JR, Wright EM (1980) Specificity of the transport system for tricarboxylic acid cycle intermediates in renal brush border. J Membr Biol 57:73–82

Wright SH, Kippen I, Wright EM (1982) Stoichiometry of Na+-succinate cotransport in renal brush border membranes. J Biol Chem 157:1773–1778

Yarchoan R, Mitsuya H, Myers CE, Broder S (1989) Clinical pharmacology of 3'-azido-2',3'-dideoxythymidine (zidovudine) and related dideoxynucleosides. N Eng J Med 321:726–738

You G, Kuze K, Kohanski RA, Amsler K, Henderson S (2000) Regulation of mOAT-mediated organic anion transport by ocadaic acid and protein kinase C in LLC-PK1 cells. J Biol Chem 275:10278–10284

Zalups RK (1993) Influence of 2,3-dimercaptopropane-1-sulfonate (DMPS) and meso-2,3-dimercaptosuccinic acid (DMSA) on the renal disposition of mercury in normal and uninephrectomized rats exposed to inorganic mercury. J Pharmacol Exp Ther 267:791–800

Zalups RK (2000) Molecular interactions with mercury in the kidney. Pharmacol Rev 52:113–143

Zhang FF, Pajor AM (2001) Topology of the Na+/dicarboxylate cotransporter: the N-terminus and hydrophilic loop 4 are located intracellularly. Biochim Biophys Acta 1511:80–89

Rev Physiol Biochem Pharmacol (2003) 146:159–216
DOI 10.1007/s10254-002-0004-7

M. Flück · H. Hoppeler

Molecular basis of skeletal muscle plasticity-from gene to form and function

Published online: 14 January 2003

Abstract Skeletal muscle shows an enormous plasticity to adapt to stimuli such as contractile activity (endurance exercise, electrical stimulation, denervation), loading conditions (resistance training, microgravity), substrate supply (nutritional interventions) or environmental factors (hypoxia). The presented data show that adaptive structural events occur in both muscle fibres (myofibrils, mitochondria) and associated structures (motoneurons and capillaries). Functional adaptations appear to involve alterations in regulatory mechanisms (neuronal, endocrine and intracellular signalling), contractile properties and metabolic capacities. With the appropriate molecular techniques it has been demonstrated over the past 10 years that rapid changes in skeletal muscle mRNA expression occur with exercise in human and rodent species. Recently, gene expression profiling analysis has demonstrated that transcriptional adaptations in skeletal muscle due to changes in loading involve a broad range of genes and that mRNA changes often run parallel for genes in the same functional categories. These changes can be matched to the structural/functional adaptations known to occur with corresponding stimuli. Several signalling pathways involving cytoplasmic protein kinases and nuclear-encoded transcription factors are recognized as potential master regulators that transduce physiological stress into transcriptional adaptations of batteries of metabolic and contractile genes. Nuclear reprogramming is recognized as an important event in muscle plasticity and may be related to the adaptations in the myosin type, protein turnover, and the cytoplasma-to-myonucleus ratio. The accessibility of muscle tissue to biopsies in conjunction with the advent of high-throughput gene expression analysis technology points to skeletal muscle plasticity as a particularly useful paradigm for studying gene regulatory phenomena in humans.

Abbreviations

Stimuli: *CLFS* Chronic low-frequency electric stimulation · *CR* Caloric restriction · *DE* Detraining · *DV* Denervation · *EE* Endurance exercise · *ER* Endurance runners · *I* Immobilization · *IN* Inactivity · *MG* Real or simulated microgravity · *RT* Resistance training · *WL* Weightlifters

M. Flück (✉) · H. Hoppeler
Institute of Anatomy, University of Bern, Bühlstrasse 26, 3000 Bern 9, Switzerland
e-mail: flueck@ana.unibe.ch · Tel.: +31-631-4637 · Fax: +31-631-3807

Structure, function: *βOX* β-Oxidation · *DEL* Deltoidus · *EC coupling* Excitation-contraction coupling · *EDL* Extensor digitorum longus · *Gls* Glycolysis · *H+* Reducing equivalents · *IMF mitochondria* Interfibrillar mitochondria · *IMCL* Intra-myocellular lipid · *KC* Krebs cycle · *M* Muscle · *NMJ* Neuromuscular junction · *SR* Sarcoplasmic reticulum · *S mitochondria* Subsarcolemmal mitochondria · *Tn* Troponin · *VL* Vastus lateralis · *VO2max* Maximal oxygen consumption

Signals, sensors and transducers: *ACTH* Corticotropin · *AMPK* 5'-AMP-activated protein kinase · *ATP* Adenosine 5'-triphosphate · *Ca2+* Intracellular calcium · *CaMKII* Ca^{2+}/CaM kinase II · *Cor* Cortisol · *EN* Epinephrine · *ERK* Extracellular signal-regulated kinase · *GH* Growth hormone · *IGFBP-3* Insulin-like growth factor binding protein 3 · *IGF-I* Insulin-like growth factor I · *JNK* c-jun N-terminal kinase · *c-jun* cellular counterpart of retroviral insert from avian sarcoma virus 17 · *Ins* Insulin · *lep* Leptin · *MAPK* Mitogen-activated (microtubule-associated) protein kinase · *NRF-1 and -2* Nuclear respiratory factor 1 and 2 · *p38* p38 MAPK · *RE* Renin · *ROS* Reactive oxygen species · *T3* Triiodothyronine · *T4* Tetraiodothyronine · *Tes* Testosterone · *TFAM* Mitochondrial transcription factor · *TSH* Thyroid stimulating hormone

Genes: *3'UTR* 3' Untranslated region · *5'UTR* 5' Untranslated region · *ACAA2* 3-ketoacyl-CoA thiolase · *ACC* Acetyl-CoA carboxylase · *AIF1* Allograft inflammatory factor 1 · *AOX1* Aldehyde oxidase · *ATP5A* ATP synthase alpha chain · *ATP5C* ATP synthase gamma chain · *ATP5G3* ATP synthase lipid-binding protein · *ATP5J* ATP synthase coupling factor 6 · *ATPB* ATP synthase beta chain · *Cat H* Cathepsin H · *CDC16Hs* Cell division cycle 16 · *c-fos* c-fos Proto-oncogene · *c-jun* c-jun Proto-oncogene · *CK* Creatine kinase · *Col* Collagen type · *CPT I and II* Carnitine *O*-palmitoyltransferases I and II · *COX* Cytochrome C oxidase subunits · *CSF-1* Colony stimulating factor · *CXCL5* Small inducible cytokine B5 · *CYP2A6* Cytochrome P450 2A6 · *CYP2B6* Cytochrome P450 2B6 · *CYP2C8* Cytochrome P450 2C8 · *CYPdb1* Cytochrome P450 db1 · *DDO* D-aspartate oxidase · *DIA1* NADH-cytochrome b5 reductase · *DNA* Deoxyribonucleic acid · *DP5* Neuronal death protein · *EGFR* Epidermal growth factor receptor · *ERF* Ets2 repressor factor · *FAP* Fibroblast activation protein · *FAT/CD36* Fatty acid translocase · *FBP2* D-fructose-1,6-bisphosphate 1-phosphohydrolase · *FN* Fibronectin · *Fum* Fumarase · *gamma 1* Interferon gamma treatment inducible mRNA · *glut-1* Glucose transporter 1 · *HIAP1* Inhibitor of apoptosis protein 1 · *HIF-1α* Hypoxia-inducible factor 1 alpha · *HPARG* Poly(ADP-ribose) glycohydrolase · *HPXEL* Peroxisomal enoyl-CoA hydratase-like protein · *HSC* Heat shock cognate · *HSP* Heat shock protein · *HSP27 and 70* Heat shock protein 27 and 70 · *HSP2* Heparan sulfate proteoglycan 2 · *IL-1, -12 and -18* Interleukin 1, 12 and 18 · *IMPDH1* Inosine-5'-monophosphate dehydrogenase 1 · *LDH* lactate dehydrogenase · *LIF* Leukaemia inhibitory factor precursor · *LN* Laminin · *LPL* Lipoprotein lipase · *LRP1* Low-density lipoprotein receptor-related protein 1 · *LTC4S* Human leucotriene C4 synthase · *Mac-2* Macrophage subpopulation-specific antigen 2 · *MARK3* MAP/microtubule affinity-regulating kinase 3 long · *MAT1* CDK-activating kinase assembly factor · *MCAD* Medium chain-specific acyl-CoA dehydrogenase · *MCT1, 2 and 4* Monocarboxylate transporters 1, 2 and 4 · *MHC* Myosin heavy chain · *MHC10* Nonmuscle type B myosin heavy chain (MYH10) · *MHC9* Nonmuscle type A myosin heavy chain (MYH9) · *mRNA* Messenger ribonucleic acid · *MSH3* DNA mismatch repair protein · *MT1B* Metallothionein-IB · *MT1F* Metallothionein I F · *MEF2* Myocyte enhancer factor 2 · *myoD* Myoblast determination protein · *MRF4* Muscle regulatory factor 4 · *myf 5 and 6* Myogenic factors 5 and 6 · *nAChR* Nicotinic acetyl choline receptor · *NADH6* Mitochondrially-encoded NADH dehydrogenase subunit · *NF-kappa B* Nuclear factor kappa B · *NCAM1* Neural cell adhesion molecule 1 · *NDUFV1* NADH ubiquinone oxidoreductase 51 kDa subunit · *NDUFV2* NADH-ubiquinone oxidoreductase 24 kDa subunit complex core protein 2 · *NF-kB p65* Nuclear factor kappa B p65 subunit · *NF-Y* Nuclear factor Y protein subunit A · *NNT* Mitochondrial NAD(P) transhydrogenase · *NRG-1* Neuregulin · *ORP150* 150 kDa Oxygen-regulated protein · *p21* Cyclin-dependent kinase inhibitor 1 · *PON2* Paraoxonase 2 · *PDK4* Pyruvate dehydrogenase kinase 4 · *PP2Ag1* Serine/threonine protein phosphatase 2A · *PPAR* Peroxisome proliferator-activated receptor · *RANTES* Regulated upon activation, normal T cell expressed and secreted · *RECQ2* RecQ-like type 2 DNA helicase · *RNA* Ribonucleic acid · *RYK* Related to receptor tyrosine kinase · *SCHAD* Short chain 3-hydroxyacyl-CoA dehydrogenase · *SDH* Succinate dehydrogenase · *SH3GL3* SH3-containing GRB2-like protein 3 · *SOD2* Manganese superoxide dismutase · *SVCT2* Sodium-dependent vitamin C transporter · *tie 2* Angiopoietin 1 receptor · *TRAP230* Thyroid hormone receptor-associated protein complex component · *TSP4* Thrombospondin 4 · *UCP* Mitochondrial uncoupling protein · *UQCRB* Ubiquinol-cytochrome C reductase · *UQCRC2* Ubiquinol-cytochrome C reductase · *UQCRH* Ubiquinol-cytochrome C reductase complex 11 kDa protein · *VEGF* Vascular endothelial growth factor

Introduction

Skeletal muscle structure and function can be changed as a consequence of stimuli that modify skeletal muscle contractile activity (endurance exercise, electrical stimulation, denervation), load on the muscle (resistance training, microgravity), substrate supply (nutritional interventions) or environmental factors such as hypoxia and thermal stress. Moreover, skeletal muscle tissue undergoes gradual changes with ageing. This malleability is an important feature of skeletal muscle tissue and allows for an economic design of muscle tissue sufficient for given conditions at a certain time but adaptable to changes imposed by muscle use and the environment (Hoppeler and Flück 2002). The phenomenon of plasticity is observed in all vertebrate species; however, a large variability in the responsiveness and degree of muscle adaptations is observed among species, individuals of a species and among different functional systems of muscle tissue such as the contractile apparatus, energy conversion and excitation-contraction coupling.

In the context of muscle malleability, the fibre and its environment appear to conspire. Recent data from rodent models indicate that skeletal muscle plasticity involves an integrated response of fibre (i.e. cytoarchitecture and composition; Chopard et al. 2001; Frenette and Tidball 1998) and fibre-associated structures (i.e. nerves, capillaries; Tyml and Mathieu-Costello 2001; Deschenes et al. 2001). Transcriptional reprogramming of nuclei is recognized to be a major event early on in this process (Wittwer et al. 2002a; Carson et al. 2002). The involvement of signalling in the sensation and transduction of stimuli into activation of specific gene expression events is widely recognized (Sakamoto and Goodyear 2002; Martineau and Gardiner 2002; Nader and Esser 2001). The present review gives an overview of the molecular basis of structure-function adaptations of skeletal muscle in rodents and humans. The current state of knowledge is summarized in a model describing how physiological stimuli are sensed at the molecular level and translated into the changes that modify skeletal muscle phenotype. Due to the rapid progress in this field, this article cannot be exhaustive but reflects the bias of the authors of what they perceive to be major factors involved in skeletal muscle plasticity. The reader is also referred to previous articles with a similar focus (Booth and Thomason 1991; Hood 2001; Baldwin and Haddad 2001; Booth and Baldwin 1995; Howald 1982).

Concepts

Muscle function

Contraction of mammalian skeletal muscle is under neuronal control. Triggering of an action potential at the neuromuscular junctions through the activity of an efferent α-motor neuron causes the synchronous contraction of the muscle fibres constituting the motor unit. The contraction process in the muscle fibre is then initiated by the release of Ca^{2+} from intracellular stores and provoking ATP-dependent interaction changes of myosin and actin in the contractile apparatus (Huxley 1988). In addition to the α-neurons responsible for the control of contractile activity, muscle contains thin myelinated (A delta or group III) and unmyelinated (C or group IV) afferents receptive to chemical, mechanical, and thermal stimuli, playing a role in the circulatory and respiratory adjustments during exercise (Kniffeki et al. 1981; Christensen and Galbo 1983).

Specific molecular diversities of the main proteins in the Ca^{2+}-dependent proteins involved in the EC coupling process largely determine the contractile properties of muscle fibres (Berchtold et al. 2000). Changes in the expression of slow (type I) and fast (type IIa, IIx and IIb) myosin heavy- and light-chain components of the sarcomere, and changes in the firing frequency of the innervating nerve are both known to contribute to differences in contraction speed of single fibres (Pette and Staron 2000; Schiaffino and Reggiani 1994, 1996).

Energy demand of skeletal muscle can be increased by two orders of magnitude by contractile activity (Hargreaves 2000; Booth and Thomason 1991). At the onset of exercise, the creatine and arginine ATP stores are self-sufficient for a few seconds. Subsequently, ATP production by glycolysis takes over. Glycolysis yields ATP at high rates but delivers the by-product lactate, which is exported into the blood stream and into adjacent muscle fibres and may hinder muscle performance (Brooks and Mercier 1994; Brooks 2000). The ATP for continued muscle work is regenerated from ADP through oxidative phosphorylation of reduction equivalents arising from the combustion of carbohydrates and fatty acids (Fig. 1). Glycolysis is an order of magnitude less efficient in terms of ATP yield per mole of glucose than the mitochondrial oxidation of carbohydrates or fatty acids (Hargreaves 2000; Weibel 2002b). During aerobic work, the balance between carbohydrate and fat metabolism is controlled by the intensity of exercise and the supply of nutrients from intra- (IMCL, glycogen) and extracellular stores (in liver and adipose tissue). Extracellular substrates need to be delivered through the capillary system whereby it is recognized that the design of the capillary network is determined by the conditions for oxygen transfer to muscle cells, as there are practically no stores of oxygen in muscle cells (Hoppeler and Weibel 2000 see Fig. 1). In general terms, combustion of carbohydrates is favoured at high-intensity exercise (over 50% of VO_2max), whereas the proportion of energy production through β-oxidation is larger during prolonged exercise and at moderate intensities (below 50% of VO_2max; Brooks and Mercier 1994).

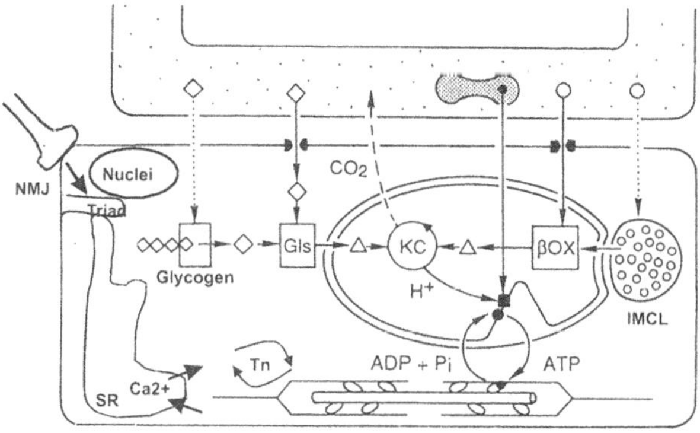

Fig. 1 Substrate pathways and muscle contraction. Model for structure-function relationships of oxygen and intracellular substrate supply to the mitochondria of skeletal muscle cells and control of muscle contraction by nerve activity. The generation of an action potential at the NMJ causes release of Ca^{2+} from the sarcoplasmic reticulum into the cytoplasmic space. Through binding to troponin, this provokes the modulation of the actin-myosin interaction, thereby shortening the sarcomere. This so-called cycle of excitation-contraction coupling is reversible and ATP-dependent. The ATP equivalents necessary for muscle contraction are mainly derived from metabolic conversion of carbohydrates and fatty acids. *Open circles* indicate fatty acids, *diamonds* indicate glucose; a *row of diamonds* indicates polymerized glycogen and *triangles* indicate acetyl-CoA. *Arrows* show the pathways of intracellular substrate breakdown from the intracellular stores to the terminal oxidase in the mitochondrial inner membrane (*black square*). Other *arrows* indicate the supply routes of oxygen (*black dot*) and substrates from the capillaries, with *dotted arrows* for the supply route to intracellular stores, temporally split from the phase of oxidation. Paired *semicircles* (*in black*) indicate transport systems facilitating glucose and fatty acid uptake. *Dashed arrow* indicates CO_2 discharge to the blood. For full-length names, the reader is referred to the list of abbreviations

The molecular composition of the sarcomere, its metabolic make-up, capillary supply and the mode of motor unit recruitment all have an influence on the functional muscle phenotype by determining the key parameters contractility and resistance to fatigue. In addition, we must recognize that muscle tissue, by virtue of its size, making up as much as 50% of body mass in some athletic species, is a key metabolic organ importantly involved in conserving the energy balance of the individual (Booth and Thomason 1991; Hoppeler and Flück 2002).

Molecular mechanisms of muscle malleability

During development, skeletal muscles are formed by fusion of mesodermal precursor cells to primary myotubes. A population of muscle precursor cells stays residual under the basal lamina (satellite cells) and can respond to various stimuli. Satellite cells may divide and proliferate, thus forming new muscle fibres (Schultz and McCormick 1994). Alternatively, they can fuse with existing muscle fibres, thereby adding to the pool of nuclei when increased need for transcriptional activity arises (Schiaffino and Reggiani 1994). The nuclear domain limitation theory predicts that nuclei can provide mRNA for only a limited myocellular volume (the nuclear domain; Allen et al. 1999). As skeletal muscle nuclei provide the necessary mRNA coding for ribosomal, structural and metabolic proteins, it was suggested that the cytoplasma-to-myonucleus ratio determines the capacity of a muscle to support protein turnover rates as well as increased synthesis of perturbed protein classes (reviewed in Booth and Baldwin 1995). The number of myonuclei and thus the size of the nuclear domains appear to play a critical role in establishing the window of adaptive potential of skeletal muscle to alter its phenotype. In these events, the recruitable satellite cell population seems to be of importance. Last but not least, selective apoptotic death of myonuclei has been proposed to occur during muscle atrophy as a mechanism for removing parts of myofibres without affecting their viability (Sandri 2002).

On a structural level, skeletal muscle plasticity involves modifications of cellular (mitochondria, myofibrils, etc.) and extracellular compartments (capillaries, nerves, connective tissue). Modulation of the proteic and to a lesser degree also of the lipidic make-up of muscle tissue has been shown to be the molecular components of muscle malleability (Booth and Baldwin 1995; Hoppeler and Weibel 1998; Bruhn et al. 1991). In particular, changes in expression of myofibrillar and metabolic proteins have often been demonstrated to be involved in skeletal muscle plasticity (Schiaffino and Reggiani 1996; Booth and Thomason 1991).

The change of protein content and of protein isoforms towards a new steady-state during an adaptive event can potentially be controlled by modifications in many steps from DNA to the assembled translation products (Fig. 2). Altered pretranslational, translational and post-translational events are all involved in the molecular regulation of the skeletal muscle phenotype. Three decades ago, solid evidence for the essential role of de novo RNA synthesis (transcription) in loading induced skeletal muscle hypertrophy was presented (Goldberg and Goodman 1969; Sobel and Kaufman 1970). These investigators showed the activity of RNA polymerase to be increased with hypertrophy. Treatment with actinomycin D, an inhibitor of DNA-dependent RNA synthesis was shown to prevent skeletal muscle hypertrophy. More recently, rapid transcriptional regulation of gene expression in human skeletal muscle during recovery from exercise has been demonstrated by nuclear run-ons (Pilegaard et al. 2000). Endurance exercise training has been shown to increase the steady-state levels of certain metabolic mRNAs in human skeletal muscle (Vogt et al. 2001; Pilegaard et al. 2000; Puntschart et al. 1995a).

The level of mRNA is determined by the rate of mRNA synthesis and the rate of mRNA decay (i.e. its chemical half-life). It is well established that the half-life time of many mRNAs can fluctuate in response to environmental stimuli such as nutrient levels, cytokines, hormones, temperature shifts and viral infections (Day and Tuite 1998). So far, only indirect evidence has been provided for control of RNA stability in muscle with exercise, by demonstrating that contractile activity modulates the level of factors interacting with a region of the 3'UTR of the cytochrome C transcript known to determine mRNA degradation (Yan et al. 1996). The observations indicate that enhanced RNA stabilization may contribute to the enhanced mRNA level and thus ultimately to the increased mitochondrial densities seen in human skeletal muscle with endurance exercise.

As expected, in general we find that cellular and molecular adaptations in human and animal models for muscle plasticity are congruent when physiologically equivalent stimuli are compared. For example, a single bout of resistance training causes similar increases in protein synthesis in humans and in rats (Phillips et al. 1999; Hernandez et al. 2000). Likewise an acute bout of endurance training causes a similar increase in VEGF expression, with a similar time course in humans and rats (Richardson et al., 1999; Breen et al. 1996). However, interspecies congruency of adaptive events is by no means absolute. Involvement of the same biological processes, but to different extents, is noted in some skeletal muscle adaptations. Differences between human and rat models for load-induced muscle hypertrophy include the extent as well as the kinetics of cellular events such as the change

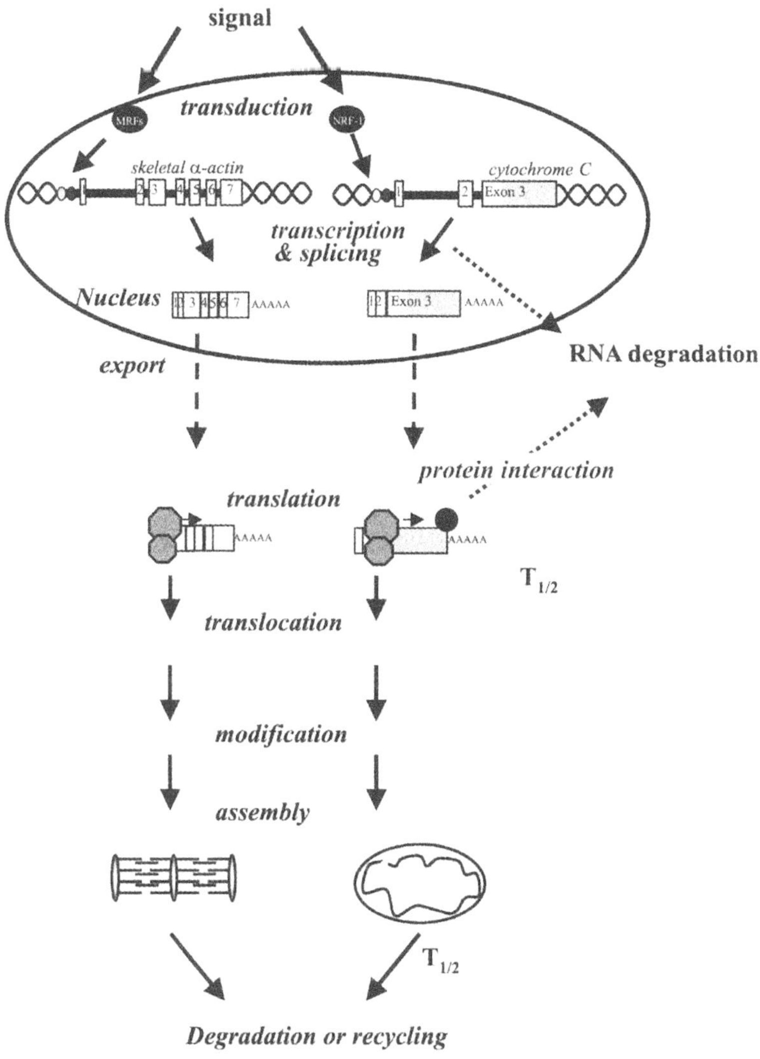

Fig. 2 Control of gene expression. Various extracellular signals initiate transcriptional, post-transcriptional, translational and post-translational responses. This influences the rate of synthesis and degradation and ultimately determines the steady-state of each expressed gene and protein. Expression steps indicated in *italics* are demonstrated to be involved in the molecular mechanism of muscle plasticity

in fibre diameter and the activation of satellite cells (Kadi and Thornell 2000; Snow 1990). These differences in response were tentatively explained by the higher relative severity of the stimuli in the rat versus the human model (Booth and Thomason 1991). Alternatively, we have to consider that observed differences in malleability and the time course of adaptations may simply reflect allometric scaling, i.e. be related to the 3/4 power relationship between basal metabolic rate and body mass (Weibel 2002a).

A survey of the molecular responses of skeletal muscle to changes in contractile activity demonstrates that as a general rule the directional change of a given mRNA is in the same direction as the directional change of the protein during its adaptation (Booth and Baldwin 1995). When RNA and translation products do not change in the same direction, in some specific cases this may be explained by high turnover or high basal concentrations of the protein (Andersen and Schiaffino 1997). An increased level of mRNA is assumed to be translated into protein and to cause a micro-adaptation in protein concentration (Day and Tuite 1998). In the absence of a proportional increase in degradation, this would increase the total amount of protein. However, for highly abundant proteins or in the presence of increased protein degradation, increases in mRNA content may not be detected as net changes in protein concentration (Fig. 2). The scenario for increased protein turnover is supported by the fact that 2 weeks of daily wheel running augmented skeletal alpha actin mRNA in fast- but not slow-twitch hindlimb muscle and increased fractional protein synthesis in slow- and fast-twitch hindlimb muscle in rats (Munoz et al. 1994; Morrison et al. 1989).

Post-translational modifications of proteins are a universal and efficient way to modulate the functional properties of enzymes and of pathways. Modifications of enzymes involved in metabolism and signalling have been reported as acute responses to several physiological stressors of skeletal muscle (Booth and Baldwin 1995). Moreover, modulation of translation efficiency has been noted to occur with muscle loading and increased contractile activity (reviewed in Booth and Baldwin 1995). Several of these adaptations are short-lived and are down-regulated by degradation or recycling of the modified protein or by removal of the post-translational modification (Bergamini 1992; Wilkinson 2000; Sorkin and Waters 1993).

The focus of this review will be on changes in muscle mRNA steady-state levels, as these can be taken as an index of changes in gene expression and currently appear to be the best described and understood molecular events that underlie muscle plasticity.

Adaptations of skeletal muscle to different stimuli

Contractile activity

The importance of contractile activity in determining the muscle phenotype is widely recognized. An important role in modulating phenotype has been assigned to the motoneuron (Gundersen 1998). It has been demonstrated that the extent and modality of neuronal activity is crucial for regulating contractile properties such as speed, strength and endurance. The signals that determine muscle phenotype are to a large extent coded in the pattern and frequency of electrical activity (Pette and Staron 2000). Large amounts of continuous low-frequency activity lead to a slowing of shortening velocity by influencing, among other things, myosin heavy-chain expression. Short bursts of high-frequency activity lead to a fast muscle phenotype (Pette 1998; Booth and Baldwin 1995).

Endurance exercise

Endurance exercise training has been the intervention of choice to explore the malleability of mitochondrial structure and function in humans. In a typical training experiment during which initially untrained subjects were trained intensely for 6 weeks (5 times/week for 30 min on a bicycle ergometer); total mitochondrial volume density was increased by 40% in

m. vastus lateralis (Hoppeler et al. 1985). The data are compatible with the contention that it is a relatively small fraction of the total muscle volume which is specifically trained with an exercise intervention and it is the mitochondria in these muscles that adapt to the stimulus and can be held responsible for the increase in VO_2max (Saltin 1986; Hoppeler et al. 1985).

Endurance exercise training increases the smaller volume fraction of subsarcolemmal mitochondria (SS mitochondria) more than the larger fraction of interfibrillar mitochondria (IMF mitochondria; Hoppeler et al. 1985). This was noted as early as 1973 (Hoppeler et al. 1973) and has since been confirmed in many studies using different separation techniques for mitochondria (Chilibeck et al. 1998; Bizeau et al. 1998; Roussel et al. 2000). Differences in mitochondrial functions such as protein import (Takahashi and Hood 1996) or protein synthesis and degradation (Connor et al. 2000) point to possible mechanisms by which the two mitochondrial subpopulations are capable of reacting differently to stress. In some studies, IMF mitochondria were shown to have higher respiration rates and higher respiratory control ratios than SS mitochondria (Lombardi et al. 2000). However, a consensus on the functional identity of SS and IMF mitochondria and the significance of the differential response to a variety of stimuli has not been reached.

Using a morphometric approach on individual histochemically typed muscle fibres Howald et al. (Howald et al. 1985) were able to demonstrate that bicycle endurance exercise training of 6 weeks duration (same study as Hoppeler et al. 1985) leads to an increase of the volume density of mitochondria in all three fibre types, the increase being larger in type IIa than in type I and type IIb/x fibres. This finding is surprising, as one would not expect type IIb/x fibres to be recruited in an endurance-type exercise such as bicycling at low intensity (approx. 20% of maximal short-term power output). Howald et al. (1985) also noted that there was a broad overlap of mitochondrial volume densities among all three fibre types before and after training. This begs the question as to the diversity of molecular phenomena, which might occur in different fibre types as a consequence of a functionally uniform intervention such as a classic endurance-type training intervention. The complex response of the individual muscle fibre types also calls for circumspection in the interpretation of molecular responses obtained on bulk extracts of muscle biopsy material. Moreover, there is evidence that aerobic exercise (4 h of exercise at 40% of VO_2max) stimulates protein turnover by increasing protein breakdown and protein synthesis in the recovery phase in human skeletal muscle (Carraro et al. 1990b). Moreover, the same type of exercise increased the fractional synthesis rate of fibronectin but not of albumin (Carraro et al. 1990b). This indicates that in response to exercise, there is also a stimulation of the synthesis of some acute-phase (extracellular matrix) proteins.

Endurance exercise training leads to a shift in skeletal muscle mitochondria towards an increased use of lipids as a substrate source (Holloszy and Booth 1976; Holloszy and Coyle 1984). We noted early (Hoppeler et al. 1973) that the intramyocellular lipid content (IMCL) in trained subjects was significantly larger than in controls. Likewise, 6 weeks of endurance exercise training approximately doubled IMCL (0.47 vs. 0.92%) of muscle fibre volume (Hoppeler et al. 1985; Howald et al. 1985; Table 1). It is generally accepted now that endurance exercise training leads to an increased use of lipids as a substrate both with regard to relative and absolute exercise intensities (van Loon et al. 2001) and augments IMCL content of muscle fibres. However, the mechanisms by which these phenomena are controlled remain to be established. Last but not least, recent observations demonstrating an increase in reduced glutathione after recurrent episodes of high-intensity exercise ses-

Table 1 Structural/functional adaptations of primate skeletal muscle to physiological stimuli (for the full-length names, the reader is referred to the list of abbreviations; blank field, to the best of our knowledge not reported; *s* seconds, *m* minutes, *h* hours, *d* days, *w* weeks, *mo* months, *y* years)

	Stimuli														
	High contractile activity-low loading					Low contractile activity-high loading			Metabolic intervention			Hypoxia		Ageing	
	EE	CLFS	DV	IN/DT	References	RT	MG	References	FA supp.	CR	References		References		References
STRUCTURAL															
Myofibrils															
Cross-sectional area															
Total muscle	–	↑mo-y	–	–	–	↑w	↓w-mo (VL)	1-7	–	–	–	↓mo	8	↓y	7
Total fibre	→y	↑mo-y	↓mo	↓w ↑w in ER	9-12	→w↑mo	↓w-mo (VL)	5, 13, 14	–	–	–	↓mo	8	↓y	7
Type I	→y	→w-mo	↓y	↑w in ER	9, 11, 12, 15-17	→w↑mo	↓mo (VL) →w-mo (DEL)	1, 5, 18-20	–	–	–	–	–	→y	16
Type IIa	→y	→w-mo	↓y	↓w in WL ↑w↓mo in ER	9, 11, 12, 15-17	→w↑mo	↓mo (VL) →w-mo (DEL)	1, 5, 18-20	–	–	–	–	–	↓y	16
Type IIx	→y	→w-mo	↓y	↓w	9, 11	→w↑mo	↓mo (VL) →w-mo (DEL)	1, 5, 18-20	–	–	–	–	–	↓y	16
Volume density															
Total fibre	–	–	–	–	–	→w↑mo	→w	2, 5, 21	–	–	22	↑mo	23	–	–
% I Fibres	↑mo-y	→mo-y		→d-w↓y	9, 12, 21, 24-30	→↓w-mo	→w-mo	1, 2, 25, 31	–	–	–	–	–	–	–
% IIa Fibres	↓y	↑mo-y	↑mo-y	→d-w	9, 10 24	↑w-y(↓mo)	→w-mo	1, 2, 19, 25, 31-33	–	–	–	–	–	↓y	34
% IIx Fibres	↓d	↓mo-y	↑mo-y	→d-w↑mo	9-12, 21, 24, 27	→↓w	→w-mo	1, 2, 19, 25, 31, 35	–	–	–	–	–	↓y	34
% Hybrid I/IIa fibres	↑y	–	–	↑mo-y	36	↓ mo	–	37	–	–	–	–	–	↑s	38
% Hybrid IIa/IIx fibres	↑y	–	–	↑mo-y	36	↓mo	–	25, 31, 37	–	–	–	–	–	↑s	38
Mitochondria (total)															
[Volume density]	↑w	–	–	–	39	↓w-y	↓mo (VL) →w-mo (DEL)	2, 13, 19	→w	–	40	↓mo	8	–	–
[Absolute volume]	–	–	–	–	–	→w	↓mo(VL)	2, 13, 19	–	–	–	–	–	–	–

Table 1 (continued)

	Stimuli														
	High contractile activity-low loading					Low contractile activity-high loading			Metabolic intervention			Hypoxia		Ageing	
	EE	CLFS	DV	IN/DT	References	RT	MG	References	FA supp.	CR	References		References		References
Muscle nuclei															
Myonuclei/fibre	–	–	–	–	–	↑mo-y	↓mo	33, 42, 43	–	–	–	–	–	–	–
Satellite cells	–	–	–	–	–	↑mo	–	33, 42	–	–	–	–	–	–	–
Capillarity															
Capillary-to-fibre ratio	↑w	→w-mo	↓y	↓w / type I fibre	11, 15, 16	→w	→w-mo	2, 5, 18	–	–	–	↓mo	8	↓y / IIa and IIx fibre	16
[Volume density]	↑w	–	–	–	11, 15, 43	→w	→w-mo	2, 5, 18	–	–	–	↑mo	8	–	–
Fuel stores															
Glycogen	↓m	–	–	–	44	–	–		–	–	–	–	–	–	–
IMCL (volume density)	↓h↑w-mo	–	–	–	27, 39	→w	→↑w	2, 5, 13, 18	↑w	–	40	→mo	8	–	–
Oxidative damage (myofibrils/lipids)	–	–	–	–	–	–	–	–	–	↓y	45	–	–	↑y	45
Motoneurons and -units	–	–	–	–	–	–	–	–	–	–	–	–	–	↓y	34, 46
FUNCTIONAL															
Regulation															
Neuronal activity	↑s-m	–	–	↓d-mo in WL	11, 47, 48	↓d↑w	↑d-w	7, 30, 49-51	–	–		–		–	
Endocrine	↑m GH, RE, EN ↓m Ins ↓h T3/T4, TSH, Ins	–	↓h Tes, GH, IGF-I	↑w GH, Tes ↓w cor in WL	11, 52-54	↑m GH, Tes, IGFBP-3 ↑w Tes ↓m lep ↓w cor	–	35, 55-58	–	–		(+EE) ↑m ACTH (+EE)↓m Ins	54	↓y Tes, IGF-I, IGFBP-3	57
Mechano	↑m-h JNK, ERK, p38 activity	–	–	–	59-63	–	–	–	–	–	–	–	–	–	–

Table 1 (continued)

	Stimuli														
	High contractile activity-low loading					Low contractile activity-high loading			Metabolic intervention			Hypoxia		Ageing	
	EE	CLFS	DV	IN/DT	References	RT	MG	References	FA supp.	CR	Refer-ences		Refer-ences		Refer-ences
Metabolic	↑m AMPKα2 activity	–	–	–	44, 64, 65	–	–	–	–	–	–	–	–	–	–
Contraction															
Maximal shortening velocity	–	↑w-mo	–	–	66	↑w	→↑w	3, 14, 67-71	–	–		-	–	↓y	7, 72
Muscle strength/ power	↓h→y	↑mo-y	–	→d-w↓mo in WL	10, 11, 73-75	↑w	↓w-mo	1, 3-6, 20, 35, 49, 76	–	–		–	–	↓y	7
Metabolism															
Myoglobin	→w	–	–	→w-mo	11, 77	–	–	–	–	–	↑y	78	–	–	
CHO metabolism	↑w	–	–	↑↓w	11, 79	↑w	→w	1, 2, 18	↓d	–	80, 81	(+EE)↑ m	82-84	↓y	85
FA oxidation (mito ox. capacity)	↑h-d-w	↑mo-y	–	↓w-mo	11, 15, 16, 65, 86, 87	–	↓w-mo (VL) →w-mo (DEL)	1-3, 18	→↑d	–	80, 81	(+EE)↓ m	83, 84	–	–
Krebs cycle (respiration)	↑w	↑w-mo	↓y	↓w-mo	11, 15, 88	↑w	→w	1-3, 18	–	–		–	–	↓y	85, 89
Energy transfer	–	–	–	→mo	11	↑w	–	1, 6	–	–	–	–	↓y	85	
Fatty acid synthesis	↑m phospho ACC	–	–	–	44, 90	–	–	–	–	–		–	–		–
Resistance to fatigue	↑w	↑mo-y	–	–	10, 24, 66, 91	↑w	–	76	↑w	–	92	–	–	↓y	9
VO_2max	↑w	↑mo-y	–	↓d-mo	9-11, 24, 39	→w	↓w-mo (VL)	2, 69	→w	–	92	–	–	–	
Synthesis rate															
Protein synthesis	↑h (mixed and FN)	–	–	–	93, 94	↑h-w (mixed and MHC)	–	7, 28, 95-99	–	–		–	–	↓y (mixed + MHC) →y (actin)	28
Protein degradation	–	–	–	–	–	↑h (mixed)	–	96, 98	–	–		–	–	–	–
RNA activity	–	–	–	–	–	↑h	–	97	–	–	–	–	–	–	

Table 1 (continued)
[1] Costill et al. 1979, [2] Ferretti et al. 1997, [3] Fitts et al. 2001, [4] Hespel et al. 2001, [5] Luethi et al. 1986, [6] VandenBorne et al. 1998, [7] Vandervoort 2002, [8] Hoppeler et al. 1990, [9] Ingjer 1979, [10] Mohr et al. 1997, [11] Mujika and Padilla 2001, [12] Round et al. 1993, [13] MacDougall et al. 1979, [14] Widrick et al. 1999, [15] Martin et al. 1992, [16] Proctor et al. 1995, [17] Schaub et al. 1989, [18] Desplanches et al. 1998, [19] Hikida et al. 2000, [20] Hortobagyi et al. 2000, [21] Lotta et al. 1991, [22] Hoppeler and Weibel 2000, [23] Hoppeler and Weibel 1998, [24] Andersen and Schiaffino 1997, [25] Andersen et al. 2000, [26] Burnham et al. 1997, [27] Howald et al. 1985, [28] Hasten et al. 2000, [29] Ricoy et al. 1998, [30] Sale 1988, [31] Andersen et al. 1994b, [32] Andersen et al. 1994a, [33] Kadi et al. 1999, [34] Carmeli and Reznick 1994, [35] Staron et al. 1994, [36] Klitgaard et al. 1990c, [37] Williamson et al. 2001, [38] Klitgaard et al. 1990a, [39] Hoppeler et al. 1985, [40] Hoppeler et al. 1999, [41] Kadi and Thornell 2000, [42] Ohira et al. 1999, [43] Zumstein et al. 1983, [44] Stephens et al. 2002, [45] Zainal et al. 2000, [46] Panenic and Gardiner 1998, [47] Ray and Gracey 1997, [48] Seals and Enoka 1989, [49] Connelly and Vandervoort 2000, [50] Ertl et al. 2002, [51] Hakkinen and Kauhanen 1989, [52] Bauman and Spungen 2000, [53] Hesse et al. 1989, [54] Kjaer et al. 1999, [55] Craig et al. 1989, [56] Hansen et al. 2001, [57] Kraemer et al. 1999, [58] Tuominen et al. 1997, [59] Aronson et al. 1997, [60] Aronson et al. 1998, [61] Boppart et al. 1999, [62] Wong and Booth 1990, [63] Yu et al. 2001, [64] Chen et al. 2000, [65] Tunstall et al. 2002b, [66] Hjeltnes et al. 1998, [67] Degens et al. 1999, [68] Larsson et al. 1996, [69] Paavolainen et al. 1999, [70] Trappe et al. 2000, [71] Widrick et al. 1997, [72] Larsson et al. 1997, [73] Bentley et al. 1998, [74] Harridge 1996, [75] Leveritt et al. 2000, [76] Lesmes et al. 1978, [77] Masuda et al. 2001, [78] Reynafarje 1962, [79] Linossier et al. 1997, [80] Carey et al. 2001, [81] Goedecke et al. 1999, [82] Green et al. 1992, [83] Parolin et al. 2000a, [84] Parolin et al. 2000b, [85] Pastoris et al. 2000, [86] Brooks and Mercier 1994, [87] Martin 1996, [88] Bengtsson et al. 2001, [89] Trounce et al. 1989, [90] Chen et al. 2000, [91] Harridge et al. 2002, [92] Hoppeler and Vogt 2001a, [93] Carraro et al. 1990a, [94] Carraro et al. 1990b, [95] Balagopal et al. 2001, [96] Biolo et al. 1995, [97] Chesley et al. 1992, [98] Phillips et al. 1997, [99] Yarasheki et al. 1993b

sions indicate that redox potential could be increased with endurance training (Rabinovich et al. 2001) and argues for expressional changes in redox metabolism.

The effects of different exercise training paradigms on myosin heavy-chain expression have recently been reviewed by Baldwin and Haddad (2001). World-class marathon runners and extreme endurance athletes have been shown to have a strong bias towards the expression of the slow type I MHC. As much as 80-90% of the MHC pool is composed of the slow type I MHC, with the remainder being type IIa MHC (Andersen et al. 2000; Table 1). In the context of these observations, it has been shown in humans that a short period of high-intensity endurance training induces a shift from fast MHC isoforms toward the slow variety within histochemically typed fibres of the m. vastus lateralis (Schaub et al. 1989). Furthermore, a report by O'Neill et al. (1999) demonstrated that 7 days of cycling exercise induces a significant down-regulation of fast IIx MHC mRNA (Table 2). Thus, although gifted athletes may have a genetic predisposition to excel in certain athletic events due to their inherent leg muscle MHC profiles, it appears that the MHC gene profile of an individual can be altered via chronic increases in contractile activity such as physical training. As indicated above, world-class marathon runners and ultra-endurance athletes are reported to have remarkably high type I fibre numbers in their trained muscle groups (Andersen et al. 2000; Ricoy et al. 1998), whereas muscles of sprinters and weightlifters predominantly consist of IIa/IIx fibres (Andersen et al. 1994b, 2000). It further appears that extreme usage (and disuse) induced alterations of the muscle MHC isoform profile result in hybrid fibres that express a combination of MHC isoforms such as type I/IIa and IIx/IIa (Klitgaard et al. 1990a). How much of these patterns of MHC gene expression is due to genetic predisposition and how much to the specificity of training (or some combination of these) is unresolved (reviewed in Baldwin and Haddad 2001).

Table 2 Expressional adaptations of primate skeletal muscle to physiological stimuli (for the full-length names, the reader is referred to the list of abbreviations; blank field, to the best of our knowledge not reported)

	Stimuli														
	High contractile activity-low loading					Low contractile activity-high loading			Metabolic intervention			Hypoxia		Ageing	
	EE	CLFS	DV	IN/ DV	Refer- ences	RT	MG/I	Refer- ences	FA supp.	CR	Refer- ences		Refer- ences		Refer- ences
FUNCTIONAL UNIT															
Contraction-structure	–	↑w MHCI	–	–	1	↑↓w-m MHCI	↓w MHCI	2-4	–	–	–	–	–	→y MHCI	2
Myosin heavy Chains	–	–	–	–	–	→↑w-m MHCIIa ↓m (aged)	→w MHCIIa	1, 3-5	–	–	–	–	–	↓y MHCIIa	2
	→m ↓d-w-mo MHCIIx	↓w MHCIIx	↑w-mo MHCIIx	↑w-mo MHCIIx	1, 4, 6-8	↓→↑w-m MHCIIx	↑w MHCIIx	1, 3, 4, 6	–	–	–	–	–	↓y MHCIIa	2
ECM and cytoskeleton	–	–	–	–	–	–	–		–	↑y Col I and III and VI and VIII, FN, HSP2, LN $\alpha 2$, desmin, tubulin, actin β, MHC9 and 10	9	–	–	↓y TSP4	9
Metabolism															
Carbohydrates															
Glycolysis	↑m-h HKII	–	–	–	10	–	–	–	–	–	–	–	–	–	–
Glucose stores	↑m-h glut-4, GYS ↑d GYS	–	–	–	10, 11	–	–	–	–	–	–	–	–	↓y glycogenin, glut-3, FBP2	9
Oxygen															
Myoglobin	–	–	–	–	–	–	–	–	–	–		↑y +EE↑w	12, 13	–	–
Fatty acids															
Import and β-oxidation	↑m-h CPTI, LPL, FAT ↑wCPTI, MCAD, VLCAD	–	–	–	10, 13-15	–	–	–	–	↓y ACAA2, HPXEL ↑yLRP1	9	–	–	↓y SCHAD	9

Table 2 (continued)

	Stimuli														
	High contractile activity-low loading					Low contractile activity-high loading			Metabolic intervention			Hypoxia		Ageing	
	EE	CLFS	DV	IN/DV	References	RT	MG/I	References	FA supp.	CR	References		References		References
ATP production and transfer															
Krebs cycle	↑m-hPDK4 ↑w SDH, Fum	–	–	–	10, 16, 17	–	–	–	↑dPDK4	↑yPDK4 ↓y MDH1	9, 29	–	–	↓y SDH, MDH1, Fum	9
El chain and ox. phos.	↑wNADH6, COXI and IV	–	–	–	17, 18	–	–	–	–	↑h-d UCP-3↓y UQCRC2 UQCRB, NDUFV2, CytC, COX IV and VIII, ATPB, NNT ↑y NDUFV1, DIA1	9, 19	–	–	↓y UQCRH, NDUFV2, ATP5A, ATP5C, ATP5G3, ATP5J, COX IV and VIII and VIa	9
Proton uncoupling	↑m-h UCP-3	–	–	–	10	–	–	–	↑w UCP-2 and 3	–	20, 29	–	–	–	–
Energy transfer	–	–	–	–	–	–	–	–	–	–	–	–	–	↓y CK	9
Protein turnover															
Turnover	–	–	–	–	–	↑d HSC/HSP70, HSP27	–	21	–	–	–	–	–	↓y Cat H	9
Folding	↑m-h HO-1, HSP-70	–	–	–	10, 22		–	–	–	–	–	–	–	↑y HSP27 and 70	9, 23
Regulation															
Signal transduction															
Mechano	–	–	–	–	–	–	–	–	–	–	–	–	–	–	–
Metabolic	↑mo PPARα	–	–	–	14	–	–	–	–	–	–	(+EE) ↑w HIF-1α	13	–	–
Endocrine response	–	–	–	–	–	–	–	–	–	–	–	–	–	↓y TRAP230	9

Table 2 (continued)

	Stimuli														
	High contractile activity-low loading					Low contractile activity-high loading			Metabolic intervention			Hypoxia		Ageing	
	EE	CLFS	DV	IN/DV	References	RT	MG/I	References	FA supp.	CR	References		References		References
Transcription	↑m-h c-fos, fosB,fra-1, c-jun, junB and D ↑wTFAM	–	–	–	17, 18, 24	↑h-d c-jun, ↑w myogenin →w MRF4, MyoD, Myf5, ↓↑h-d EGR-1	→w myogenin, MRF4, MyoD, Myf5	23, 25	–	–	–	–	–	↓y ERF	9
Cell cycle															
Angiogenesis	↑m-hVEGF	–	–	–	26, 27	↑h-d VEGF	–	23	–	–	–	+EE ↑w VEGF	13	–	–
Satellite cell proliferation	–	–	–	–	–	↑w-m Leu19	–	4	–	–	–	–	–	–	–
Myonuclei incorporation	–	–	–	–	–	↑y dMHC	–	4	–	–	–	–	–	–	–
Muscle cell growth arrest	–	–	–	–	–	–	–	–	–	–	–	–	–	↑y p21	9
Cell proliferation	–	–	–	–	–	–	–	–	–	↑y RYK, FAP	9	–	–	↓y IGF-BP 5 and 7, MAT1, IMPDH1, EGFR, Mac-2, PP2Ag1, MARK3	9
(Neuronal) cell death	–	–	–	–	–	–	–	–	–	–	–	–	–	↑y SH3GL3, DP5, granzyme A	9
Neuronal remodelling	–	–	–	–	28	–	–	–	–	–	–	–	–	↑y NRG-1, NCAM1, reelin, phogrin	9
Repair															
(Oxidative) Stress/ damage	–	–	–	–	–	–	–	–	–	–	–	–	–	↑y MT1F, ORP150, NF-kB p65	9, 23
DNA repair	–	–	–	–	–	–	–	–	–	–	–	-	–	↓y XRCC1, HPARG, MSH3, RECQ2	9

Table 2 (continued)

	Stimuli														
	High contractile activity-low loading					Low contractile activity-high loading			Metabolic intervention			Hypoxia		Ageing	
	EE	CLFS	DV	IN/ DV	Refer- ences	RT	MG/I	Refer- ences	FA supp.	CR	Refer- ences		Refer- ences		Refer- ences
Detoxification	–	–	–	–	–	–	–	–	–	–	9	–	–	↓y SOD2, PON2 ↑y AOX1, CYP2B6, CYP2A6, CYP2C8	9
Inflammation	–	–	–	–	–	↑h-d RANTES, IL-1β	–	23	–	↓y RING6 OTF-2	9	–	–	↑y granulin, gamma 1, JAK1, LTC4S, CSF-1, chitotriosidase, IL-18, CXCL5	9

[1] Harridge et al. 2002, [2] Balagopal et al. 2001, [3] Hortobagyi et al. 2000, [4] Kadi and Thornell 1999, [5] Andersen et al. 1994a, [6] Andersen and Aagaard 2000, [7] Andersen et al. 2000, [8] O'Neill et al. 1999, [9] Kayo et al. 2001, [10] Pilegaard et al. 2000, [11] Kraniou et al. 2000, [12] Reynafarje 1962, [13] Vogt et al. 2001, [14] Horowitz et al. 2000, [15] Tunstall et al. 2002b, [16] Martin et al. 1992, [17] Puntschart et al. 1995a, [18] Bengtsson et al. 2001, [19] Tunstall et al. 2002a, [20] Schrauwen et al. 2001a, [21] Thompson et al. 2001, [22] Puntschart et al. 1996, [23] Jozsi et al. 2000, [24] Puntschart et al. 1998, [25] Hespel et al. 2001, [26] Gustafsson et al. 1999, [27] Richardson et al. 1999, [28] Panenic and Gardiner 1998, [29] Peters et al. 2001

The effects of endurance exercise such as running on the MHC profile appear to be both muscle-specific and dose-dependent. For example, in rodents, when animals are trained to run at moderate to high intensity (~30 m/min; ~20% incline at ~75% VO_2max) for several weeks, the running effects on the MHC profile of the soleus are manifested only when animals run for longer durations (60 and 90 min/day; Demirel et al. 1999). In mixed fast muscles (red vastus and gastrocnemius), which have a composition bias to type IIx and IIa MHC expression, both the type IIa and IIx MHCs are up-regulated relative to the sedentary state, whereas the IIb MHC is significantly down-regulated under these training conditions (Demirel et al. 1999). If running is extended for longer durations, it is even possible to induce increased expression of the type I MHC in fast muscles (Demirel et al. 1999; Green et al. 1984).

It has been shown that systematic endurance exercise training increases the steady-state level of a number of mRNAs encoding mitochondrial proteins in proportion to the increase in mitochondrial volume density (Puntschart et al. 1995a). Moreover, acute bouts of endurance exercise cause a significant rapid (0-4 h after exercise) transient increase in the mRNA level of several proteins involved in regulating mitochondrial functions (Pilegaard et al. 2000). Additionally, exercise seems to rapidly affect cellular trafficking, i.e. glut-4 and possibly small G proteins (Kraniou et al. 2000) (reviewed in Booth and Baldwin 1995). Endurance exercise training further causes a fine-tuning of the mRNA level of MCAD involved in beta oxidation of medium chain fatty acids (Vogt et al. 2001). This observation is compatible with the functional observation that endurance exercise shifts the functional capacity of mitochondria towards an increased use of lipids as a substrate (Holloszy and Booth 1976; Holloszy and Coyle 1984). Bicycling exercise also affects expression of genes involved in the modulation of muscle substrate supply, i.e. to an induction of key components of fatty acid transport (FAT/CD36, FABPpm, LPL) and β-oxidation (CPTI) in human m. vastus lateralis during the recovery phase (Pilegaard et al. 2000; Tunstall et al. 2002b). Data from animal experiments support that LPL expression is at least in part controlled by local contractile activity (Hamilton et al. 1998). Activation of gene expression therefore appears to be the main mechanism for the accumulation of post-transcriptional micro-adaptations responsible for the subsequent structural and biochemical adaptations of the mitochondrial compartment in exercised skeletal muscle.

Looking at mitochondria in more detail, we find that the mitochondrial organelle incorporates numerous proteins involved in beta oxidation, the Krebs cycle and the mitochondrial respiratory chain, in the inner and outer membrane and its matrix. Only a few of these proteins are coded on the mitochondrial genome while most are coded on the nuclear genome. Therefore, proteins from nuclear-encoded genes, as well as phospholipids, must be synthesized in the cytosol, imported into mitochondria and incorporated into their appropriate locations for mitochondrial biogenesis to be successful (Hood 2001). The steady-state level of mitochondrially, as well as nuclear-encoded mRNAs of the mitochondrial respiratory chain and the mitochondrial DNA concentration is increased in m. vastus lateralis of athletes versus untrained subjects. By contrast, genomic DNA was not increased in athletes while mitochondrial DNA was increased (Puntschart et al. 1995a). These observations indicate a specific way in which the expression of mitochondrial proteins from the nuclear and from the mitochondrial genomes is coordinated. For nuclear encoded genes, we find an increase in transcription leading the observed increase in mRNA of downstream genes. For mitochondrially encoded species, the rate of transcription is unchanged and the total quantity of mitochondrial DNA is up-regulated. Thus, the proportion of mitochondrial RNA to mitochondrial DNA remains constant. This is achieved because endurance exercise causes an increase in the level of the mitochondrial transcription factor (TFAM) in proportion to the increased abundance of mitochondrial- and nuclear-encoded mitochondrial proteins. A transient increase in TFAM and nuclear respiratory factor 1 and 2 (NRF-1 and NRF-2) mRNA has also been documented for rat muscle (Bengtsson et al. 2001; Gordon et al. 2001a; Murakami et al. 1998). Therefore, post-exercise transient expression of NRF-1 through controlling TFAM expression may be a key factor for coordinating nuclear and mitochondrial gene transcription for mitochondrial biogenesis in exercised skeletal muscle.

The increase in oxidative capacity of skeletal muscle with endurance training in human (and rats) is accompanied by a similar increase in muscle capillarity. Using morphometric techniques, the capillary-to-fibre ratio is significantly higher in orienteers (2.70) than in untrained men (2.07) and in untrained women (1.37) (Zumstein et al. 1983). To our knowledge, the molecular events leading to concerted capillary growth in human exercise training have yet to be studied in detail. In human skeletal muscle, exercise affects expression of the *VEGF* gene, which is involved in formation and permeability of blood vessels (Richardson et al. 1999; Gustafsson et al. 1999).

Data from animal studies further indicate the involvement of other molecules such as angiotensin II, basic fibroblast growth factor and transforming growth factor $\beta 1$, but not of nitric oxide, in capillary remodelling induced by exercise (Amaral et al. 2001b; Breen et al. 1996; Lloyd et al. 2001; Richardson et al. 1999). Blocking of angiotensin II-signalling by angiotensin-converting enzyme inhibitors and angiotensin II type I receptor blockers completely blocked VEGF-dependent short-term (3 days) capillary remodelling in rat skeletal hindlimb muscle induced by intense running exercise or 7 days of electrical stimulation (Amaral et al. 2001a, 2001b). The involvement of the renin-angiotensin system in control of VEGF expression and vessel density support for an important role of microcirculation in capillary remodelling of skeletal muscle through eventual modulation of vascular resistance and shear stress (Hudlicka 1998).

Using microarray analysis (Atlas array no. 7740-1) we recently analysed the level of distinct mRNAs involved in regulatory processes in human m. vastus lateralis and the difference in their expression between untrained subjects (mean $VO_2max=39$ ml/min/kg; $n=7$) and professional cyclists (mean $VO_2max=72$ ml/min/kg; $n=7$). High expression of genes involved in inflammation (*IL-1*, *IL-13*) and stress response (*HSP-27*, *c-jun*, UV excision repair protein RAD23A) was detected pointing to an eventual role in skeletal muscle homeostasis while others with well-established muscle regulatory function were often less expressed (Fig. 3). Using the Mann-Whitney U-test, we identified only 15 out of 408 expressed genes to be differently expressed between untrained subjects and professional cyclists. At the 0.05% significance level, this number of transcripts is within the range of randomly expected events in a population of 408. We have therefore refrained from a biological interpretation of these results. It is assumed that a highly fluctuating gene expression between individuals could mask adaptations in regulatory gene expression (Wittwer et al. unpublished observations).

Inactivity/detraining

The adaptability of skeletal muscle to variable levels of functional demands is well underlined by the observation that muscular detraining occurs during periods of insufficient

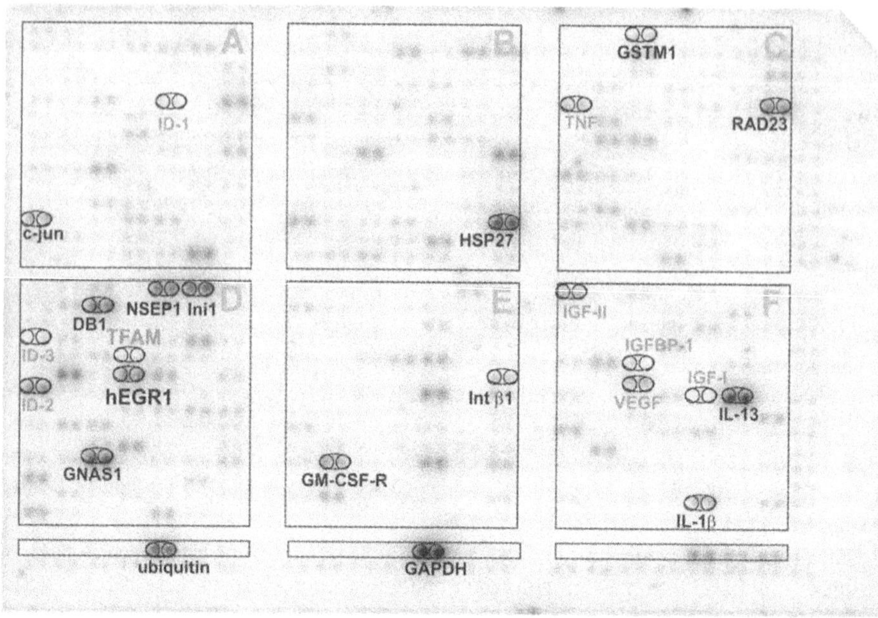

Fig. 3A-F Regulatory gene profile in human m. vastus lateralis. Autoradiogram of a microarray that was hybridized with radiolabelled target cDNA synthesized from RNA isolated from untrained human m. vastus lateralis. The *boxes* denote sectors that were doubly spotted with probes of genes belonging to related functional categories: **A** cell cycle regulation; **B** signal transduction; **C** cell death and DNA interaction; **D** transcription; **E** cell surface receptors; **F** cell-cell communication. The spots assigned with *black and grey text* correspond to the ten genes which were on average most highly expressed and to genes with established muscle regulatory function

training stimulus. The underlying structural-functional changes to detraining have been recently reviewed by Mujika and Padilla (2001). Fibre distribution remains unchanged during the initial weeks of inactivity, but oxidative fibres may decrease in endurance athletes and increase in strength-trained athletes within 8 weeks after training was stopped. Muscle fibre cross-sectional area declines rapidly in strength and sprint athletes, and in recently endurance-trained subjects, whereas it may increase slightly in endurance athletes. Force production declines slowly and in relation to decreased EMG activity. Strength performance in general is readily maintained for up to 4 weeks of inactivity, but highly trained athletes' eccentric force and sport-specific power, and recently acquired isokinetic strength, may decline significantly. Detraining may be characterized by a decreased capillary density, which could take place within 2-3 weeks of inactivity. Arterial-venous oxygen difference declines if training arrest continues beyond 3-8 weeks. Moreover, rapid and progressive reductions were noted within 3-46 weeks of detraining in oxidative and Krebs cycle enzyme activities (β-hydroxyacyl-CoA dehydrogenase, citrate synthase, succinate dehydrogenase, malate dehydrogenase) to bring about a reduced mitochondrial ATP production. The above changes are related to the reduction in VO_2max observed during long-term training cessation. In contrast, myoglobin concentration was found unchanged in the m. gastrocnemius of endurance-trained runners and cyclists after 3 and 12 weeks of training cessation. Moreover, glycolytic enzyme activities (phosphofructokinase, lactate dehydrogenase, phosphorylase) in muscles of trained and untrained individuals show nonsystematic changes during periods of training cessation (Table 1). These muscular characteristics remain above sedentary values in the detrained athlete but usually return to baseline values in recently trained individuals. Interestingly, increases were observed the concentration of growth hormone and testosterone while cortisol was decreased in the plasma concentrations of weightlifters after 12 weeks of detraining. The complete pattern of adaptations indicates that the qualitative changes in fibre types may depend on the training status of the individuals. To the best of our knowledge, no data are available on the expressional events that underlie the phenomenon of muscular detraining in humans.

Cross-reinnervation

Cross-reinnervation (cross-union) experiments performed by Buller, Eccles and Eccles on cats (Buller et al. 1960) were at the origin of the paradigm of skeletal muscle phenotype. This kind of experiment demonstrated that the type of motor neuron innervation is determining many muscle characteristics. When a denervated muscle exhibiting fast-twitch properties is ectopically reinnervated by a slow nerve, the muscle adopts most slow-twitch properties, and vice versa (Buller et al. 1960).

The early cross-reinnervation experiments are supplemented now with studies leading to the identification of molecular events involved in muscle reprogramming. A particular focus has been the study of the myogenic transcription factors myogenin, myoD, MRF4/herculin/myf-6 and myf5. Together, these factors play an important role in myogenesis through binding to consensus Ebox recognition sites, which exist in numerous muscle-specific genes (Olson 1990). Cross-reinnervation of the slow rat soleus muscle with a fast nerve decreased myogenin mRNA expression in regions usually expressing fast myosin heavy chain (Hughes et al. 1993). Moreover, thyroid hormone treatment resulted in activation of the myoD and fast myosin heavy-chain gene expression in the slow soleus while no significant change in either slow MHC, MRF4 or myogenin mRNA accumulation was detected. Based on the observations that the levels of myogenin are higher in slow-twitch muscles, while myoD is associated with fast-type fibres, it was proposed that myogenin and myoD may function as intermediaries to selectively regulate fibre-type-dependent contractile gene expression. Specifically, because myogenin controls expression of a battery of muscle-specific genes, including slow MHC expression (Olson 1990), it was suggested that myoD is involved in controlling fast glycolytic and that myogenin would control slow *MHC* gene expression (Hughes et al. 1993). Consequently, an increase in the myogenin:myoD ratio would preferentially drive expression of slow-type contractile proteins in skeletal muscle (Fig. 4).

Electric stimulation

Chronic low-frequency stimulation (CLFS) of mammalian skeletal muscle causes specific changes in muscle properties (Tables 1, 2). Much of the early work on the effect of chronic stimulation was done in animal models (Pette and Vrbova 1992). These experiments have led to the conclusion that the firing frequency and activity pattern of motor-neurons are the key determinants of the contractile characteristics of muscle fibres.

CLFS activates all motor units of the target muscle synchronously and with the same impulse pattern. It is thus possible to attain much higher levels of muscle activity over

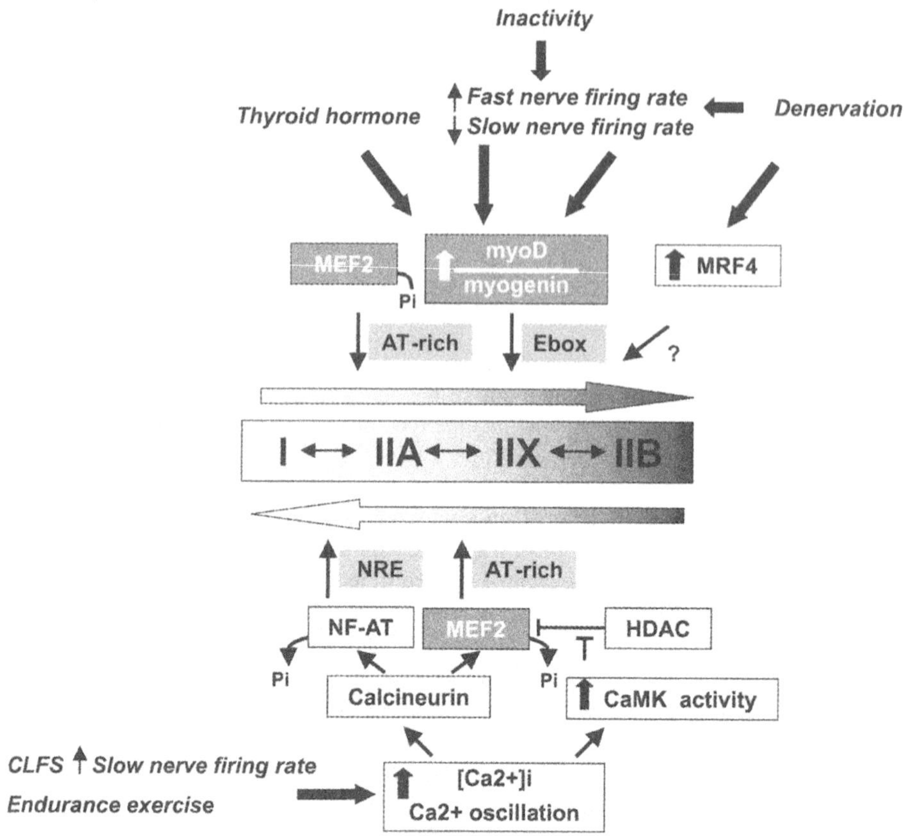

Fig. 4 Molecular control of fibre-type-specific MHC expression. Model indicating the molecular factors that have been implied in control of myosin heavy-chain expression as a consequence of different stimuli in (rodent) skeletal muscle myonuclei. Reduced rate of slow nerve firing, denervation, thyroid hormone and eventually inactivity cause an increase in the myoD/myogenin ratio of myogenic transcription factors and phosphorylated MEF2 drives expression of fast type myosin II (Hughes et al. 1993). Conversely, through increases in intracellular calcium $[Ca^{2+}]i$, chronic low-frequency stimulation (*CLFS*) and endurance exercise cause the activation of the phosphatase calcineurin and the protein kinase CaMKII. Calcineurin causes dephosphorylation of transcription factor NF-AT and MEF2 and initiates transcription of myogenin. CaMKII provokes the liberation of MEF2 from the complex with the transcriptional repressor histone deacetylase (*HDAC*) (Olson and Williams 2000). These events promote the expression of slow type myosins (Chin et al. 1998; Wu et al. 2001, 2002)

time than with any voluntary exercise regimen. CLFS thus challenges muscle to its full adaptive potential (Pette 2001). The major adaptations observed with CLFS affect the contractile apparatus and its energy supply. Chronic stimulation of fast-type animal muscle increases time to peak, twitch tension and half relaxation time while the maximal rate of tetanic tension development is decreased (Pette and Vrbova 1992). Moreover, increased resistance to fatigue and increased aerobic-oxidative capacity are consistent findings observed with CLFS of animal skeletal muscle (Pette and Vrbova 1992; Reichmann et al. 1985; Table 1). The observed functional changes are explained by a drop in muscle fibre diameter, changes in the quality of the myofibrillar apparatus and the Ca^{2+} regulatory system towards the slow muscle type as well as an increase in capillarity (reviewed in Pette and Vrbova 1992). The increase in activities of enzymes of the aerobic-oxidative metabolism is held responsible for the increased fatigue resistance. A linear correlation is shown to exist between the increase in citrate synthase activity, total mitochondrial volume and aerobic-oxidative capacity (Reichmann et al. 1985). Recent data indicate that the functional changes provoked by increasing the contractile activity of skeletal muscle of spinal cord-injured humans by chronic low-frequency stimulation are qualitatively similar to those seen with endurance exercise (Tables 1, 2). Exercise training of spinal cord-injured individuals (30 min/day, two to three times per week for 1 year) on an ergometer under electrical stimulation reverses inactivity-associated performance changes in skeletal muscle (Mohr et al. 1997). Resistance to fatigue, work output, oxygen uptake, muscle cross-sectional area were all increased, while the number of MHC-type IIa fibres was increased and the type IIb (IIx) fibre number was reduced. Similar changes in MHC expression with an augmentation of MHC type IIa and a drop in MHC type IIa fibres were seen after 6 and 12 months of functional electric stimulation (Andersen et al. 1996). Intriguingly, in both studies, MHC type I fibres did not increase.

In the clinical setting, electrical stimulation therapy either of individual muscles or in a coordinated functional manner is occasionally used to counteract atrophy, pressure sores, bone loss, and glucose intolerance (Mohr et al. 1997). Improvements in muscle metabolic properties, insulin sensitivity, and oxidation capacity may be closely related to the transition of fibre types within the fast phenotypes (i.e. from MHC IIx to MHC IIa) (Andersen et al. 1996; Burnham et al. 1997; Mohr et al. 1997).

It has been demonstrated early on that low-frequency (10 Hz, 12 h/day) electric stimulation of fast-twitch rabbit muscles induces the tissue yield of the total polyadenylated RNA, polyadenylated RNAs specifically translatable in vitro, total ribosomes whereby an increase in monosomes precedes the increase in the polysomes (Seedorf et al. 1986). The increase in lactate dehydrogenase isoenzymes and citrate synthase was found to be due to altered transcription levels as well as post-transcriptional control mechanisms. The latter are held responsible for the early rise in CS concentration. Later, data from human and animal models indicated that numerous pretranslational changes contribute to adaptations of skeletal muscle to chronic electrical stimulation. For example, continuous indirect electrical stimulation of tibialis anterior muscle increases CPT II mRNA (Yan et al. 1995). Recent data from paraplegic humans indicate that the up-regulation of mRNA for MHC-I isoform and down-regulation of the MHC-IIx isoform both contribute to changes in the contractile apparatus and hence to the reduction in speed of contraction with chronic electrical stimulation (Harridge et al. 2002). Last but not least, expression of the myogenic factors myogenin and MyoD were found to increase as a consequence of electric stimulation while myogenin is implicated in the control of the increase in nAChR expression with CLFS (reviewed in Buonanno et al. 1998).

Denervation

Denervation due to spinal cord injury leads to drastic adaptations of human skeletal muscle. Paralyzed muscles of individuals with chronic traumatic spinal cord injury (SCI) are characterized by a massive reduction of the size of all muscle fibres, by a high incidence of fast muscle fibres and in some instances even by a complete loss of slow fibres in what would normally be mixed fibre-type muscles (Burnham et al. 1997; Lotta et al. 1991; Martin et al. 1992; Round et al. 1993) (Fig. 5; Tables 1, 2). For both type I and type II fibres,

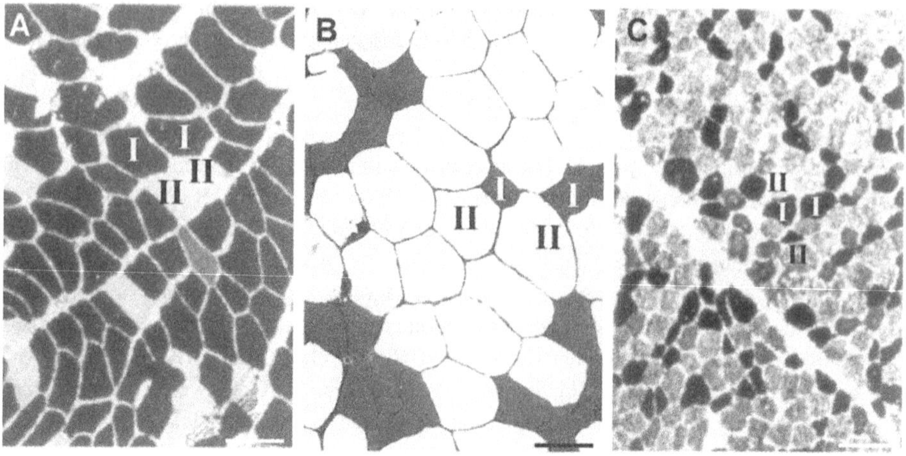

Fig. 5A-C . Plasticity of human skeletal muscle fibre types. Histochemical pictures showing adaptations of fibre types in human m. vastus lateralis with endurance exercise (**A**), resistance training (**B**) and denervation subsequent to spinal cord injury (**C**). Slow- (*I*) and fast-type (*II*) fibres are indicated. Bar, 100 μM

the cross-sectional area, activities of succinate dehydrogenase, and the capillary-to-fibre ratio are significantly reduced in paralyzed muscles (Martin et al. 1992). Moreover, the activity of the primary enzyme for conversion of circulating lipoprotein triglyceride into (membrane-permeable) free fatty acids, lipoprotein lipase (LPL), is decreased in denervated rat soleus muscle, supporting the concept that contractile activity is importantly involved in the regulation of the metabolic muscle phenotype (Smol et al. 2001).

The myosin heavy-chain (MHC) isoform expression appears to be altered towards an increase in MHC II. Furthermore, SCI leads not only to a general relative increase in MHC II expression, but also to a relative increase in the expression of the fastest of the human MHC isoforms, MHC IIx (Burnham et al. 1997; Lotta et al. 1991; Round et al. 1993). These changes are thus opposite to the adaptations seen with increased motoneuron activity such as during CLFS (Tables 1, 2). This observation supports the suggestion that human MHC IIx (formerly called IIb) is the default *MHC* gene (Goldspink et al. 1991) and that the high expression of this isoform in paraplegic muscle is an adaptive response to a lack of usage, specifically of the loss of neural input and mechanical loading (Andersen et al. 1996; Harridge 1996; Talmadge 2000).

In animal models, the myogenic factors myogenin, myoD, MRF4/herculin/myf-6 and myf5 have been demonstrated to undergo rapid nuclear accumulation in mature myofibres after denervation (Weis et al. 2000; Eftimie et al. 1991; Walters et al. 2000). There are data indicating that the expression of the myogenic regulatory factors in denervated muscles depend on the muscle phenotype (Walters et al. 2000). Fast muscles respond very quickly to denervation by increasing the level of MRF transcripts MyoD and myogenin while slow muscles did not show significant increases in expression up to 48 h after denervation (Walters et al. 2000; Fig. 4). MRF4 has been suggested to regulate genes encoding adult contractile proteins and acetylcholine receptor subunits, indicating that MRF4 may have important roles in the gene programs activated after denervation and during muscle regeneration (Weis et al. 2000). Overall, the data indicate that muscle-specific transcription factors of the helix-loop-helix family such as myoD and myogenin are important for regulating genes related to metabolic profile, fibre size and possibly myosin heavy-chain expression in response to changes in nerve-dependent contractile activity (Gundersen 1998; Buonanno et al. 1998).

Experiments with rats show that denervation-induced transient increases in myogenin and sustained up-regulation of levels of myoD cause an increase in MHC IIx expression in soleus and EDL but no atrophy of these muscles (Dupont-Versteegden et al. 1998). These findings indicate that the mechanisms underlying maintenance of muscle mass are distinct from those controlling myosin heavy-chain expression. The results support the contention that altered myogenin and myoD levels are involved in control of MHC expression. It was shown that MyoD was preferentially expressed in satellite cells in soleus muscle but was more readily detectable in myofibre nuclei in EDL. By contrast, myogenin was readily detected in both myofibre and satellite cell nuclei, suggesting varied functions of these myogenic factors in different muscles (Dupont-Versteegden et al. 1998). These findings are compatible with a scenario whereby new satellite cells have to be recruited and to fuse with type I fibres of slow-twitch solei to express type II MHC in these former type I fibres. This suggests that the recruited myonuclei substitute during this fibre transformation for a lack of myonuclei which switch back to or re-express the default-type type II MHC. In the fast-twitch EDL muscle, however, myonuclei can be induced to express Myo D and type II MHC isoforms (Fig. 4).

However, there are a number of negative correlations between the activity-dependent regulation of MyoD and myogenin and genes encoding contractile proteins that challenge the model of Hughes et al. (1993), which predicts that the myogenin:myoD ratio determines expression of slow versus fast-type contractile proteins in skeletal muscle. An alternative model was proposed that postulates that motor neuron innervation alone induces transcription regulatory mechanism(s) that activate the expression of numerous genes encoding contractile proteins and metabolic enzymes (Buonanno et al. 1998; Olson and Williams 2000; Fig. 4). Future experiments are now necessary to identify to what extent regulatory factors in addition to the myogenin:myoD ratio are needed to explain the observed plasticity of MHC expression, contractile activity and metabolic pathways observed as a consequence of denervation in rodent and human skeletal muscle (Talmadge 2000).

Muscle loading

Several manipulations which change the mechanical load to which muscle is exposed have been shown to have pronounced effects on muscle structure and function. In the following sections, we discuss the structural and functional changes as well as the molecular events associated with various modalities for increasing or decreasing muscle loading.

Resistance training

Using high-intensity, low-repetitive (strength-type) exercise, human skeletal muscles show marked gains in strength that are due both to neuronal adaptations and to an increase in muscle cross-sectional area (see Sale1988). There is consensus that the gain in muscle cross-sectional area is mainly due to an increase in myofibrillar volume (i.e. contractile protein; MacDougall et al. 1979; Luethi et al. 1986). The cross-sectional area of all fibre types is increased following resistance training, with a tendency for larger increases in type II than type I fibres (Hortobagyi et al. 2000; Table 1). Volume densities of mitochondria are found to be reduced and may be as low as 2% in some elite athletes. This decrease is thought to be due to a dilution of a constant mitochondrial volume in larger muscle fibres (MacDougall et al. 1979; Luethi et al. 1986). Marked structural abnormalities such as central nuclei and atrophied fibres were noted in elite power lifters and body builders (MacDougall et al. 1982). These fibre abnormalities have been assumed to be related to the use of anabolic steroids in elite strength athletes. For further structural and functional characteristics of strength-trained human muscle, see also Alway et al. (1988) and Prince et al. (1981). The fact that strength training seems to leave the mitochondrial compartment relatively unaffected is supported by an unchanged distribution of subsarcolemmal versus interfibrillar mitochondria demonstrated by quantitative SDH stain of mitochondria (Chilibeck et al. 1999). Neuronal adaptations within the earlier weeks of resistance training have been suggested (Hakkinen and Kauhanen 1989; see also Sale 1988). However, some studies have not been able to demonstrate changes in the control properties of the nervous system after 8 weeks of resistance training (Rich and Cafarelli 2000). Conversely, the pronounced effect of resistance training in elderly subjects (strength gain up to 100%) has been shown to reside more in neuronal adaptations of the motor control than in the moderate increase in muscle cross-sectional area of 5-10% (Vandervoort 2002; Connelly and Vandervoort 2000).

Numerous experiments point to the importance of changes in protein turnover for the hypertrophic process (Wong and Booth 1990; Table 1). Resistance training is known to cause a (maximal) increase in fractional synthesis rate and fractional breakdown rate within 3 h after exercise that are maintained up to 48 h after the training in humans (Phillips et al. 1999; Biolo et al. 1995). Mixed and MHC synthesis rates remain enhanced as a consequence of continued resistance training of young and old subjects (Yarasheski et al. 1993b; Balagopal et al. 2001; Hasten et al. 2000). Due the lack of pertinent expression data, it is currently not known whether the absence of mitochondrial adaptations with strength training is due to transcriptional regulatory events (Chesley et al. 1992) or to enhanced mitochondrial turnover.

In humans, resistance training has been shown to up-regulate fast IIa MHC and eventually the slow type MHCI isoforms while expression of the fast type IIx MHC may be down-regulated (Adams et al. 1993; Andersen et al. 1994b; Klitgaard et al. 1990b). An increased mRNA level probably contributes to the increased size of type I and IIa fibre types. Recent data indicate that the increase in pure type MHC IIa-expressing fibres can be attributed to a decrease in the number of fibres that are hybrid with regard to the expression of slow and fast MHC isoforms (Williamson et al. 2001). The eventual drop in total MHC IIx expression with resistance training is probably due to a reduction in MHC IIx expression as a consequence of differentiation of IIx-expressing hybrid fibres towards pure IIa-expressing fibres. The increase in size of pure MHC IIx-expressing fibres also observed seems to be of lesser importance.

If subjects are detrained for several weeks, it appears that this process is reversed and the re-expression of the fast type IIx in (hybrid) fibres appears to be greater than in the pretraining period (Andersen and Aagaard 2000). However, even though resistance training increases type IIa MHC expression, it is uncertain how these molecular changes relate to the functional changes seen with strength training, in particular in the initial stages of strength training (Carroll et al. 1998).

In animal models, it has been clearly established that satellite cell activation is involved and may be a prerequisite for fibre hypertrophy (Schultz and McCormick 1994). Using a marker for satellite cells it has been documented that expression of early markers of myogenesis is activated in satellite cells and muscle fibres in response to resistance training in humans (Kadi and Thornell 1999). Satellite cells are believed to proliferate and fuse with existing fibres, thereby contributing to an increase in myonuclei per muscle fibre. Another study indicated that the number of myonuclei was increased with fibre hypertrophy and positively correlated with the increased number of satellite cells (Kadi and Thornell 2000). This suggests that the acquisition of additional myonuclei is required to support the enlargement of muscle cells during strength training. Moreover, hypertrophy of skeletal muscle of untrained elderly men induced by high-intensity resistance training did not result in significant changes in the cytoplasma-to-myonucleus ratio despite the increased cross-sectional size of all fibre types and transition of type IIx towards IIa fibres (Hikida et al. 2000). The cellular changes occurring during hypertrophic adaptations thus essentially confirm the nuclear domain theory, suggesting that the cytoplasma-to-myonucleus ratio is a function of the myosin type and the amount of protein turnover (see Booth and Baldwin 1995). Taken together, the available evidence suggests that satellite cells are recruited during hypertrophy of muscle fibres in order to maintain the cytoplasma-to-myonucleus ratio.

Rodent models have provided a great deal of evidence on the molecular events responsible for muscle plasticity with changes in the loading conditions. Goldberg and Goodman (1969) used actinomycin D (a transcription inhibitor) to block load-induced muscle hypertrophy, demonstrating the importance of mRNA changes for muscle plasticity. The advent of microarray techniques now allows for studying the regulation of hundreds of genes simultaneously and therefore provides the opportunity for the detection of concerted regulation of gene clusters (synexpression groups; Niehrs and Pollet 1999). Recent gene profiling data document that work-induced muscle hypertrophy is an integrated transcriptional response whereby genes related to carbohydrate and protein metabolism, autocrine/paracrine factors, extracellular matrix proteins, transcription factors and cell regulatory factors change together (Carson et al. 2002). It has been proposed that the pronounced increase in the muscle regulatory factor myogenin in overloaded muscle supports its involvement as a major controller of the complex fast-to-slow transformation process (Carson et al. 2002). Another important player in muscle hypertrophy seems to be the *ski* gene (Sutrave et al. 1990). It has been demonstrated that type II fast fibres undergo selective hypertrophy in muscles of transgenic mice that overexpress the *ski* gene. Supporting evidence for the role of the *ski* gene in muscle hypertrophy comes from experiments indicating a threefold increase in *ski* mRNA after injury (Soeta et al. 2001). MRF4 has been identified as an additional factor of potential importance for muscle hypertrophy (Hespel et al. 2001). This role is suggested by an increase in MRF4 expression during recovery of muscle fibre diameter from disuse atrophy after 10 weeks of knee-extension exercise under oral creatine monohydrate supplementation in humans (Hespel et al. 2001). Recently it has been demonstrated that a single bout of resistance exercise in young male subjects has significant effects on the transcriptome (Jozsi et al. 2000). The expression of mRNAs involved in the stress/damage (HSP27) and inflammatory response, the angiogenic factor VEGF and several transcription factors were found changed in biopsies of m. vastus lateralis 24 h after a single bout of resistance exercise. Similarly, large increases in the level of the heat-shock proteins (HSPs) HSP27 and HSC/HSP70 were identified 48 h after a single bout of resistance exercise in human m. biceps brachii (Thompson et al. 2001). HSPs fulfil diverse and distinct functions in protein folding and stabilization (Liu and Steinacker 2001), indicating that their increase after exercise may be related to the longer-lasting changes in protein turnover observed after resistance exercise (see Table 1).

Real and simulated microgravity (immobilization)

The adaptations of skeletal muscle tissue to microgravity have recently been reviewed (Fitts et al. 2001). Adaptations to microgravity conditions have been studied in biopsies from astronauts before and after space flights of various durations as well as in experiments involving prolonged bedrest (simulated microgravity). As far as skeletal muscle tissue is concerned, both interventions lead to similar structural and functional adaptations. After 42 days of a head-down tilt bedrest (with no countermeasures), we found a 17% reduction in muscle cross-sectional area and a similar reduction in volume density of mitochondria in biopsies of m. vastus lateralis (Ferretti et al. 1997). The total loss of oxidative capacity in this muscle thus amounted to 28.5%, with a concomitant loss of 22% of total capillary length. Biopsies obtained in the same subject from m. deltoideus showed no loss of muscle cross-sectional area or fibre size, no difference in muscle mitochondrial volume and no decrease in capillarity (Desplanches et al. 1998). Space flight of 6 months duration leads to a decrease in maximal voluntary contraction of the plantar flexors of the foot of 20-48% (see review by Fitts et al. 2001). This loss of muscle strength is due both to muscle atrophy as well as to selective loss of contractile proteins, leading to a decrease in force per cross-sectional area. Already after 17 days of space flight, type I fibres in m. soleus of astronauts were found to manifest less average peak Ca^{2+}-activated force. This was attributed to a reduction in fibre diameter and/or force per cross-sectional area (Widrick et al. 1999). Conversely, mean velocity during unloaded contractions was greater but could not be explained by alterations in myosin heavy- or light-chain composition. After microgravity exposure there is an increase in the muscle fatigability and a shift away from lipid towards glucose metabolism in rats (reviewed in Fitts et al. 2001). However, the results of muscle inactivity after real space flights must be interpreted with caution, as the observed changes occur despite vigorous counter measures designed to minimize loss of muscle structure and function in orbit.

Comparing real with simulated microgravity, the results in m. vastus lateralis are found to be similar under both conditions. However, the results of m. deltoideus remain puzzling. It is unclear whether upper and lower limb muscles react differentially to inactivation or whether m. deltoideus escaped atrophy in the bedrest experiment because of an increased circumstantial use of this muscle under these particular circumstances. This could also be the case in real space flight where an increased use of upper body muscles for stabilization is likely.

The molecular mechanisms responsible for microgravity adaptations have mainly been studied using earth-bound animal models. Hindlimb suspension of rats and immobilization are two experimental interventions that reproduce many of the gross functional and structural adaptations in skeletal muscle observed with space flight. Hindlimb suspension leads to fibre atrophy in m. soleus with a selective loss of myofibrils, a shift towards higher speeds of contraction, a reduced resistance to fatigue and a reduced oxidative capacity (Fitts et al. 2001; Thomason and Booth 1990). The hindlimb suspension model has produced evidence for a reduced protein synthesis combined with increased protein degradation, a shift towards expression of fast myosin heavy-chain isoforms and a shift towards a preferred utilization of carbohydrates over fatty acids as substrates (Fitts et al. 2001; Ste-

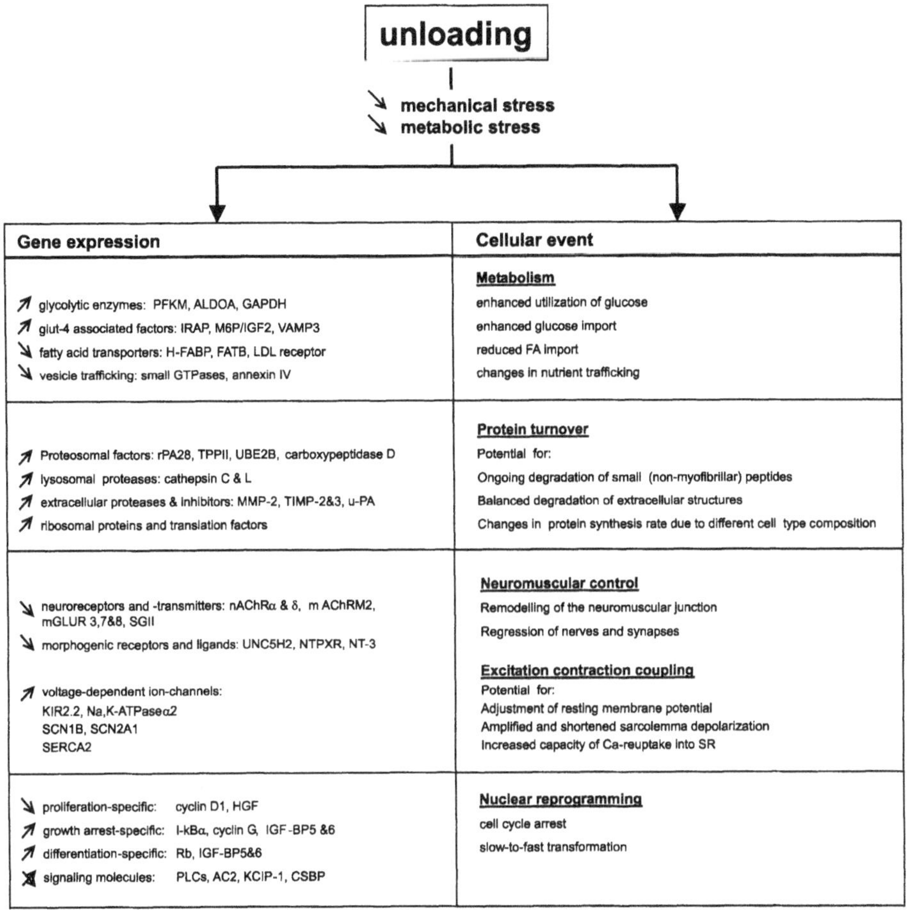

Fig. 6 Effect of simulated microgravity on the rat M. soleus gene profile. Schematic diagram summarizing the expressional changes in functional categories in hindlimb suspended rat m. soleus and their interpretation in terms of affected cellular event (Wittwer et al. 2002a). Unloading is expected to cause these expressional adaptations through a reduction in mechanical as well as metabolic stress

vens et al. 1999; Baldwin and Haddad 2001; Booth and Baldwin 1995). The latter is due to post-transcriptional (enzymatic) events. Moreover, it has been demonstrated that activation of the ubiquitin-proteasome pathway, the NF-kappa B pathway and apoptosis is involved in muscle loss with rodent m. soleus atrophy (Hunter et al. 2002; Ikemoto et al. 2001; Allen et al. 1997; Sandri 2002). Recently, we have demonstrated that prolonged unloading of rat soleus muscle causes massive adaptations of the gene profile (Wittwer et al. 2002). Expression levels of many genes involved in fibre transformation, metabolism, nutrient trafficking, protein turnover and cell regulation were changed. Additionally, expressional changes of extracellular proteases, of genes involved in nerve-muscle interaction and of EC coupling were identified to change with hindlimb suspension-induced soleus atrophy (Fig. 6). The results support a modulation of nutrient trafficking by glut-4 associated factors and small G proteins. Many of the changes of the transcriptome observed in this study were described for the first time. The data on molecular markers indicating changes at the neuromuscular junction with microgravity are supported by structural evidence demonstrating that space flight modifies the morphology of the neuromuscular junction (Deschenes et al. 2001). We would assume that the widespread changes in the transcriptome elicited by hindlimb suspension in rats is typical for disuse atrophy and we suspect that prolonged microgravity exposition would lead to similar changes in human muscles.

Nutritional interventions

High-fat diet

It has long been known that adequate glycogen stores are necessary for optimal muscle performance (Bergstroem et al. 1967). However, it has only recently been recognized that nutritional interventions manipulating the macronutrient composition of the diet can have a significant impact on muscle structure and function (Helge 1996; Helge et al. 1998; Hoppeler et al. 1999). As indicated above, endurance exercise training leads to an increased use of fat as a substrate. This adaptational feature has been identified as an important means of sparing glycogen during long-term exercise (Holloszy and Coyle 1984). Experiments in rats have suggested that high-fat diets alone can enhance endurance performance by increasing muscle tissue capacity for oxidative metabolism, in particular the ability to oxidize fat (Miller et al. 1984). It was also found that these changes were additive to endurance exercise training (Simi et al. 1991). Likewise, in humans a sizable number of studies report a decrease in the respiratory exchange ratio as a consequence of a high-fat diet (Carey et al. 2001; Jansson and Kaijser 1982; Phinney et al. 1983; Lambert et al. 1994), indicating an increased reliance of muscle tissue on fat as a substrate. High-fat diets gradually increase fat oxidation both at rest and during exercise, while there is no change in rate of appearance of fatty acids and plasma-derived fatty acid oxidation (Schrauwen et al. 2000). However, unlike in animals, a high-fat diet in humans has no effect on muscle oxidative capacity (Goedecke et al. 1999; Hoppeler et al. 1999). Of the enzymes involved in beta-oxidation, only carnitine acyl transferase activity was increased, while citrate synthase and 3-hydroxyl-acyl-CoA dehydrogenase was not changed, in human m. vastus lateralis after 10 days of high-fat diet (Goedecke et al. 1999; Peters et al. 2001; Table 1). Likewise, muscle mitochondrial content seems to be unaffected by a high-fat diet in humans, while intramyocellular lipid (IMCL) content of muscle cells appears to be elevated by this diet intervention (Hoppeler 1999; Vogt et al., unpublished observations). Interestingly, the reports on performance gains in human athletes after high-fat diets remain controversial (Vogt et al., unpublished observations). An increase in endurance capacity seems to be most likely for ultraendurance events when care is taken that the muscle glycogen pool is fully loaded (Lambert et al. 2001).

Little is known on the molecular events underlying these metabolic adaptations of the muscle cell. The available data support the notion that the expression of enzymes involved in regulation of mitochondrial function, i.e. proton uncoupling (UCP-2 and -3) and the entry of carbohydrate-derived pyruvate into the Krebs cycle (PDK4), is affected (Schrauwen et al. 2001a; Peters et al. 2001). The induction of PDK4 mRNA and its activity 1 day after onset with the high-fat diet indicates enhanced entry of carbohydrate-derived pyruvate into the Krebs cycle (Peters et al. 2001). The up-regulation of UCP-2 and 3 mRNA in m. vastus lateralis of subjects after 10 weeks of high-fat diet was more pronounced in humans with high proportions of type IIa fibres, suggesting a role for UCPs in lipid utilization (Schrauwen et al. 2001a). This eventually indicates protection of mitochondria against accumulation of non-esterified fatty acids inside the mitochondria (Schrauwen et al. 2001b). An important adaptive mechanism is the activation of the key regulator fat metabolizing enzyme expression, peroxisome proliferator-activated receptor (PPAR), through lipid products. Activation of PPARα and PPARγ may be related to the induction of UCP-2 and -3 mRNA expression (Vidal-Puig et al. 1997; Pedraza et al. 2000).

Recently, it has been demonstrated that metabolic interventions affect the regulation of metabolism via transcriptional and post-translational events. It has been realized that amino acids are not only substrates of chemical reactions but also act as signalling molecules in control of protein synthesis in skeletal muscle (Kimball 2002). The branched-chain amino acid leucine mediates signalling, leading to increased initiation of mRNA translation as well as to an increase in the phosphorylation status of the translational repression protein 4E-BP1 and the ribosomal protein S6 kinase (S6K1) (Greiwe et al. 2001). Conjugated linoleic acid isoforms have been demonstrated to improve glucose transport and glycogen synthase activity and to increase UCP-2 mRNA in skeletal muscle of obese rats (Ryder et al. 2001). A challenge for future studies is to identify whether the adaptations identified in rodents play a role in explaining the metabolic changes following diet interventions in humans.

Caloric restriction

Caloric restriction (CR) has been recommended to control weight and the blood lipid profile, particularly in the elderly (Schlienger et al. 1995). Animal experimentation shows that caloric restriction reduces fibre loss and mitochondrial abnormalities in aged rat muscle (Aspnes et al. 1997). To our knowledge, no data have been published on the effect of caloric restriction on structural or functional adaptations in human skeletal muscle. Recent data from experiments with rodents indicate a significant modulation of gene expression under caloric restriction (Weindruch et al. 2002). Caloric restriction increased expression of genes involved in carbohydrate and fatty acid metabolism and decreased expression of genes involved in stress-response/DNA repair and detoxification processes. Notably, these data suggest that caloric restriction could reverse age-related changes in gene expression.

It is obvious that nutritional preferences and requirements vary widely between humans and mice. Nevertheless, there are some indications of similar transcriptional adaptations occurring in both species as a consequence of caloric restriction. An important role of caloric restriction in regulation of muscle plasticity in humans is indicated by the recent observation that caloric restriction results in an up-regulation of cytoskeletal protein-encoding genes and a decrease in the expression of genes involved in mitochondrial bioenergetics in m. vastus lateralis of rhesus monkeys (Kayo et al. 2001). Moreover, caloric restriction in primates is found to reduce oxidative damage (Zainal et al. 2000). These results indicate that the induction of an oxidative stress-induced transcriptional response may be a common feature of ageing in skeletal muscle of rodents and primates. However, the extent to which caloric restriction modifies these responses may be species-specific.

Hypoxia

Intermittent and continuous exposure to hypoxia causes many adaptive events in human skeletal muscle tissue (see Hoppeler and Vogt 2001a). A significant 10% loss in thigh muscle cross-sectional area in 14 subjects after return from mountaineering expeditions to the Himalayas has been reported (Hoppeler et al. 1990). Post-expedition volume density of mitochondria was reduced by 23%, such that the total loss of oxidative capacity of m. vastus lateralis was close to 30%. The muscle capillary bed was not much affected by high-altitude exposure, with the result that a practically unchanged capillary network supplied a smaller muscle oxidative capacity after return from expedition. The results of a substantial loss of muscle oxidative capacity as a consequence of permanent exposure to hypoxia is substantiated by biochemical results obtained from the same subjects (Howald et al. 1990) but also from results of Operation Everest II, a simulation of an ascent to Mt. Everest in the pressure chamber (MacDougall et al. 1991; Green et al. 1989). The combined structural and functional evidence suggests that lowlanders do not tolerate long-term severe hypoxia well. This view is supported by the observation that the quantity of lipofuscin present in muscle fibres is increased by over threefold after returning from an expedition (Martinelli et al. 1990). Lipofuscin is a degradation product formed by lipid peroxidation and characterizes cytological damage possibly incurred by radical formation in muscle cells in hypoxia (Radak et al. 1994). Hypoxia training also initiates a shift in muscle metabolism towards more carbohydrate oxidation (Vogt et al. 2001).

Changes in gene expression in mammalian skeletal muscles as a consequence of hypoxia exposure have not yet received much attention. Training in intermittent hypoxia has been shown to improve oxygen transfer to muscle mitochondria by increasing muscle capillarity and muscle myoglobin content (Vogt et al. 2001). These changes are believed to be related to hypoxia-inducible factor 1 alpha (HIF-1α), which is a transcription factor controlling the expression of many genes making up the hypoxia response (Semenza 1999). HIF-1α is constitutively expressed but degraded rapidly under normoxic conditions. Hypoxia has been demonstrated to stabilize HIF-1α and thus make it available for transcription regulation (Jewell et al. 2001) (see also "Metabolic factors"). We were also able to show recently that HIF-1α mRNA is induced in human skeletal muscle after endurance exercise training in hypoxia (Vogt et al. 2001). Recent experiments with rats hint at the molecular changes possibly responsible for the functional adaptations in skeletal muscle to chronic hypoxia. Protein expression of the mitochondria-associated Bcl-2 was inversely related to the oxidative character of muscles and markedly induced in rat muscles after 3 weeks' exposure to chronic hypoxia. This change paralleled the oxidative character of the muscles (heart > soleus > gastrocnemius). No sign of apoptosis was detected in these experiments (Riva et al. 2001). It was suggested that the adaptations were an anti-apoptotic mechanism allowing protection against the lack of oxygen in oxidative muscles. It has recently been shown that protein expression of the enzymes and transporters of lactate, lactate dehydrogenase (LDH) and monocarboxylate transporters (MCT1, 2 and 4) is affected in a tissue-specific manner by long-term exposure to hypobaric hypoxia (McClelland and Brooks 2002).

It is currently not known whether the adaptability of the lactate shuttle (LDH and MCTs) contributes to the changes in the control of lactate metabolism that relate to the paradoxical observation that maximal lactate concentration at VO_2max is reduced after adaptation to altitude (Hochachka et al. 2002).

The involvement of transcriptional mechanisms in skeletal muscle adaptations to hypoxia have been thoroughly studied in fish (see Gracey et al. 2001). These gene expression profiling data document a very rapid shut-down of energy-requiring processes such as protein synthesis and locomotion in skeletal muscle and a concurrent suppression of cell growth and proliferation involved in hypoxia survival mechanisms. Future studies will have to provide the evidence on whether similar expressional changes may be responsible for the observed reduction in fibre size and shift in energy metabolism observed in human skeletal muscle with stays at high altitudes.

Ageing

Age-related changes in muscle mass and functional properties are the result of a complex hierarchical system of regulation of cell ageing processes and corresponding cell adaptive responses (Navarro et al. 2001; Tables 1, 2). It is generally accepted that there is a decrease in muscle mass and a decrease in muscle oxidative capacity, in particular after the age of 70 in humans. It is not clear, however, how much of the observed changes are due to inactivity and how much are due to the ageing process *per se*. It has been demonstrated that old muscle retains its ability to enlarge during resistance training (Booth and Baldwin 1995; Balagopal et al. 2001; Hikida et al. 2000) and to increase oxidative capacity after endurance training (Kiessling et al. 1973). Numerous aspects of skeletal muscle plasticity thus do not seem to be compromised by the ageing process. Ageing muscle has been shown to be different in motor-unit innervation, post-synaptic properties (sarcolemmal channel alterations and depolarization properties), EC coupling and calcium homeostasis (Weindruch1995; Navarro et al. 2001; Klitgaard et al., 1990c). Old muscle also responds differently to growth hormone and testosterone release as a consequence of an acute bout of resistance exercise than young muscle (Craig et al. 1989). Old muscle has a higher proportion of type I/IIa and IIa/IIx hybrid fibres (Klitgaard et al. 1990a), exhibits reduced state III mitochondrial respiration rates (Trounce et al. 1989) and a decrease in citrate synthase activity (Pastoris et al. 2000). Data from rhesus monkeys demonstrate age-dependent accretion of signs of oxidative damage in myofibrils (Zainal et al. 2000). These changes may be related to impaired mitochondrial respiration rates, as the O_2 that is incompletely reduced by mitochondria yields reactive oxygen species (Chandel and Schumacker 2000). The fractional synthesis rate of MHC and mixed protein is reduced in aged muscle (Hasten et al. 2000). As contributing factors to muscle ageing hormonal factors, modulation by caloric restriction and decreased exercise activities have been recognized. Gene profiling data imply that specific expressional changes are associated with the ageing process in mouse skeletal muscle (Weindruch et al. 2002). In particular, it was noted that expression of genes involved in stress response and neuronal growth were increased while expression of genes responsible for protein turnover, calcium handling and energy metabolism were reduced. The observations of differences in the expression of factors involved in nerve morphogenesis are supportive evidence for a process of denervation and reinnervation in ageing human muscle (Vandervoort 2002).

In rhesus monkeys, experiments taking advantage of high-density oligonucleotide arrays indicate a selective up-regulation of transcripts involved in inflammation and oxidative stress and a down-regulation of genes involved in mitochondrial electron transport and oxidative phosphorylation with ageing (Kayo et al. 2001). These data are broadly compatible with the changes of the transcriptome described in human skeletal muscle with ageing. A recent gene profiling study comparing skeletal muscle tissue from healthy elderly (62-75 years old) to young (20-34 years old) men demonstrated elevated expressions of genes typically involved in stress and damage response and a decrease in expression of a gene encoding a DNA repair/cell cycle checkpoint protein (Jozsi et al. 2000; Table 2). Several genes with important roles in the adaptation of muscle to exercise, specifically the processes of angiogenesis and cell proliferation, showed a similar response to exercise in older and younger subjects. Other responses, however, such as those seen after resistance exercise were different between the two age groups. Jozsi et al. (2000) concluded from these results that there may be a molecular basis for the difference in the response of skeletal muscle tissue to resistance training in elderly people.

Thermal stress

Very little is known concerning the adaptations of skeletal muscle to extreme thermal conditions. During exercise, multiple afferent signals including body temperature feed back on sympathetic centres and in turn influence cardiovascular, hormonal, and metabolic responses, electrolyte homeostasis and muscular performance (Christensen and Galbo 1983). Using a combination of indirect calorimetry and stable isotope methodology, it has also been demonstrated recently that exposure for 2 h at 10°C stimulates heat production by involuntary muscle contraction (cold shivering) and increased the oxidation of plasma glucose, muscle glycogen and of lipids by more than twofold (Haman et al. 2002). These data indicate that lipids and muscle glycogen provide most of the energy for heat production. Reversible ultrastructural modifications possibly related to muscle temperature have been described in horses exercising at higher intensities (McCutcheon et al. 1992). Heat-stress is known to partially prevent hindlimb suspension-induced atrophy through a process involving heat-shock proteins (Naito et al. 2000). Moreover, it has been shown for rat gastrocnemius muscle that 6 weeks of uphill running (5 days/week, 60 min/day) on a treadmill at cold temperature (4° or 8°C), concomitant with a drop in exercising core temperature, causes a greater increase in cytochrome-c oxidase activity than running at ambient temperature of 23°C (Mitchell et al. 2002). The data indicate that decreasing exercising core temperature may enhance mitochondrial biogenesis. However, there seems to be practically no further data in the literature on mammalian skeletal muscle response to a temperature challenge. Muscle tissue in certain fish species exposed to water temperature changes with season exhibits a marked and well-described molecular plasticity (Johnston and Temple 2002). However, the changes occurring in fish muscle as a consequence of temperature adaptations are outside the scope of this article.

Signals, sensors and transducers

What could be the sensing mechanisms responsible for the specific adaptational events induced in skeletal muscle by exercise and changes in the environment? For example, strength training causes a selective increase in the myofibrillar compartment, thus increasing the capacity for force production. By contrast, endurance exercise predominantly increases the mitochondrial and capillary compartment, giving rise to an improvement of the capacity for aerobic energy turnover. The unique and specific muscle cellular responses to strength- and endurance-type exercise make it likely that different sensing systems are involved with high-intensity low-repetitive (strength) training versus low-intensity high-repetitive (endurance) exercise training. The massive and selective perturbations of cellular homeostasis with exercises of different intensity and duration suggests that both metabolic *and* mechanical factors can be sensed separately and need to be integrated into complex transcriptional responses. The consequence of these transcriptional regulations is seen as the malleability of cellular structure. The modified structures then enable the transformed muscle to better withstand an imposed physiological stress (Booth and Thomason 1991; Booth and Baldwin 1995). Apart from metabolic and mechanical signals, there is ample evidence for the involvement of hormonal (growth hormone, IGFs, ACTH, thyroid state and testosterone) as well as neuronal factors (via intracellular Ca^{2+}) playing important roles in the plastic response of muscle tissue.

For the sake of simplicity, this article will focus on the concept that mechanical, metabolic, hormonal and neuronal factors are the key signals distinctly sensed by skeletal muscle tissue (Fig. 7). These signals will be discussed separately. Finally we will focus on the integration of these signals into the transcriptional adaptations that are seen in skeletal muscle.

Mechanical factors

The role of mechanical factors is best supported by animal model studies demonstrating that mechanical factors stretch or modulate transcriptional induction of the *c-jun* and *c-fos* genes within 1 h (Dawes et al. 1996). In situ hybridization experiments of human m. vastus lateralis show that an increase in c-fos and c-jun mRNA after running occurs in a patchy expression pattern not corresponding to fibre type distribution. This has been taken to support the notion that factors other than metabolic products related to fibre recruitment contribute to the observed massive up-regulation of *c-fos* gene transcription in exercised human skeletal muscle (Puntschart et al. 1998).

The contention that mechanical factors are involved in regulation of the mRNA concentration with exercise training is further supported by the observations in mechanically stressed skeletal muscle fibres that mechanisms known to contribute to pre-transcriptional control are affected concomitantly with the transcription of downstream target genes. In particular, exercise can induce all three limbs of the MAP-kinase pathways (ERK1/2, JNK and p38) in human m. vastus lateralis (Aronson et al. 1997, 1998; Yu et al. 2001; Boppart et al. 1999; Widegren et al. 2001; Table 1). Activation of these kinases is known to affect transcription factors of the jun/fos and ATF/CREB family (Hunter and Karin 1992; Table 2). The latter are known to bind to the promoter of many genes, e.g. *c-jun*, *c-fos* and cytochrome C, that are affected simultaneously by exercise (Hood 2001). In situ studies delineated that mechanical factors, i.e. the degree of tension in rat muscle fibres, control JNK and ERK1/2 differently and directly (Martineau and Gardiner 2001). In contrast, the p38 pathway in skeletal muscle is only responsive when mechanical stress (stretch, exercise) is applied in vivo (Boppart et al. 2001). This can be taken as an indication that p38 is indirectly controlled by mechanical factors, possibly through endocrine mechanisms. Furthermore, the activation of the p70S6 kinase has also been linked to mechanical or metabolic stimuli, which may induce muscle hypertrophy by influencing translation (factors) (Nader and Esser 2001). The mechanically induced release and production of growth factors (IGF-I, the IGF-I splice form MGF, IL-6) has been recognized as another signalling route in skeletal muscle exercise, potentially representing an important link between contracting

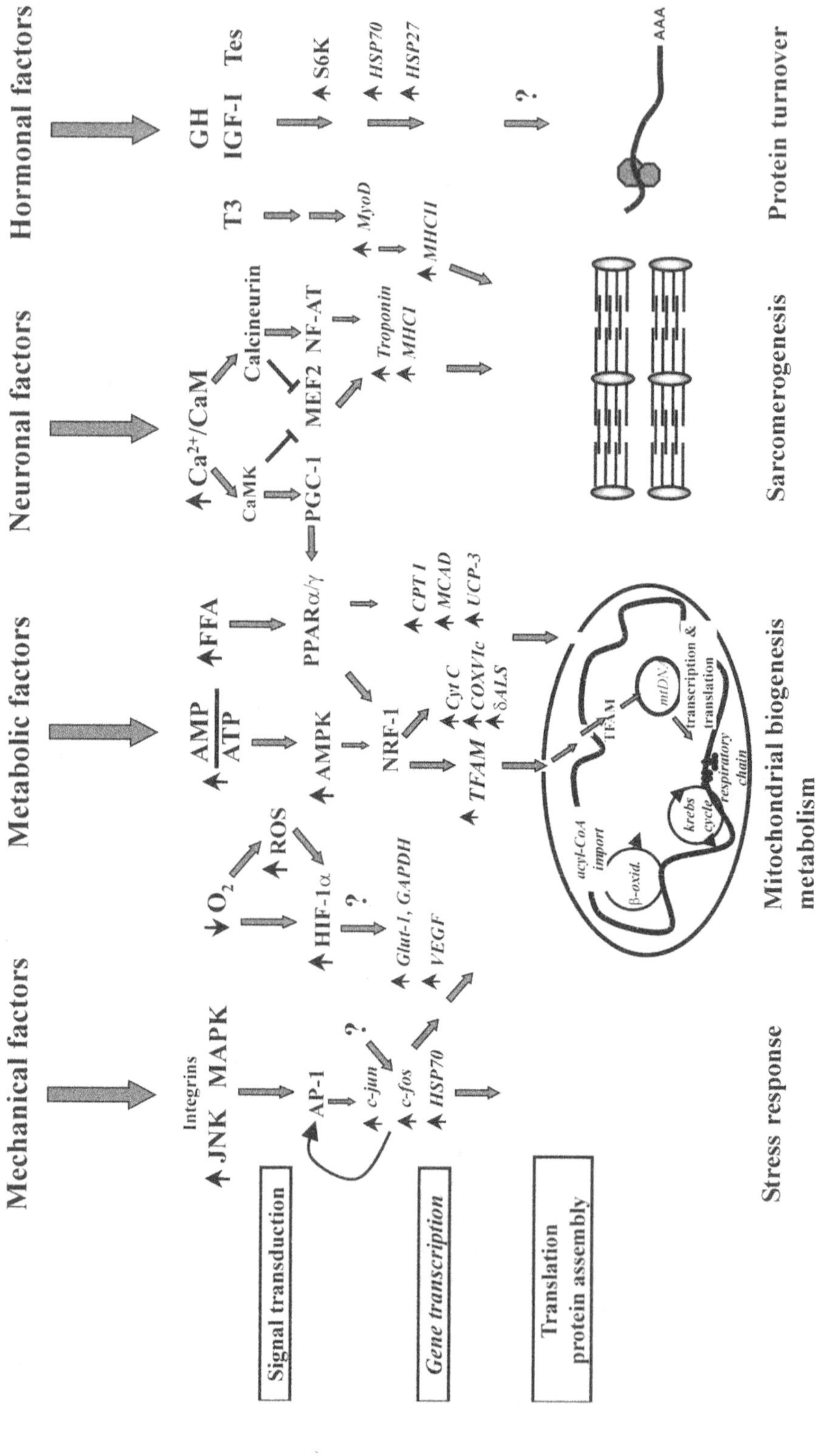
Mechanical factors
Metabolic factors
Neuronal factors
Hormonal factors
Signal transduction
Gene transcription
Translation
protein assembly
Integrins
JNK MAPK
AP-1
c-jun
c-fos
HSP70
O2
ROS
HIF-1α
Glut-1, GAPDH
VEGF
AMP
ATP
FFA
AMPK
PPARα/γ
NRF-1
CPT I
MCAD
UCP-3
Cyt C
COXVIc
δALS
TFAM
Ca2+/CaM
CaMK
Calcineurin
PGC-1
MEF2
NF-AT
Troponin
MHCI
MyoD
MHCII
T3
GH
IGF-1
Tes
S6K
HSP70
HSP27
acyl-CoA
import
β-oxid.
krebs
cycle
respiratory
chain
TFAM
mtDNA
transcription &
translation
AAA
Stress response
Mitochondrial biogenesis
metabolism
Sarcomerogenesis
Protein turnover

skeletal muscles and exercise-related metabolic changes (Goldspink 1999; Pedersen et al. 2001; Adams et al. 1999). Current thinking suggests that the activation of all these mechano-transduction events is ultimately linked to mechano-sensation via integrins and associated kinases (Gordon et al. 2001b; Carson and Wei 2000; Chiquet and Flück 2001).

Metabolic factors

Several metabolic perturbations that occur in severely exercised skeletal muscle, such as a drop in pH and oxygen tension and an increase in AMP concentration and free radical production (to name only a few), may be sensed and could serve to control changes in gene expression. The acute chemical perturbations of muscle cells are likely linked in multiple ways to signal transduction events that ultimately influence gene expression. A drop in the ATP/AMP ratio, activating 5'-AMP-activated protein kinase (AMPK) (Bergeron et al. 2001) is believed to be an important sensing mechanisms that links metabolic changes to gene expression (Aronson et al. 1998; Hood 2001). The sensing of cellular energy charge by the AMPK cascade supports the energy charge hypothesis first proposed in 1960 by Daniel Atkinson (in Hardie and Hawley 2001). Unsaturated long-chain fatty acids are released from adipose tissue during exercise and can directly stimulate PPARα-activated gene transcription (Escher and Wahli 2000; Kliewer et al. 1997; Huss et al. 2001). Unsaturated long-chain fatty acids must thus be considered to be paracrine factors modulating the capacity of muscle cells to transport and metabolize fatty acids (Fig. 7).

Ever since Reynafarje (1962), local tissue hypoxia has been assumed to be an important stimulus for muscle tissue adaptations. Using ^{1}H nuclear magnetic resonance spectroscopy, it was shown that myoglobin desaturation occurs within 20 s of onset of exercise in human quadriceps muscle (Richardson et al. 1995). The finding of a low muscle oxygen tension during exercise is supported by the earlier studies of Gayeski and Honig (1978). Using the same technique, Richardson et al. (2001) have shown recently that oxygen saturation in human skeletal muscle, calculated from myoglobin desaturation, is reduced with onset of exercise, to a plateau. This has been taken to indicate that maximal muscle local hypoxic conditions may prevail even during low-intensity exercise in normoxia. This reasoning is based on the assumption that oxygen is homogenously distributed within and among muscle fibres. This may not be the case. Mitochondria, the oxygen sink, are more prevalent in oxidative fibres by a factor of at least three and even within fibres they are clustered in the fibre periphery (Howald et al. 1985). As a consequence, we would assume that the average ^{1}H nuclear magnetic resonance deoxy signal does not account for local hypoxia conditions within individual fibres. However, it is these local conditions that are likely to be relevant for the molecular response of muscle fibres (Lindstedt and Wells 1988; Gayeski and Honig 1978).

The observation of a decrease in muscle oxidative capacity with long-term hypoxia exposure came as a surprise (MacDougall et al. 1991; Kayser and Hoppeler 1991). A number

Fig. 7 Integration of signals into changes of skeletal muscle gene expression. Mechanical, metabolic, neuronal and hormonal factors are transduced via complex pathways into adaptations in gene expression. A prominent increase in activity of several kinases and downstream transcription factors (for details see text) is observed within minutes after application of stress to muscle tissue. This may link to induced transcription of downstream genes and functional adaptations. Recent data indicate that convergence between multiple pathways exists

of studies have been designed to maximize hypoxic stress on muscle (and to minimize muscle deterioration) by exposing working subjects to normobaric or hypobaric hypoxia for the period of the exercise training sessions only (short-term hypoxia; Terrados et al. 1988, 1990; Desplanches et al. 1993; Emonson et al. 1997; Melissa et al. 1997; Vogt et al. 2001). The consensus seems to be that training in short-term hypoxia produces training effects similar to, but not identical with the effects seen after training in normoxia (see Hoppeler and Vogt 2001b). The reason why long-term hypoxia is damaging for muscle cells is not yet fully understood (Hoppeler and Vogt 2001b).

Hypoxia has been demonstrated to instantaneously stabilize the transcription factor HIF-1α, which appears to be a master gene involved in the hypoxia response of most mammalian cells (Semenza 1999). In HeLaS3 cell culture, 0.5% O_2 provokes an increase in HIF-1α from undetectable levels within 2 min (Jewell et al. 2001). Continuous hypoxia induces HIF-1α with maximal levels after 1 h and gradually decreasing thereafter in various organs of mice (Stroka et al. 2001). The increase in HIF-1α is mediated by reducing hydroxylation of a proline and in the oxygen-dependent degradation domain of HIF-1α. This is achieved by interference with a hypoxia-inducible proline hydroxylase; as in normoxia, HIF-1α is targeted for destruction via the ubiquitin pathway (Wenger and Bauer 2001; Zhu and Bunn 1999; Kietzmann et al. 2000). Recently, hydroxylation of an asparagine residue within the HIF-1α C-terminal transactivation domain has been noted to block its association with co-activators (Lando et al. 2002; Mahon et al. 2001).

The increase in HIF-1α and its dimer formation with HIF-1β (ARNT) subsequently induces expression of a number of hypoxia-activated genes (Semenza 1999). The heterodimeric complex (HIF-1) drives transcription of a variety of hypoxia-inducible genes including angiogenic factors, glucose transporters and glycolytic enzymes through binding to cis-acting hypoxia-response elements (HREs) in the promoters (Semenza 1999; Pages et al. 2000). Recently, direct evidence was provided that in rat skeletal muscle, oxidant production is increased during prolonged exercise, with as potential sources for oxidants both the mitochondrial respiratory chain and the NADPH oxidase (Bejma and Ji 1999). Radicals (reactive oxygen species, ROS) produced in this process are known to activate multiple pathways that influence gene expression and have been shown to be linked to the hydroxylation of HIF-1α (Kietzmann et al. 2000). In this regard it is interesting to note that mitochondria-derived ROS are required to initiate HIF-1α stabilization during hypoxia (Chandel et al. 2000; Chandel and Schumacker 2000) and that enzymic and non-enzymic oxygen free radical-generating systems have been implicated to control hydroxylation of prolyl and arginyl residues (Stadtman 1990). These lines of evidence indicate that the ubiquitous transcription factor, HIF-1α, could be a major molecular component that integrates the events that translate increased muscle activity into plastic remodelling of muscle with local tissue hypoxia as a signal.

Neuronal factors

It is well known that transient increases in intracellular Ca^{2+} are involved in fibre transformation. This is believed to occur through modifying Ca^{2+}-calmodulin-dependent processes (Sreter et al. 1987). Recent data provide indirect evidence for neuronal modulation of muscle function by Ca^{2+}-dependent processes and indicate that calmodulin may in fact function as a sensor (Talmadge 2000). Ca^{2+} is periodically released from the SR during contractions, thereby increasing cytoplasmic Ca^{2+} concentration by nearly two magnitudes before Ca^{2+} is pumped back into the SR via Ca-ATPAse(s) (Berchtold et al. 2000). Such periodic increases in intracellular Ca^{2+} might trigger the activation of Ca^{2+}/calmodulin-dependent enzymes acting as calcium sensors (Talmadge 2000). This scenario is supported by the observation that Ca^{2+}/calmodulin kinase II is a decoder for Ca^{2+} oscillations (Dupont and Goldbeter 1998) and that Ca^{2+}-independent CaMK(II) activity is increased in exercised rodent muscle (Flück et al. 2000). The latter observation could eventually be related to an elevation of intracellular free calcium to levels high enough to activate CaMK. This would then be similar to the situation observed in chronically stimulated muscles (Chin and Allen 1996, reviewed in Talmadge 2000). An increase in autonomous CaMK activity may in turn favour mitochondrial biogenesis, as CaMK seems to be involved in mitochondrial biogenesis by causing up-regulation of a master regulator of mitochondrial biogenesis, PGC-1 which resides upstream of PPARs (Wu et al. 2002). Much attention has also been drawn to results that implicate the Ca^{2+}/calmodulin-dependent calcineurin:NF-AT and CaMK pathways in the regulation of muscle-specific gene expression (Talmadge 2000; Wu et al. 2002). Calcineurin:NF-AT has also been found to influence the expression of slow muscle fibre specific contractile proteins (i.e. troponin) as well as to help to control transformation from type II towards type I fibres (Chin et al. 1998; Figs. 4,7).

Recently, molecular correlates for neuronal reprogramming as an important mechanism in muscle plasticity have been identified by the demonstration that synapses (Deschenes et al. 2001), elements of the neuromuscular junction (Wittwer et al. 2002a) and motor units (Leterme and Falempin 1996) are modified in animal models of microgravity. The observations that serotonin receptors adapt to hindlimb unloading (Wittwer et al. 2002a) suggest that afferent metabolic receptors in afferent nerves are also modulated by changes in muscle loading conditions. Taken together, the available evidence suggests that neuronal factors are involved in muscle plasticity either by interfering with contractile activity or by modulating mechanical and metabolic signals, on several levels of hierarchical organization.

Hormonal factors

It is well documented that hormones have a pronounced effect on muscle phenotype (Frisch 1999; Everts 1996; Florini et al. 1991). There is considerable evidence that serum levels of many hormones are influenced by exercise, muscle loading and ageing. For example, endurance exercise acutely increases serum growth hormone, endorphin and renin levels while insulin levels are reduced (Kjaer et al. 1999). Also, a single resistance exercise session (45-60 min) involving all major muscle groups increases serum GH by more than fourfold but had no effect on testosterone levels (trained or untrained people) (Craig et al. 1989). By contrast, an acute high-intensity weightlifting protocol increased the levels of serum testosterone, cortisol and growth hormone in junior elite male weightlifters (Kraemer et al. 1992). Total serum testosterone levels were increased in both men and women after 6 and 8 weeks of heavy resistance training while the increase in serum growth hormone levels seen after exercise did not result in a change in basal GH concentration (Kraemer et al. 1998). Resistance training was found to cause alterations in the acid-labile subunit of the circulating IGF system in humans (Nindl et al. 2001). Such an increase in protein could also be seen with overloading of rat muscle potentially preceding increases in IGF mRNA (Adams et al. 1993). Space flight lowers thyroxine and triiodothyronine levels in human, causes a loss of insulin sensitivity, increases the activity of the sympathetic system but leaves growth hormone levels unaffected (Strollo 1999). The level of 13 hormones including thyroid stimulating hormone (TSH) and growth hormone were dramatically increased after a triathlon event where dehydration did not significantly contribute to the observed endocrine changes (Malarkey et al. 1993). In well-trained young subjects, long-distance runs (45-75 km) were shown to selectively increase T4 levels, whereas in older and less high-performance athletes, the same stimulus provoked a significant decrease of the levels of T4, T3 and TSH (Hesse et al. 1989; Table 1). Differences in the response of GH levels with lower-body resistance training involving a combination of concentric and eccentric muscle work indicate that serum GH levels are sensitive to the type of muscle action (Kraemer et al. 2001). Subjects with a spinal cord injury were shown to have a depression of the anabolic hormones, testosterone and growth hormone/IGF-I and decreased serum T3 levels (Bauman and Spungen 2000). In this context, it may also be noted that hypothyroidism is characterized by exercise intolerance (McAllister et al. 1997).

Evidence for involvement of hormones in muscle plasticity is indicated by observations on the influence of the thyroid state on MHC expression in rodents (Booth and Baldwin 1995). Skeletal muscle is one of the major target organs for thyroid hormones. Although thyroid hormones are essential during growth, both their excess and their deficiency cause muscle wasting by yet unknown mechanisms. The muscles most commonly affected are those used during prolonged effort (slow-twitch muscles). Most of the peripheral effects of the thyroid hormones can be ascribed to the action of triiodothyronine (T3), which is produced by de-iodination of thyroxine (T4) in liver and kidney. From the plasma, T3 is actively transported into skeletal muscle (Everts 1996). Thyroid hormone (T3) plays an important role in the regulation of muscle plasticity. Hyperthyroidism and doses of the thyroid hormones T3, cause an up-regulation of IIa and possibly IIx fibres in 30-40% of rat soleus fibres (Baldwin and Haddad 2001). In contrast, hypothyroidism is known to cause IIa fibres to shift to express only type I MHC. The thyroid state and particularly a single T3 injection affects Ca^{2+} ATPase and Na+, K+ ATPase as early as after 24 h. In humans, a linear correlation between the Na+, K+ ATPase concentration of skeletal muscle and the free T4 index was established (Everts 1996). Current evidence indicates that T3 largely exerts its effect by first interacting with receptor complexes in the nucleus. The subcellular action of the thyroid hormone-thyroid receptor interaction is thought to be mediated via their combined interaction with DNA sequences located in the promoter (regulatory) sequence of thyroid-responsive genes designated as TRE elements (reviewed in Booth and Baldwin 1995).

Excess doses of the thyroid hormone triiodothyronine have been shown in young subjects to increase β-adrenergic receptor density in all types of skeletal muscle fibres (Martin 1993). Thyroid exposure results in alterations of the myogenin/myoD mRNA expression pattern and MHC (Hughes et al. 1993). This latter observation is in general agreement with the finding that triiodothyronine up-regulates expression of genes involved in cellular functions including transcriptional control, mRNA maturation, protein turnover, signal transduction, cellular trafficking, and energy metabolism in skeletal muscle of healthy men (Clement et al. 2002). Non-genomic actions of thyroid hormone independent of nuclear receptors for the hormone have been described. These actions include alterations in solute transport (Ca^{2+}, Na+, glucose), changes in activities of several kinases, including protein kinase C, cAMP-dependent protein kinase and pyruvate kinase M2, effects on efficiency of specific mRNA translation and mRNA t1/2, modulation of mitochondrial respiration, and regulation of actin polymerization (Davis and Davis 1996). All these effects could influcncc musclc phenotype.

A link has been established between β-adrenergic receptor level density and oxidative capacity in skeletal muscle. The administration of β-adrenergic agonist to humans during long-term bedrest experiments has been shown to maintain or increase the activity of CS and SDH in skeletal muscle (reviewed in Booth and Baldwin 1995).

Hormones with an anabolic action (growth hormone, insulin-like growth factors, and testosterone) stimulate human muscle growth mainly by increasing protein synthesis or by decreasing protein breakdown (insulin) (Rooyackers and Nair 1997). The stress hormones (glucagon, glucocorticoids, and catecholamines) cause muscle catabolism when up-regulated together; however, the effects of the individual hormones on human muscle and their mechanisms of action remain to be defined more clearly. Moreover, it has been demonstrated that the adipocyte-derived hormone leptin increases glucose and fatty acid metabolism in skeletal muscle (Ceddia et al. 2001). Despite the evidence that heavy, glycogen-depleting exercise lowers serum leptin concentrations in correlation with serum triglyceride concentration (Tuominen et al. 1997), the involvement of increased secretion of this adipocyte-derived factor in control of exercise-induced muscle plasticity and fuel homeostasis in humans remains to be established (Bradley et al. 2001).

Testosterone has pronounced effects on muscle size and strength in normal men, especially when combined with strength training (Bhasin et al. 1996). This implies that the increased serum testosterone levels seen after resistance exercise may contribute to increased protein synthesis (Table 1). Similarly, doses of GH administration increase fasting serum IGF-I but may increase the rate of muscle protein synthesis in normal adults (Gamrin et al. 2000; Fryburg et al. 1992; Yarasheski et al. 1993a, 1993b) but not in experienced weightlifters (Yarasheski et al. 1993a). Growth hormone stimulates growth of skeletal muscle and connective tissue, increases rate of protein synthesis, and decreases rate of glucose use, thereby counteracting insulin action on lipid and glucose metabolism (Dominici and Turyn 2002; Howrie 1987). Recently, GH administration alone was demonstrated to decrease fat mass, increase fat-free mass, and induce MHC IIx in healthy elderly men whereas resistance training did overrule the changes in MHC (the MHC IIx isoform was decreased) composition induced by GH administration alone (Lange et al. 2002). This parallels the observation that supplementation of resistance training with GH dose not further enhance muscle anabolism and function (Yarasheski et al. 1992). Secretion of most of the hormones mentioned above is under the control of the central nervous system. This points to the important role of the brain for muscle plasticity. Growth hormone (somatotropin) is secreted by the anterior pituitary gland in response to various stimuli, including exercise, hypoglycaemia, and arginine.

Taken together, many hormones have been shown to modulate specific aspects of muscle phenotype. An integrated view of muscle plasticity therefore needs to account for the hormonal milieu experienced by the muscle during any adaptive event.

Signal integration

All factors discussed in the preceding paragraph are likely to interact with each other and to affect skeletal muscle gene expression at multiple levels. It is likely that many sig-

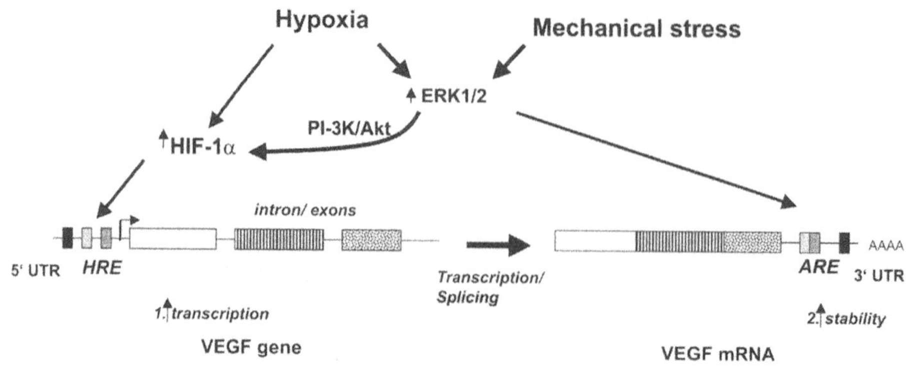

Fig. 8 Possible cooperation of oxygen and mechanical signal into changed VEGF mRNA level. The level of VEGF mRNA synthesis and degradation is controlled through cis-elements. In normoxia, VEGF mRNA is rapidly degraded through a mechanism involving two instability elements in its 3'UTR (adenylate-uridylate-rich elements, *ARE*). A drop in oxygen tension (pO_2) stabilizes the transcription factor HIF-1α that in turn interacts through hypoxia response elements (*HRE*) with the VEGF 5'UTR, which induces VEGF transcription. Moreover, a drop in oxygen tension and mechanical stress induce the MAPK pathway which increases in the half-life time of VEGF mRNA through modulating the binding of a hypoxia-inducible protein to a stability element in the VEGF 3'UTR. Finally, phosphorylation of HIF-1α by ERK1/2 has been shown to enhances HIF-1-dependent transcriptional activation of VEGF. Thus mechanical factors and metabolic factors (oxygen tension) can possibly interact at the transcriptional and post-transcriptional level to control VEGF mRNA in skeletal muscle through HIF-1α and MAPK

nalling pathways exert different and superimposed effects on promoter activity as well as on RNA stabilization, degradation and translation.

So far, no study reports on the complex task of monitoring the effect of exercise on multiple levels of expression control for one single gene. However, how signal integration could work is illustrated by the regulation of VEGF transcription and mRNA stability by hypoxia (Fig. 8). The VEGF mRNA falls into a class of labile messengers which codes for transiently expressed proteins, induced by acute stress signals. The prime mechanism for hypoxia-mediated regulation of VEGF mRNA synthesis acts through cis-elements (Hypoxia response elements, HRE) within its 5'UTR (Pages et al. 2000). The decay of VEGF mRNA is determined through two instability elements (adenylate-uridylate-rich elements, ARE) in the 3'UTR, which promote VEGF mRNA degradation in normoxia, and a stability element, which increases the VEGF mRNA half-life time in hypoxia. In culture, it has been shown that hypoxia increases the steady-state levels of VEGF mRNA through inducing the VEGF promoter activity via binding of the transcription factor HIF-1 (hypoxia-inducible factor 1) to HREs (Forsythe et al. 1996). HIF-1 is a heterodimer of the HIF-1α and HIF-1β proteins and its activity in acutely regulated on the post-translational level by stabilization of HIF-1α in hypoxia (Semenza 1999). Moreover, under low oxygen tension HIF-1α is directly phosphorylated via the mechano-sensitive and hypoxia-sensitive ERK1/2 MAPK pathway and enhances HIF-1-dependent transcriptional activation of VEGF (Berra et al. 2000; Seko et al. 1997; Fig. 8). Hypoxia also enhances the stability of the VEGF mRNA through binding of a hypoxia-inducible protein to the 3'UTR (Levy et al. 1996). Furthermore, the tyrosine kinase inhibitor genistein, at doses known to inhibit ERK phosphorylation (Tang et al. 1998), inhibits hypoxia-induced protein binding to the VEGF 3'UTR and blocks the hypoxia-induced stabilization of VEGF via the 3'UTR (Levy et al. 1996). Finally, dominant-negative mutant forms of c-Src or of Raf-1, both of which can

activate the MAPK pathway, blocked the hypoxia-induced stabilization of VEGF 3'UTR transcripts (Mukhopadhyay et al. 1995). Thus mechanical factors and metabolic factors (oxygen tension) can interact through HIF-1α and MAPK at transcriptional and post-transcriptional levels to control VEGF mRNA in skeletal muscle.

Concerning the changes in mitochondrial mass or volume in human skeletal muscle, it is apparent that the factors that orchestrate mitochondrial biosynthesis with endurance training are manifold. Factors that are involved in the coordination of mitochondrial protein expression from the nuclear and mitochondrial genome include the nuclear respiratory factors 1 and 2 (NRF-1 and NRF-2) and the mitochondrial transcription factor A (TFAM). Expression of TFAM, and several nuclear encoded proteins (cytochrome C), is promoted by the nuclear-encoded transcription factors NRF-1 and NRF-2 (Virbasius and Scarpulla 1994; Evans and Scarpulla 1990). TFAM is essentially involved in general mitochondrial biogenesis (Larsson et al. 1998). It translocates into the mitochondria where it induces transcription and replication of mitochondrial DNA (Shadel and Clayton 1997; Fig. 7).

The involvement of factors other than TFAM and NRF-1 in exercise induced mitochondrial adaptations is suggested by the observations that selective transcriptional up-regulation of enzymes involved in beta oxidation of fatty acids with exercise is not explained by NRF-1 or NRF-2 alone (Lehman et al. 2000 and references therein).

Candidate signalling molecules that emerge as additional transducers of exercise stimuli to increased expression of mitochondrial proteins include the transcription factor complex activator protein-1 (AP-1), 5'AMP-activated protein kinase (AMPK) and peroxisome proliferator-activated receptor-alpha and gamma (PPARα/γ). Treadmill running and ergometer training, for 30 min above the aerobic threshold, induce expression of the jun/fos family of transcription factors in human m. vastus lateralis (Puntschart et al. 1998; Aronson et al. 1998). Dimers of the jun/fos family bind as AP-1 complex to the promoter, and/or influence the promoter activity of several nuclear encoded mitochondrial proteins, i.e. cytochrome C, carnitine palmitoyltransferase (CPT) and mitochondrial malic enzyme, some of which are affected acutely by exercise (Hood 2001; Hunter and Karin 1992; Butta et al. 2001; Brady et al. 1992; Pilegaard et al. 2000). AMPK activation is involved in the regulation of skeletal muscle metabolism during exercise (Winder et al. 2000; Zhou et al. 2000) and the $\alpha 2$ isoform of AMPK has been demonstrated to be activated by ergometer exercise in an intensity-dependent manner in human m. vastus lateralis (Wojtaszewski et al. 2000; Fig. 7). AMPK activation in rat muscle was held responsible for increasing cytochrome C and delta-aminolevulinic acid synthase protein levels, the activity of mitochondrial enzymes citrate synthase, succinate dehydrogenase, and malate dehydrogenase, and the skeletal muscle mitochondria-specific uncoupling protein-3 mRNA (Winder et al. 2000; Zhou et al. 2000). Moreover, Ca^{2+}-calodulin-dependent kinases recently have been implicated in regulation of mitochondrial biogenesis of skeletal muscle by a process that involves induced expression of the master regulator of mitochondrial biogenesis, the peroxisome proliferator-activated receptor γ co-activator PGC-1 (Wu et al. 2002). PGC-1 leads to a powerful induction of *NRF-1* and *NRF-2* gene expression and co-activates the transcriptional function of NRF-1 on the promoter of the mitochondrial transcription factor TFAM (Wu et al. 1999). A role for Ca^{2+} and CaMK in control of mitochondrial biogenesis during contractile activity is further supported by the finding that electric stimulation of cardiac cell cultures induces the mRNA of the transcription factor NRF-1 (Xia et al. 1997).

The inverse relationship between the cellular oxygen concentration and transcription factor HIF-1α protein levels indicates that HIF-1α is a main molecular sensor integrating increased muscle activity and oxygen availability into muscular remodelling. Recent evidence indicates that also phosphorylation of HIF-1α, through hypoxia-activated mitogen-activated protein kinases (MAPK), contributes to induced HIF-1α activity and promotes expression of its target VEGF (Pages et al. 2000; Minet et al. 2000; Semenza1999). There is increasing evidence that phosphorylation of HIF-1α activity, by a MAPK-dependent mechanism, is also important in gene regulation under normoxic conditions in response to growth factors in culture (Richard et al. 2000; Zelzer et al. 1998). On the other hand, it is found that exercise, presumably due to increased mechanical stress during contractile activity (Zou et al. 1998), also induces MAPK activity (Sherwood et al. 1999). Moreover, functional cooperation of HIF-1α with the MAPK-dependent transcription factor c-jun in hypoxia-induced gene transcription has been demonstrated (Alfranca et al. 2002). Last but not least, HIF-1α accumulates in response to mechanical stress (haemodynamic overload) in the nuclei of cardiac myocytes in a manner dependent on the concomitantly activated phosphatidylinositol 3-kinase (PI-3K)-Akt pathway and links to induced VEGF mRNA expression (Kim et al. 2002; Sodhi et al. 2001). A picture is therefore suggested whereby integration of exercise signals into transcription of angiogenic factors in skeletal muscle could be achieved by a cooperation and/or competition of mechano- and (local) hypoxia-sensitive pathways at the level of HIF-1α (Fig. 8).

The results on the expressional co-regulation of myofibrillar proteins MHC and the myogenic factors (foremost myoD and myogenin) suggest that myogenic factors integrate nerve activity and thyroid state into remodelling of the contractile apparatus. However, these factors do not appear involved in the mechanisms underlying maintenance of muscle mass (Dupont-Versteegden et al. 1998).

Conclusions

The present data suggest that exercise and environmental stress acting on skeletal muscle cause adaptive events in muscle fibres and associated structures and contribute to muscle plasticity. Changes in the gene profile are recognized as an interface that integrates the physiological perturbations by providing the necessary instructive information for the remodelling process of muscle structure and function.

The enormous and well-described plasticity of muscle tissue and its accessibility to biopsies makes muscle the organ of choice to study gene regulatory phenomena in humans. With the advent of technology to monitor thousands of genes at once, the skeletal muscle plasticity is a paradigm whose exploration is of greatest potential to understand the adaptations of human biological systems to physiological stimuli in the tissue context of the living. With the appropriate molecular techniques it has been demonstrated in the past 10 years that rapid changes in mRNA occur with exercise in human and rodent species (Pilegaard et al. 2000; Neufer and Dohm 1993; Puntschart et al. 1998). Recent expression profile data demonstrate that transcriptional adaptations in muscle due to changes in loading involve co-incident adaptations of genes in several functional categories. The co-directional expressional changes of genes belonging to functional categories (synexpression groups) indicates that expression of a battery of genes is controlled by a master transcriptional regulators. Signalling pathways involving the nuclear-encoded transcription factors

NRF-1, TFAM, AP-1, PPAR and AMPK, HIF-1α, and myogenic regulatory factors are implicated in transducing and integrating physiological stress into transcriptional adaptations of metabolic and contractile genes. These changes are matched to structural/functional adaptations and enzyme activity known to occur with corresponding stimuli. Nuclear reprogramming is recognized as an important event in muscle plasticity and may be related to the adaptations in the myosin type, protein turnover, and the cytoplasma-to-myonucleus ratio, as predicted by nuclear domain theory.

Future experiments will have to aim at understanding the acute regulatory phenomena occurring as a consequence of a single exercise intervention. Acute regulatory phenomena will have to be linked to the steady state changes in mRNA levels of structural genes ultimately responsible for the gross and subtle structural and functional modifications described as skeletal muscle plasticity. To this end, future approaches will make use of high-throughput technology such as protein and DNA arrays to identify the characteristic molecular adaptations in animal and human models.

Moderate exercise has a beneficial effect of body weight, metabolism and insulin resistance and energy expenditure and can counteract the syndrome of physical frailty (Evans 1995; Booth and Baldwin 1995). Understanding the molecular key to muscle plasticity therefore may be of importance for understanding the development of diseases such as obesity, NIDDM, hypertension and hyperlipidaemia and will give insight into these important clinical processes (Booth et al. 2002).

References

Adams GR, Hather BM, Baldwin KM, Dudley GA (1993) Skeletal muscle myosin heavy chain composition and resistance training. J Appl Physiol 74:911-915

Adams GR, Haddad F, Baldwin KM (1999) Time course of changes in markers of myogenesis in overloaded rat skeletal muscles. J Appl Physiol 87:1705-1712

Alfranca A, Gutierrez MD, Vara A, Aragones J, Vidal F, Landazuri MO (2002) c-Jun and hypoxia-inducible factor 1 functionally cooperate in hypoxia-induced gene transcription. Mol Cell Biol 22:12-22

Allen DL, Linderman JK, Roy RR, et al (1997) Apoptosis: a mechanism contributing to remodeling of skeletal muscle in response to hindlimb unweighting. Am J Physiol 273:C579-C587

Allen DL, Roy RR, Edgerton VR (1999) Myonuclear domains in muscle adaptation and disease. Muscle Nerve 22:1350-1360

Alway SE, MacDougall JD, Sale DG, Sutton JR, McComas AJ (1988) Functional and structural adaptations in skeletal muscle of trained athletes. J Appl Physiol 64:1114-1120

Amaral SL, Linderman JR, Morse MM, Greene AS (2001a) Angiogenesis induced by electrical stimulation is mediated by angiotensin II and VEGF. Microcirculation. 8:57-67

Amaral SL, Papanek PE, Greene AS (2001b) Angiotensin II and VEGF are involved in angiogenesis induced by short-term exercise training. Am J Physiol Heart Circ Physiol 281:H1163-H1169

Andersen JL, Aagaard P (2000) Myosin heavy chain IIX overshoot in human skeletal muscle. Muscle Nerve 23:1095-1104

Andersen JL, Klitgaard H, Bangsbo J, Saltin B (1994a) Myosin heavy chain isoforms in single fibres from m-vastus lateralis of soccer players-effects of strength-training. Acta Physiol Scand 150:21-26

Andersen JL, Klitgaard H, Saltin B (1994b) Myosin heavy chain isoforms in single fibres from m vastus lateralis of sprinters: Influence of training. Acta Physiol Scand 151:135-142

Andersen JL, Mohr T, Biering-Sorensen F, Galbo H, Kjaer M (1996) Myosin heavy chain isoform transformation in single fibres from m. vastus lateralis in spinal cord injured individuals: effects of long-term functional electrical stimulation (FES). Pflugers Arch 431:513-518

Andersen JL, Schiaffino S (1997) Mismatch between myosin heavy chain mRNA and protein distribution in human skeletal muscle fibers. Am J Physiol 272:C1881-C1889

Andersen JL, Schjerling P, Saltin B (2000) Muscle, genes and athletic performance. Sci Am 283:48-55

Aronson D, Violan MA, Dufresne SD, Zangen D, Fielding RA, Goodyear LJ (1997) Exercise stimulates the mitogen-activated protein kinase pathway in human skeletal muscle. J Clin Invest 99:1251-1257

Aronson D, Boppart MD, Dufresne SD, Fielding RA, Goodyear LJ (1998) Exercise stimulates c-Jun NH2 kinase activity and c-Jun transcriptional activity in human skeletal muscle. Biochem Biophys Res Commun 251:106-110

Aspnes LE, Lee CM, Weindruch R, Chung SS, Roecker EB, Aiken JM (1997) Caloric restriction reduces fiber loss and mitochondrial abnormalities in aged rat muscle. FASEB J 11:573-581

Balagopal P, Schimke JC, Ades P, Adey D, Nair KS (2001) Age effect on transcript levels and synthesis rate of muscle MHC and response to resistance exercise. Am J Physiol Endocrinol Metab 280:E203-E208

Baldwin KM, Haddad F (2001) Plasticity in skeletal, cardiac, and smooth muscle invited review: effects of different activity and inactivity paradigms on myosin heavy chain gene expression in striated muscle. J Appl Physiol 90:345-357

Bauman WA, Spungen AM (2000) Metabolic changes in persons after spinal cord injury. Phys Med Rehabil Clin N Am 11:109-140

Bejma J, Ji LL (1999) Aging and acute exercise enhance free radical generation in rat skeletal muscle. J Appl Physiol 87:465-470

Bengtsson J, Gustafsson T, Widegren U, Jansson E, Sundberg CJ (2001) Mitochondrial transcription factor A and respiratory complex IV increase in response to exercise training in humans. Pflugers Arch 443:61-66

Bentley DJ, Zhou S, Davie AJ (1998) The effect of endurance exercise on muscle force generating capacity of the lower limbs. J Sci Med Sport 1:179-188

Berchtold MW, Brinkmeier H, Muntener M (2000) Calcium ion in skeletal muscle: its crucial role for muscle function, plasticity, and disease. Physiol Rev 80:1215-1265

Bergamini E (1992) Protein degradation and modification. Introduction and overview. Ann N Y Acad Sci 663:43-47

Bergeron R, Ren JM, Cadman KS, et al (2001) Chronic activation of AMP kinase results in NRF-1 activation and mitochondrial biogenesis. Am J Physiol Endocrinol Metab 281:E1340-E1346

Bergstroem J, Hermansen L, Hultman E, Saltin B (1967) Diet muscle glycogen and physical performance. Acta Physiol Scand 71:172-179

Berra E, Pages G, Pouyssegur J (2000) MAP kinases and hypoxia in the control of VEGF expression. Cancer Metastasis Rev 19:139-145

Bhasin S, Storer TW, Berman N, et al (1996) The effects of supraphysiologic doses of testosterone on muscle size and strength in normal men. N Engl J Med 335:1-7

Biolo G, Maggi SP, Williams BD, Tipton KD, Wolfe RR (1995) Increased rates of muscle protein turnover and amino acid transport after resistance exercise in humans. Am J Physiol Endocrinol Metabol 31:E514-E520

Bizeau ME, Willis WT, Hazel JR (1998) Differential responses to endurance training in subsarcolemmal and intermyofibrillar mitochondria. J Appl Physiol 85:1279-1284

Booth FW, Baldwin KM (1995) Muscle plasticity: energy demanding and supply processes. In: Rowell LB, Shepherd JT (eds). Handbook of physiology. Oxford University Press, New York, pp 1076-1121

Booth FW, Thomason DB (1991) Molecular and cellular adaptation of muscle in response to exercise-perspective of various models. Physiol Rev 71:541-585

Booth FW, Chakravarthy MV, Gordon SE, Spangenburg EE (2002) Waging war on physical inactivity: using modern molecular ammunition against an ancient enemy. J Appl Physiol 93:3-30

Boppart MD, Aronson D, Gibson L, Roubenoff R, Abad LW, Bean J, Goodyear LJ, Fielding RA (1999) Eccentric exercise markedly increases c-Jun NH(2)-terminal kinase activity in human skeletal muscle. J Appl Physiol 87:1668-1673

Boppart MD, Hirshman MF, Sakamoto K, Fielding RA, Goodyear LJ (2001) Static stretch increases c-Jun NH2-terminal kinase activity and p38 phosphorylation in rat skeletal muscle. Am J Physiol Cell Physiol 280:C352-C358

Bradley RL, Cleveland KA, Cheatham B (2001) The adipocyte as a secretory organ: mechanisms of vesicle transport and secretory pathways. Recent Prog Horm Res 56:329-358

Brady PS, Park EA, Liu JS, Hanson RW, Brady LJ (1992) Isolation and characterization of the promoter for the gene coding for the 68 kDa carnitine palmitoyltransferase from the rat. Biochem J 286:779-783

Breen EC, Johnson EC, Wagner H, Tseng HM, Sung LA, Wagner PD (1996) Angiogenic growth factor mRNA responses in muscle to a single bout of exercise. J Appl Physiol 81:355-361

Brooks GA (2000) Intra- and extra-cellular lactate shuttles. Med Sci Sports Exerc 32:790-799

Brooks GA, Mercier J (1994) Balance of carbohydrate and lipid utilization during exercise: the crossover concept. J Appl Physiol 76:2253-2261

Bruhn H, Frahm J, Gyngell ML, Merboldt KD, Hanicke W, Sauter R (1991) Localized proton NMR spectroscopy using stimulated echoes: applications to human skeletal muscle in vivo. Magn Reson Med 17:82-94

Buller AJ, Eccles JC, Eccles RM (1960) Interactions between motoneurons and muscles in respect of the characteristic speeds of their responses. J Physiol 150:417-439

Buonanno A, Cheng J, Venepally P, Weis J, Calvo S (1998) Activity-dependent regulation of muscle genes: repressive and stimulatory effects of innervation. Acta Physiol Scand 163:S17-S26

Burnham R, Martin T, Stein R, Bell G, MacLean I, Steadward R (1997) Skeletal muscle fibre type transformation following spinal cord injury. Spinal Cord 35:86-91

Butta N, Gonzalez-Manchon C, Arias-Salgado EG, Ayuso MS, Parrilla R (2001) Cloning and functional characterization of the 5' flanking region of the human mitochondrial malic enzyme gene. Regulatory role of Sp1 and AP-2. Eur J Biochem 268:3017-3027

Carey AL, Staudacher HM, Cummings NK, et al (2001) Effects of fat adaptation and carbohydrate restoration on prolonged endurance exercise. J Appl Physiol 91:115-122

Carmeli E, Reznick AZ (1994) The physiology and biochemistry of skeletal muscle atrophy as a function of age. Proc Soc Exp Biol Med 206:103-113

Carraro F, Hartl WH, Stuart CA, Layman DK, Jahoor F, Wolfe RR (1990a) Whole body and plasma protein synthesis in exercise and recovery in human subjects. Am J Physiol 258:E821-E831

Carraro F, Stuart CA, Hartl WH, Rosenblatt J, Wolfe RR (1990b) Effect of exercise and recovery on muscle protein synthesis in human subjects. Am J Physiol 259:E470-E476

Carroll TJ, Abernethy PJ, Logan PA, Barber M, McEniery MT (1998) Resistance training frequency: strength and myosin heavy chain responses to two and three bouts per week. Eur J Appl Physiol 78:270-275

Carson JA, Wei L (2000) Integrin signaling's potential for mediating gene expression in hypertrophying skeletal muscle. J Appl Physiol 88:337-343

Carson JA, Nettleton D, Reecy JM (2002) Differential gene expression in the rat soleus muscle during early work overload-induced hypertrophy. FASEB J 16:207-209

Ceddia RB, William WN, Jr, Curi R (2001) The response of skeletal muscle to leptin. Front Bio Sci 6:D90-D97

Chandel NS, Schumacker PT (2000) Cellular oxygen sensing by mitochondria: old questions, new insight. J Appl Physiol 88:1880-1889

Chandel NS, McClintock DS, Feliciano CE, Wood TM, Melendez JA, Rodriguez AM, Schumacker PT (2000) Reactive oxygen species generated at mitochondrial complex III stabilize hypoxia-inducible factor-1alpha during hypoxia: a mechanism of O2 sensing. J Biol Chem 275:25130-25138

Chen ZP, Mcconell GK, Michell BJ, Snow RJ, Canny BJ, Kemp BE (2000) AMPK signaling in contracting human skeletal muscle: acetyl-CoA carboxylase and NO synthase phosphorylation. Am J Physiol Endocrinol Metab 279:E1202-E1206

Chesley A, MacDougall JD, Tarnopolsky MA, Atkinson SA, Smith K (1992) Changes in human muscle protein synthesis after resistance exercise. J Appl Physiol 73:1383-1388

Chilibeck PD, Bell GJ, Socha T, Martin T (1998) The effect of aerobic exercise training on the distribution of succinate dehydrogenase activity throughout muscle fibres. Res Can J Appl Physiol 23:74-86

Chilibeck PD, Syrotuik DG, Bell GJ (1999) The effect of strength training on estimates of mitochondrial density and distribution throughout muscle fibres. Res Eur J Appl Physiol 80:604-609

Chin ER, Allen DG (1996) The role of elevations in intracellular [Ca2+] in the development of low frequency fatigue in mouse single muscle fibres. Res J Physiol 491:813-824

Chin ER, Olson EN, Richardson JA, Yang Q, Humphries C, Shelton JM, Wu H, Zhu W, Bassel-Duby R, Williams RS (1998) A calcineurin-dependent transcriptional pathway controls skeletal muscle fiber type. Genes Dev 12:2499-2509

Chiquet M, Flück M (2001) Early responses to mechanical stress: from signals at the cell surface to altered gene expression, cell and molecular responses to stress. In: Storey KB, Storey JM (eds) Protein adaptations and signal transduction. Elsevier Science, pp 97-110

Chopard A, Pons F, Marini JF (2001) Cytoskeletal protein contents before and after hindlimb suspension in a fast and slow rat skeletal muscle. Am J Physiol Regul Integr Comp Physiol 280:R323-R330

Christensen NJ, Galbo H (1983) Sympathetic nervous activity during exercise. Annu Rev Physiol 45:139-153

Clement K, Viguerie N, Diehn M, et al (2002) In vivo regulation of human skeletal muscle gene expression by thyroid hormone. Genome Res 12:281-291

Connelly DM, Vandervoort AA (2000) Effects of isokinetic strength training on concentric and eccentric torque development in the ankle dorsiflexors of older adults. J Gerontol A Biol Sci Med Sci 55:B465-B472

Connor MK, Bezborodova O, Escobar CP, Hood DA (2000) Effect of contractile activity on protein turnover in skeletal muscle mitochondrial subfractions . J Appl Physiol 2000. 88:1601-1606
Costill DL, Coyle EF, Fink WF, Lesmes GR, Witzmann F (1979) Adaptations in skeletal muscle following strength training. J Appl Physiol 46:96-99
Craig BW, Brown R, Everhart J (1989) Effects of progressive resistance training on growth hormone and testosterone levels in young and elderly subjects. Mech Age Dev 49:159-169
Davis PJ, Davis FB (1996) Nongenomic actions of thyroid hormone. Thyroid 6:497-504
Dawes NJ, Cox VM, Park KS, Nga H, Goldspink DF (1996) The induction of c-fos and c-jun in the stretched latissimus dorsi muscle of the rabbit: responses to duration, degree and re-application of the stretch stimulus. Exp Physiol 81:329-339
Day DA, Tuite MF (1998) Post-transcriptional gene regulatory mechanisms in eukaryotes: an overview. J Endocrinol 157:361-371
Degens H, Soop M, Hook P, Ljungqvist O, Larsson L (1999) Post-operative effects on insulin resistance and specific tension of single human skeletal muscle fibres. Res Clin Sci (Lond) 97:449-455
Demirel HA, Powers SK, Naito H, Hughes M, Coombes JS (1999) Exercise-induced alterations in skeletal muscle myosin heavy chain phenotype: dose-response relationship. J Appl Physiol 86:1002-1008
Deschenes MR, Britt AA, Gomes RR, Booth FW, Gordon SE (2001) Recovery of neuromuscular junction morphology following 16 days of spaceflight. Synapse 42:177-184
Desplanches D, Hoppeler H, Linossier MT, Denis C, Claassen H, Dormois D, Lacour JR, Geyssant A (1993) Effects of training in normobaric hypoxia on human muscle ultrastructure. Pflugers Arch 425:263-267
Desplanches D, Hoppeler H, Mayet MH, Denis C, Claassen H, Ferretti G (1998) Effects of bedrest on deltoideus muscle morphology and enzymes. Acta Physiol Scand 162:135-140
Dominici FP, Turyn D (2002) Growth hormone-induced alterations in the insulin-signaling system. Exp Biol Med (Maywood.) 227:149-157
Dupont G, Goldbeter A (1998) CaM kinase II as frequency decoder of Ca2+ oscillations. BioEssays 20:607-610
Dupont-Versteegden EE, Houle JD, Gurley CM, Peterson CA (1998) Early changes in muscle fiber size and gene expression in response to spinal cord transection and exercise. Am J Physiol 275:C1124-C1133
Eftimie R, Brenner HR, Buonanno A (1991) Myogenin and MyoD join a family of skeletal muscle genes regulated by electrical activity. Proc Natl Acad Sci U S A 88:1349-1353
Emonson DL, Aminuddin AH, Wight RL, Scroop GC, Gore CJ (1997) Training-induced increases in sea level VO2max and endurance are not enhanced by acute hypobaric exposure. Eur J Appl Physiol 76:8-12
Ertl AC, Diedrich A, Biaggioni I, et al (2002) Human muscle sympathetic nerve activity and plasma noradrenaline kinetics in space. J Physiol 538:321-329
Escher P, Wahli W (2000) Peroxisome proliferator-activated receptors: insight into multiple cellular functions. Mutat Res 448:121-138
Evans MJ, Scarpulla RC (1990) NRF-1: a trans-activator of nuclear-encoded respiratory genes in animal cells. Genes Dev 4:1023-1034
Evans WJ (1995) Effects of exercise on body composition and functional capacity of the elderly. J Gerontol A Biol Sci Med Sci 50 [Spec No]:147-150
Everts ME (1996) Effects of thyroid hormones on contractility and cation transport in skeletal muscle. Acta Physiol Scand 156:325-333
Ferretti G, Antonutto G, Denis C, et al (1997) The interplay of central and peripheral factors in limiting maximal O2 consumption in man after prolonged bed rest. J Physiol 501:677-686
Fitts RH, Riley DR, Widrick JJ (2001) Functional and structural adaptations of skeletal muscle to microgravity. J Exp Biol 204:3201-3208
Florini JR, Ewton DZ, Magri KA (1991) Hormones, growth factors, and myogenic differentiation. Annu Rev Physiol 53:201-216
Flück M, Waxham MN, Hamilton MT, Booth FW (2000) Skeletal muscle Ca(2+)-independent kinase activity increases during either hypertrophy or running. J Appl Physiol 88:352-358
Forsythe JA, Jiang BH, Iyer NV, et al (1996) Activation of vascular endothelial growth factor gene transcription by hypoxia-inducible factor 1. Mol Cell Biol 16:4604-4613
Frenette J, Tidball JG (1998) Mechanical loading regulates expression of talin and its mRNA, which are concentrated at myotendinous junctions. Am J Physiol 275:C818-C825
Frisch H (1999) Growth hormone and body composition in athletes. J Endocrinol Invest 22:106-109

Fryburg DA, Louard RJ, Gerow KE, Gelfand RA, Barrett EJ (1992) Growth hormone stimulates skeletal muscle protein synthesis and antagonizes insulin's antiproteolytic action in humans. Diabetes 41:424-429

Gamrin L, Essen P, Hultman E, McNurlan MA, Garlick PJ, Wernerman J (2000) Protein-sparing effect in skeletal muscle of growth hormone treatment in critically ill patients. Ann Surg 231:577-586

Gayeski TEJ, Honig (1978) Myoglobin saturation and calculated PO2 in single cells of resting gracilis muscle. Adv Exptl Biol Med 94:77-84

Goedecke JH, Christie C, Wilson G, et al (1999) Metabolic adaptations to a high-fat diet in endurance cyclists. Metabolism 48:1509-1517

Goldberg AL, Goodman HM (1969) Amino acid transport during work-induced growth of skeletal muscle. Am J Physiol 216:1111-1115

Goldspink G (1999) Changes in muscle mass and phenotype and the expression of autocrine and systemic growth factors by muscle in response to stretch and overload. J Anat 194:323-334

Goldspink G, Scutt A, Martindale J, Jaenicke T, Turay L, Gerlach GF (1991) Stretch and force generation induce rapid hypertrophy and myosin isoform gene switching in adult skeletal muscle. Biochem Soc Trans. 19:368-373

Gordon JW, Rungi AA, Inagaki H, Hood DA (2001a) Effects of contractile activity on mitochondrial transcription factor A expression in skeletal muscle. J Appl Physiol 90:389-396

Gordon SE, Flück M, Booth FW (2001b) Selected Contribution. Skeletal muscle focal adhesion kinase, paxillin, and serum response factor are loading dependent. J Appl Physiol 90:1174-1183

Gracey AY, Troll JV, Somero GN (2001) Hypoxia-induced gene expression profiling in the euryoxic fish Gillichthys mirabilis. Proc Natl Acad Sci U S A 98:1993-1998

Green HJ, Klug GA, Reichmann H, Seedorf U, Wiehrer W, Pette D (1984) Exercise-induced fiber type transition with regard to myosin, parvalbumin, and sarcoplasmic reticulum in muscles of the rat. Pfluegers Arch 400:432-438

Green HJ, Sutton JR, Cymerman A, Young PM, Houston CS (1989) Operation Everest II: adaptations in human skeletal muscle. J Appl Physiol 66:2454-2461

Green HJ, Sutton JR, Wolfel EE, Reeves JT, Butterfield GE, Brooks GA (1992) Altitude acclimatization and energy metabolic adaptations in skeletal muscle during exercise. J Appl Physiol 73:2701-2708

Greiwe JS, Kwon G, McDaniel ML, Semenkovich CF (2001) Leucine and insulin activate p70 S6 kinase through different pathways in human skeletal muscle. Am J Physiol Endocrinol Metab 281:E466-E471

Gundersen K (1998) Determination of muscle contractile properties: the importance of the nerve. Acta Physiol Scand 162:333-341

Gustafsson T, Puntschart A, Kaijser L, Jansson E, Sundberg CJ (1999) Exercise-induced expression of angiogenesis-related transcription and growth factors in human skeletal muscle. Am J Physiol 276:H679-H85

Hakkinen K, Kauhanen H (1989) Daily changes in neural activation, force-time and relaxation-time characteristics in athletes during very intense training for one week. Electromyogr Clin Neuro Physiol 29:243-249

Haman F, Perronet F, Kenny GP, et al (2002) Effect of cold exposure on fuel utilization in humans: plasma glucose, muscle glycogen, and lipids. J Appl Physiol 93:77-84

Hamilton MT, Etienne J, McClure WC, Pavey BS, Holloway AK (1998) Role of local contractile activity and muscle fiber type on LPL regulation during exercise. Am J Physiol 275:E1016-E1022

Hansen S, Kvorning T, Kjaer M, Sjogaard G (2001) The effect of short-term strength training on human skeletal muscle: the importance of physiologically elevated hormone levels. Scand J Med Sci Sports 11:347-354

Hardie DG, Hawley SA (2001) AMP-activated protein kinase: the energy charge hypothesis revisited. Bioessays 23:1112-1119

Hargreaves M (2000) Skeletal muscle metabolism during exercise in humans. Clin Exp Pharmacol Physiol 27:225-228

Harridge SD (1996) The muscle contractile system and its adaptation to training. In: Marconnet P, Saltin B, Komi P, Poortmans J (eds) Human muscular function during dynamic exercise. Karger, Basel, pp 82-94

Harridge S, Magnusson G, Saltin B (1997) Life-long endurance-trained elderly men have high aerobic power, but have similar muscle strength to non-active elderly men. Aging (Milano) 9:80-87

Harridge SD, Andersen JL, Hartkopp A, Zhou S, Biering-Sorensen F, Sandri C, Kjaer M (2002) Training by low-frequency stimulation of tibialis anterior in spinal cord-injured men. Muscle Nerve 25:685-694

Hasten DL, Pak-Loduca J, Obert KA, Yarasheski KE (2000) Resistance exercise acutely increases MHC and mixed muscle protein synthesis rates in 78-84 and 23-32 yr olds. Am J Physiol Endocrinol Metab 278:E620-E626

Helge JW (1996) Interaction of diet and training on endurance performance and substrate utilization. PhD Thesis, University of Copenhagen
Helge JW, Wulff B, Kiens B (1998) Impact of a fat rich diet on endurance performance in man. Role of the dietary period. Med Sci Sports Exerc 30:456-461
Hernandez JM, Fedele MJ, Farrell PA (2000) Time course evaluation of protein synthesis and glucose uptake after acute resistance exercise in rats. J Appl Physiol 88:1142-1149
Hespel P, Op't Eijnde B, Van Leemputte M, et al (2001) Oral creatine supplementation facilitates the rehabilitation of disuse atrophy and alters the expression of muscle myogenic factors in humans. J Physiol 536:625-633
Hesse V, Vilser C, Scheibe J, Jahreis G, Foley T (1989) Thyroid hormone metabolism under extreme body exercises. Exp Clin Endocrinol 94:82-88
Hikida RS, Staron RS, Hagerman FC, et al (2000) Effects of high-intensity resistance training on untrained older men. II. Muscle fiber characteristics and nucleo-cytoplasmic relationships. J Gerontol A Biol Sci Med Sci 55:B347-B354
Hjeltnes N, Galuska D, Bjornhom M, et al (1998) Exercise-induced overexpression of key regulatory proteins involved in glucose uptake and metabolism in tetraplegic persons: molecular mechanism for improved glucose homeostasis. FASEB J 12:1701-1712
Hochachka PW, Beatty CL, Burelle Y, Trump ME, McKenzie DC, Matheson GO (2002) The lactate paradox in human high-altitude physiological performance. News Physiol Sci 17:122-126
Holloszy JO, Booth FW (1976) Biochemical adaptation to endurance exercise in muscle. Annu Rev Physiol 38:273-291
Holloszy JO, Coyle EF (1984) Adaptations of skeletal muscle to endurance exercise and their metabolic consequences. J Appl Physiol 56:831-838
Hood DA (2001) Invited Review. Contractile activity-induced mitochondrial biogenesis in skeletal muscle. J Appl Physiol 90:1137-1157
Hoppeler H (1999) Skeletal muscle substrate metabolism. Int J Obes Relat Metab Disord 23 [Suppl 3]:S7-10
Hoppeler H, Flück M (2002) Normal mammalian skeletal muscle and its phenotypic plasticity. J Exp Biol 205:2143-2152
Hoppeler H, Vogt M (2001a) Hypoxia training for sea-level performance. Training high-living low. Adv Exp Med Biol 502:61-73
Hoppeler H, Vogt M (2001b) Muscle tissue adaptations to hypoxia. J Exp Biol 204:3133-3139
Hoppeler H, Weibel ER (1998) Limits for oxygen and substrate transport in mammals. J Exp Biol 201:1051-1064
Hoppeler H, Weibel ER (2000) Structural and functional limits for oxygen supply to muscle. Acta Physiol Scand 168:445-456
Hoppeler H, Luethi P, Claassen H, Weibel ER, Howald H (1973) The ultrastructure of the normal human skeletal muscle-a morphometric analysis on untrained men, women, and well-trained orienteers. Pfluegers Arch 344:217-232
Hoppeler H, Howald H, Conley K, Lindstedt SL, Claassen H, Vock P, Weibel ER (1985) Endurance training in humans: aerobic capacity and structure of skeletal muscle. J Appl Physiol 59:320-327
Hoppeler H, Kleinert E, Schlegel C, Claassen H, Howald H, Cerretelli P (1990) Muscular exercise at high altitude: II. Morphological adaptation of skeletal muscle to chronic hypoxia. Int J Sport Med 11:S3-S9
Hoppeler H, Billeter R, Horvath PJ, Leddy JJ, Pendergast DR (1999) Muscle structure with low- and high-fat diets in well-trained male runners. Int J Sports Med 20:522-526
Horowitz JF, Leone TC, Feng W, Kelly DP, Klein S (2000) Effect of endurance training on lipid metabolism in women: a potential role for PPARalpha in the metabolic response to training. Am J Physiol Endocrinol Metab 279:E348-E355
Hortobagyi T, Dempsey L, Fraser D, et al (2000) Changes in muscle strength, muscle fibre size and myofibrillar gene expression after immobilization and retraining in humans. J Physiol 524:293-304
Howald H (1982) Training-induced morphological and functional changes in skeletal muscle. Int J Sport Med 3:1-12
Howald H, Hoppeler H, Claassen H, Mathieu O, Straub R (1985) Influences of endurance training on the ultrastructural composition of the different muscle fiber types in humans. Pfluegers Arch 403:369-376
Howald H, Pette D, Simoneau JA, Uber A, Hoppeler H, Cerretelli P (1990) Muscular exercise at high altitude: III. Effects of chronic hypoxia on muscle enzymes. Int J Sports Med 11:S10-S14
Howrie DL (1987) Growth hormone for the treatment of growth failure in children. Clin Pharm 6:283-291
Hudlicka O (1998) Is physiological angiogenesis in skeletal muscle regulated by changes in microcirculation? Microcirculation 5:5-23

Hughes SM, Taylor JM, Tapscott SJ, Gurley CM, Carter WJ, Peterson CA (1993) Selective accumulation of MyoD and Myogenin mRNAs in fast and slow adult skeletal muscle is controlled by innervation and hormones. Develop 118:1137-1147

Hunter RB, Stevenson E, Koncarevic A, Mitchell-Felton H, Essig DA, Kandarian SC (2002) Activation of an alternative NF-kappaB pathway in skeletal muscle during disuse atrophy. FASEB J 16:529-538

Hunter T, Karin M (1992) The regulation of transcription by phosphorylation. Cell 70:375-387

Huss JM, Levy FH, Kelly DP (2001) Hypoxia inhibits the peroxisome proliferator-activated receptor alpha/ retinoid X receptor gene regulatory pathway in cardiac myocytes: a mechanism for O2-dependent modulation of mitochondrial fatty acid oxidation. J Biol Chem 276:27605-27612

Huxley A (1988) Muscular contraction. Annu Rev Physiol 50:1-16

Ikemoto M, Nikawa T, Takeda S, et al (2001) Space shuttle flight (STS-90) enhances degradation of rat myosin heavy chain in association with activation of ubiquitin-proteasome pathway. FASEB J 15:1279-1281

Ingjer F (1979) Capillary supply and mitochondrial content of different skeletal muscle types in untrained and endurance-trained men. A histochemical and ultrastructural study. Eur J Appl Physiol 40:197-209

Jansson E, Kaijser L (1982) Effect of diet on the utilization of blood-borne and intramuscular substrates during exercise in man. Acta Physiol Scand 115:19-30

Jewell UR, Kvietikova I, Scheid A, Bauer C, Wenger RH, Gassmann M (2001) Induction of HIF-1alpha in response to hypoxia is instantaneous. FASEB J 15:1312-1314

Johnston IA, Temple GK (2002) Thermal plasticity of skeletal muscle phenotype in ectothermic vertebrates and its significance for locomotory behaviour. J Exp Biol 205:2305-2322

Jozsi AC, Dupont-Versteegden EE, Taylor-Jones JM, et al (2000) Aged human muscle demonstrates an altered gene expression profile consistent with an impaired response to exercise. Mech Ageing Dev 120:45-56

Kadi F, Thornell LE (1999) Training affects myosin heavy chain phenotype in the trapezius muscle of women. Histochem Cell Biol 112:73-78

Kadi F, Thornell LE (2000) Concomitant increases in myonuclear and satellite cell content in female trapezius muscle following strength training. Histochem Cell Biol 113:99-103

Kadi F, Eriksson A, Holmner S, Butler-Browne GS, Thornell LE (1999) Cellular adaptation of the trapezius muscle in strength-trained athletes. Histochem Cell Biol 111:189-195

Kayo T, Allison DB, Weindruch R, Prolla TA (2001) Influences of aging and caloric restriction on the transcriptional profile of skeletal muscle from rhesus monkeys. Proc Natl Acad Sci U S A 98:5093-5098

Kayser B, Hoppeler H (1991) Performance capacity and muscle structure of Himalayan sherpas. J Appl Physiol 70:1938-1942

Kiessling KH, Pilstroem L, Karlsson J, Piehl K (1973) Mitochondrial volume in skeletal muscle from young and old physically untrained and trained healthy men and from alcoholics. Clin Sci 44:547-554

Kietzmann T, Fandrey J, Acker H (2000) Oxygen radicals as messengers in oxygen-dependent gene expression. News Physiol Sci 15:202-208

Kim CH, Cho YS, Chun YS, Park JW, Kim MS (2002) Early expression of myocardial HIF-1alpha in response to mechanical stresses: regulation by stretch-activated channels and the phosphatidylinositol 3-kinase signaling pathway. Circ Res 90:E25-E33

Kimball SR (2002) Regulation of global and specific mRNA translation by amino acids. J Nutr 132:883-886

Kjaer M, Hanel B, Worm L, et al (1999) Cardiovascular and neuroendocrine responses to exercise in hypoxia during impaired neural feedback from muscle. Am J Physiol 277:R76-R85

Kliewer SA, Sundseth SS, Jones SA, et al (1997) Fatty acids and eicosanoids regulate gene expression through direct interactions with peroxisome proliferator-activated receptors alpha and gamma. Proc Natl Acad Sci U S A 94:4318-4323

Klitgaard H, Bergman O, Betto R, Salviati G, Schiaffino S, Clausen T, Saltin B (1990a) Co-existence of myosin heavy chain I and IIa isoforms in human skeletal muscle fibres with endurance training. Pflugers Arch 416:470-472

Klitgaard H, Zhou M, Richter EA (1990b) A myosin heavy chain composition of single fibres from m biceps brachii of male body builders. Acta Physiol Scand 140:175-180

Klitgaard H, Zhou M, Schaffino S, Betto R, Salviati G, Sal (1990c) Ageing alters the myosin heavy chain composition of single fibres from human skeletal muscle. Acta Physiol Scand 140:55-62

Kniffeki KD, Mense S, Schmidt RF (1981) Muscle receptors with fine afferent fibers which may evoke circulatory reflexes. Circ Res 48:I25-I31

Kraemer WJ, Fry AC, Warren BJ, Stone MH, Fleck SJ, Kearney JT, Conroy BP, Maresh CM, Weseman CA, Triplett NT, et al (1992) Acute hormonal responses in elite junior weightlifters. Int J Sports Med 13:103-109

Kraemer WJ, Staron RS, Hagerman FC, Hikida RS, Fry AC, Gordon SE, Nindl BC, Gothshalk LA, Volek JS, Marx JO, Newton RU, Hakkinen K (1998) The effects of short-term resistance training on endocrine function in men and women. Eur J Appl Physiol 78:69-76

Kraemer WJ, Hakkinen K, Newton RU, Nindl BC, Volek JS, McCormick M, Gotshalk LA, Gordon SE, Fleck SJ, Campbell WW, Putukian M, Evans WJ (1999) Effects of heavy-resistance training on hormonal response patterns in younger vs. older men. J Appl Physiol 87:982-992

Kraemer WJ, Dudley GA, Tesch PA, Gordon SE, Hather BM, Volek JS, Ratamess NA (2001) The influence of muscle action on the acute growth hormone response to resistance exercise and short-term detraining. Growth Horm IGF Res 11:75-83

Kraniou Y, Cameron-Smith D, Misso M, Collier G, Hargreaves M (2000) Effects of exercise on GLUT-4 and glycogenin gene expression in human skeletal muscle. J Appl Physiol 88:794-796

Lambert EV, Speechly DP, Dennis SC, Noakes TD (1994) Enhanced endurance in trained cyclists during moderate intensity exercise following 2 weeks adaptation of a high-fat diet. Eur J Appl Physiol 69:287-293

Lambert EV, Goedecke JH, Zyle C, Murphy K, Hawley JA, Dennis SC, Noakes TD (2001) High-fat diet versus habitual diet prior to carbohydrate loading: effects of exercise metabolism and cycling performance. Int J Sport Nutr Exerc Metab 11:209-225

Lando D, Peet DJ, Gorman JJ, Whelan DA, Whitelaw ML, Bruick RK (2002) FIH-1 is an asparaginyl hydroxylase enzyme that regulates the transcriptional activity of hypoxia-inducible factor. Genes Dev 16:1466-1471

Lange KH, Andersen JL, Beyer N, et al (2002) GH administration changes myosin heavy chain isoforms in skeletal muscle but does not augment muscle strength or hypertrophy, either alone or combined with resistance exercise training in healthy elderly men. J Clin Endocrinol Metab 87:513-523

Larsson L, Li X, Berg HE, Frontera WR (1996) Effects of removal of weight-bearing function on contractility and myosin isoform composition in single human skeletal muscle cells. Pflugers Arch 432:320-328

Larsson L, Li X, Frontera WR (1997) Effects of aging on shortening velocity and myosin isoform composition in single human skeletal muscle cells. Am J Physiol 272:C638-C649

Larsson NG, Wang J, Wilhelmsson H, Oldfors A, Rustin P, Lewandoski M, Barsh GS, Clayton DA (1998) Mitochondrial transcription factor A is necessary for mtDNA maintenance and embryogenesis in mice. Nat Genet 18:231-236

Lehman JJ, Barger PM, Kovacs A, Saffitz JE, Medeiros DM, Kelly DP (2000) Peroxisome proliferator-activated receptor gamma coactivator-1 promotes cardiac mitochondrial biogenesis. J Clin Invest 106:847-856

Lesmes GR, Costill DL, Coyle EF, Fink WJ (1978) Muscle strength and power changes during maximal isokinetic training. Med Sci Sports 10:266-269

Leterme D, Falempin M (1996) Contractile properties of rat soleus motor units following 14 days of hindlimb unloading. Pflugers Arch 432:313-319

Leveritt M, MacLaughlin H, Abernethy PJ (2000) Changes in leg strength 8 and 32 h after endurance exercise. J Sports Sci 18:865-871

Levy AP, Levy NS, Goldberg MA (1996) Post-transcriptional regulation of vascular endothelial growth factor by hypoxia. J Biol Chem 271:2746-2753

Lindstedt SL, Wells DJ (1988) Skeletal muscle mitochondria: the aerobic gate? Adv Exp Med Biol 227:207-213

Linossier MT, Dormois D, Perier C, Frey J, Geyssant A, Denis C (1997) Enzyme adaptations of human skeletal muscle during bicycle short-sprint training and detraining. Acta Physiol Scand 161:439-445

Liu Y, Steinacker JM (2001) Changes in skeletal muscle heat shock proteins: pathological significance. Front Bio Sci 6:D12-D25

Lloyd PG, Yang HT, Terjung RL (2001) Arteriogenesis and angiogenesis in rat ischemic hindlimb: role of nitric oxide. Am J Physiol Heart Circ Physiol 281:H2528-H2538

Lombardi A, Damon M, Vincent A, Goglia F, Herpin P (2000) Characterisation of oxidative phosphorylation in skeletal muscle mitochondria subpopulations in pig: a study using top-down elasticity analysis. FEBS Lett 475:84-88

Lotta S, Scelsi R, Alfonsi E, et al (1991) Morphometric and neurophysiological analysis of skeletal muscle in paraplegic patients with traumatic cord lesion. Paraplegia 29:247-252

Luethi JM, Howald H, Claassen H, Roesler K, Vock P, Hoppeler H (1986) Structural changes in skeletal muscle tissue with heavy-resistance exercise. Int J Sports Med 7:123-127

MacDougall JD, Sale DG, Moroz JR, Elder GCB, Sutton JR, Howald H (1979) Mitochondrial volume density in human muscle following heavy resistance training. Med Sci Sport Exercise 11:164-166

MacDougall JD, Sale DG, Elder GC, Sutton JR (1982) Muscle ultrastructural characteristics of elite powerlifters and bodybuilders. Eur J Appl Physiol 48:117-126

MacDougall JD, Green HJ, Sutton JR, Coates G, Cymerman AYP, Houston CS (1991) Operation Everest-II-Structural adaptations in skeletal muscle in response to extreme simulated altitude. Acta Physiol Scand 142:421-427

Mahon PC, Hirota K, Semenza GL (2001) FIH-1: a novel protein that interacts with HIF-1alpha and VHL to mediate repression of HIF-1 transcriptional activity. Genes Dev 15:2675-2686

Malarkey WB, Hall JC, Rice RR Jr, et al (1993) The influence of age on endocrine responses to ultraendurance stress. J Gerontol 48:M134-M139

Martin TP, Stein RB, Hoeppner PH, Reid DC (1992) Influence of electrical stimulation on the morphological and metabolic properties of paralyzed muscle. J Appl Physiol 72:1401-1406

Martin WH III (1993) Triiodothyronine, beta-adrenergic receptors, agonist responses, and exercise capacity. Ann Thorac Surg 56:S24-S34

Martin WH III (1996) Effects of acute and chronic exercise on fat metabolism. Exerc Sport Sci Rev 24:203-231

Martineau LC, Gardiner PF (2001) Insight into skeletal muscle mechanotransduction: MAPK activation is quantitatively related to tension. J Appl Physiol 91:693-702

Martineau LC, Gardiner PF (2002) Skeletal muscle is sensitive to the tension-time integral but not to the rate of change of tension, as assessed by mechanically induced signaling. J Biomech 35:657-663

Martinelli M, Winterhalder R, Cerretelli P, Howald H, Hoppeler H (1990) Muscle lipofuscin content and satellite cell volume is increased after high altitude exposure in humans. Experientia 46:672-676

Masuda K, Okazaki K, Kuno S, Asano K, Shimojo H, Katsuta S (2001) Endurance training under 2500-m hypoxia does not increase myoglobin content in human skeletal muscle. Eur J Appl Physiol 85:486-490

McAllister RM, Delp MD, Laughlin MH (1997) A review of effects of hypothyroidism on vascular transport in skeletal muscle during exercise. Can J Appl Physiol 22:1-10

McClelland GB, Brooks GA (2002) Changes in MCT 1, MCT 4, and LDH expression are tissue-specific in rats after long-term hypobaric hypoxia. J Appl Physiol 92:1573-1584

McCutcheon LJ, Byrd SK, Hodgson DR (1992) Ultrastructural changes in skeletal muscle after fatiguing exercise. J Appl Physiol 72:1111-1117

Melissa L, MacDougall JD, Tarnopolsky MA, Cipriano N, Green HJ (1997) Skeletal muscle adaptations to training under normobaric hypoxic versus normoxic conditions. Med Sci Sports Exerc 29:238-243

Miller WC, Bryce GR, Conlee RK (1984) Adaptations to a high-fat diet that increase exercise endurance in male rats. J Appl Physiol 56:78-83

Minet E, Arnould T, Michel G, et al (2000) ERK activation upon hypoxia: involvement in HIF-1 activation. FEBS Lett 468:53-58

Mitchell CR, Harris MB, Cordaro AR, Starnes JW (2002) Effect of body temperature during exercise on skeletal muscle cytochrome c oxidase content. J Appl Physiol 93:526-530

Mohr T, Andersen JL, Biering-Sorensen F, et al (1997) Long-term adaptation to electrically induced cycle training in severe spinal cord injured individuals. Spinal Cord 35:1-16

Morrison PR, Biggs RB, Booth FW (1989) Daily running for 2 wk and mRNAs for cytochrome c and alpha-actin in rat skeletal muscle. Am J Physiol 257:C936-C939

Mujika I, Padilla S (2001) Muscular characteristics of detraining in humans. Med Sci Sports Exerc 33:1297-1303

Mukhopadhyay D, Tsiokas L, Zhou XM, Foster D, Brugge JS, Sukhatme VP (1995) Hypoxic induction of human vascular endothelial growth factor expression through c-Src activation. Nature 375:577-581

Munoz KA, Aannestad A, Tischler ME, Henriksen EJ (1994) Skeletal muscle protein content and synthesis after voluntary running and subsequent unweighting. Metabolism 43:994-999

Murakami T, Shimomura Y, Yoshimura A, Sokabe M, Fujitsuka N (1998) Induction of nuclear respiratory factor-1 expression by an acute bout of exercise in rat muscle. Biochim Biophys Acta 1381:113-122

Nader GA, Esser KA (2001) Intracellular signaling specificity in skeletal muscle in response to different modes of exercise. J Appl Physiol 90:1936-1942

Naito H, Powers SK, Demirel HA, Sugiura T, Dodd SL, Aoki J (2000) Heat stress attenuates skeletal muscle atrophy in hindlimb-unweighted rats. J Appl Physiol 88:359-363

Navarro A, Lopez-Cepero JM, Sanchez del Pino MJ (2001) Skeletal muscle and aging. Front Biol Sci 6:D26-D44

Neufer PD, Dohm GL (1993) Exercise Induces a transient increase in transcription of the GLUT-4 gene in skeletal muscle. Am J Physiol 265:C1597-C1603

Niehrs C, Pollet N (1999) Synexpression groups in eukaryotes. Nature 402:483-487

Nindl BC, Kraemer WJ, Marx JO, et al (2001) Overnight responses of the circulating IGF-I system after acute, heavy-resistance exercise. J Appl Physiol 90:1319-1326

O'Neill DS, Zheng D, Anderson WK, Dohm GL, Houmard JA (1999) Effect of endurance exercise on myosin heavy chain gene regulation in human skeletal muscle. Am J Physiol 276:R414-R419

Ohira Y, Yoshinaga T, Ohara M, et al (1999) Myonuclear domain and myosin phenotype in human soleus after bed rest with or without loading. J Appl Physiol 87:1776-1785

Olson EN (1990) MyoD family: a paradigm for development? Genes Dev 4:1454-1461

Olson EN, Williams RS (2000) Calcineurin signaling and muscle remodeling. Cell 101:689-692

Paavolainen L, Hakkinen K, Hamalainen I, Nummela A, Rusko H (1999) Explosive-strength training improves 5-km running time by improving running economy and muscle power . J Appl Physiol 86:1527-1533

Pages G, Milanini J, Richard DE, et al (2000) Signaling angiogenesis via p42/p44 MAP kinase cascade. Ann N Y Acad Sci 902:187-200

Panenic R, Gardiner PF (1998) The case for adaptability of the neuromuscular junction to endurance exercise training. Can J Appl Physiol 23:339-360

Parolin ML, Spriet LL, Hultman E, Hollidge-Horvat MG, Jones NL, Heigenhauser GJ (2000a) Regulation of glycogen phosphorylase and PDH during exercise in human skeletal muscle during hypoxia. Am J Physiol Endocrinol Metab 278:E522-E534

Parolin ML, Spriet LL, Hultman E, Matsos MP, Hollidge-Horvat MG, Jones NL, Heigenhauser GJ (2000b) Effects of PDH activation by dichloroacetate in human skeletal muscle during exercise in hypoxia. Am J Physiol Endocrinol Metab 279:E752-E761

Pastoris O,Boschi F, Verri M, Baiardi P, Felzani G, Vecchiet J, Dossena M, Catapano M(2000) The effects of aging on enzyme activities and metabolite concentrations in skeletal muscle from sedentary male and female subjects. Exp Gerontol 35:95-104

Pedersen BK, Steensberg A, Fischer C, Keller C, Ostrowski K, Schjerling P (2001) Exercise and cytokines with particular focus on muscle-derived IL-6. Exerc Immunol Rev 7:18-31

Pedraza N, Solanes G, Carmona MC, et al (2000) Impaired expression of the uncoupling protein-3 gene in skeletal muscle during lactation: fibrates and troglitazone reverse lactation-induced downregulation of the uncoupling protein-3 gene. Diabetes 49:1224-1230

Peters SJ, Harris RA, Wu P, Pehleman TL, Heigenhauser GJ, Spriet LL (2001) Human skeletal muscle PDH kinase activity and isoform expression during a 3-day high-fat/low-carbohydrate diet. Am J Physiol Endocrinol Metab 281:E1151-E1158

Pette D (1998) Training effects on the contractile apparatus. Acta Physiol Scand 162:367-376

Pette D (2001) Historical perspectives: plasticity of mammalian skeletal muscle. J Appl Physiol 90:1119-1124

Pette D, Staron RS (2000) Myosin isoforms, muscle fiber types, and transitions. Microsc Res Tech 50:500-509

Pette D, Vrbova G (1992) Adaptation of mammalian skeletal muscle fibers to chronic electrical stimulation. Rev Physiol Biochem Pharmacol 120:115-202

Phillips SM, Tipton KD, Aarsland A, Wolf SE, Wolfe RR (1997) Mixed muscle protein synthesis and breakdown after resistance exercise in humans. Am J Physiol 273:E99-E107

Phillips SM, Tipton KD, Ferrando AA, Wolfe RR (1999) Resistance training reduces the acute exercise-induced increase in muscle protein turnover. Am J Physiol 276:E118-E124

Phinney SD, Bistrian BR, Evans WJ, Gervino E, Blackburn GL (1983) The human metabolic response to chronic ketosis without caloric restriction: preservation of submaximal exercise capability with reduced carbohydrate oxidation. Metabolism 32:769-777

Pilegaard H, Ordway GA, Saltin B, Neufer PD (2000) Transcriptional regulation of gene expression in human skeletal muscle during recovery from exercise. Am J Physiol Endocrinol Metab 279:E806-E814

Prince FP, Hikida RS, Hagerman FC, Staron RS, Allen WH (1981) A morphometric analysis of human muscle fibers with relation to fiber types and adaptations to exercise. J Neurol Sci 49:165-179

Proctor DN, Sinning WE, Walro JM, Sieck GC, Lemon PW (1995) Oxidative capacity of human muscle fiber types: effects of age and training status. J Appl Physiol 78:2033-2038

Puntschart A, Claassen H, Jostarndt K, Hoppeler H, Billeter R (1995a) mRNAs of enzymes involved in energy metabolism and mtDNA are increased in endurance trained athletes. Am J Physiol 269:C619-C625

Puntschart A, et al (1995b) Immediate-early-(IEG) and heat-shock-protein-(HSP) gene expression after a single bout of exercise in humans. USGEB, Experientia 51:A22(S05-S25)

Puntschart A, Vogt M, Widmer HR, Hoppeler H, Billeter R (1996) Hsp70 expression in human skeletal muscle after exercise. Acta Physiol Scand 157:411-417

Puntschart A, Wey E, Jostarndt K, Vogt M, Wittwer M, Widmer HR, Hoppeler H, Billeter R(1998) Expression of fos and jun genes in human skeletal muscle after exercise. Am J Physiol 43:C129-C137

Rabinovich RA, Ardite E, Troosters T, et al (2001) Reduced muscle redox capacity after endurance training in patients with chronic obstructive pulmonary disease. Am J Respir Crit Care Med 164:1114-1118

Radak Z, Lee K, Choi W, et al (1994) Oxidative stress induced by intermittent exposure at a simulated altitude of 4000 m decreases mitochondrial superoxide dismutase content in soleus muscle of rats. Eur J Appl Physiol 69:392-395

Ray CA, Gracey KH (1997) Augmentation of exercise-induced muscle sympathetic nerve activity during muscle heating. J Appl Physiol 82:1719-1725

Reichmann H, Hoppeler H, Mathieu-Costello O, von Bergen F, Pette D (1985) Biochemical and ultrastructural changes of skeletal muscle mitochondria after chronic electrical stimulation in rabbits. Pfluegers Arch 404:1-9

Reynafarje B (1962) Myoglobin content and enzymatic activity of muscle and altitude adaptation. J Appl Physiol 17:301-305

Rich C, Cafarelli E (2000) Submaximal motor unit firing rates after 8 wk of isometric resistance training. Med Sci Sports Exerc 32:190-196

Richard DE, Berra E, Pouyssegur J (2000) Nonhypoxic pathway mediates the induction of hypoxia-inducible factor 1alpha in vascular smooth muscle cells. J Biol Chem 275:26765-26771

Richardson RS, Noyszewski EA, Kendrick KF, Leigh JS, Wagner PD (1995) Myoglobin O2 desaturation during exercise. Evidence of limited O2 transport. J Clin Invest 96:1916-1926

Richardson RS, Wagner H, Mudaliar SR, Henry R, Noyszewski EA, Wagner PD (1999) Human VEGF gene expression in skeletal muscle: effect of acute normoxic and hypoxic exercise. Am J Physiol 277:H2247-H2252

Richardson RS, Newcomer SC, Noyszewski EA (2001) Skeletal muscle intracellular PO(2) assessed by myoglobin desaturation: response to graded exercise. J Appl Physiol 91:2679-2685

Ricoy JR, Encinas AR, Cabello A, Madero S, Arenas J (1998) Histochemical study of the vastus lateralis muscle fibre types of athletes. J Physiol BioChem 54:41-47

Riva C, Chevrier C, Pasqual N, Saks V, Rossi A (2001) Bcl-2/Bax protein expression in heart, slow-twitch and fast-twitch muscles in young rats growing under chronic hypoxia conditions. Mol Cell BioChem 226:9-16

Rooyackers OE, Nair KS (1997) Hormonal regulation of human muscle protein metabolism. Annu Rev Nutr 17:457-485

Round JM, Barr FM, Moffat B, Jones DA (1993) Fibre areas and histochemical fibre types in the quadriceps muscle of paraplegic subjects. J Neurol.Sci 116:207-211

Roussel D, Lhenry F, Ecochard L, Sempore B, Rouanet JL, Favier R (2000) Differential effects of endurance training and creatine depletion on regional mitochondrial adaptations in rat skeletal muscle. Biochem J 350:547-553

Ryder JW, Chibalin AV, Zierath JR (2001) Intracellular mechanisms underlying increases in glucose uptake in response to insulin or exercise in skeletal muscle. Acta Physiol Scand 171:249-257

Sakamoto K, Goodyear LJ (2002) Invited Review. Intracellular signaling in contracting skeletal muscle. J Appl Physiol 93:369-383

Sale DG (1988) Neural adaptation to resistance training. Med Sci Sports Exerc 20:S135-S145

Saltin B (1986) Physiological adaptation to physical conditioning Old problems revisited. Acta Med Scand S711:11-24

Sandri M (2002) Apoptotic signaling in skeletal muscle fibers during atrophy. Curr Opin Clin Nutr Metab Care 5:249-253

Schaub MC, Brunner UT, von Schulthess C, Neidhart M, Baumann H (1989) Adaptation of contractile proteins in human heart and skeletal muscles. BioMed Biochim Acta 48:S306-S312

Schiaffino S, Reggiani C (1994) Myosin isoforms in mammalian skeletal muscle. J Appl Physiol 77:493-501

Schiaffino S, Reggiani C (1996) Molecular diversity of myofibrillar proteins: gene regulation and functional significance. Physiol Rev 76:371-423

Schlienger JL, Pradignac A, Grunenberger F (1995) Nutrition of the elderly: a challenge between facts and needs. Horm Res 43:46-51

Schrauwen P, Wagenmakers AJ, Marken Lichtenbelt WD, Saris WH, Westerterp KR (2000) Increase in fat oxidation on a high-fat diet is accompanied by an increase in triglyceride-derived fatty acid oxidation. Diabetes 49:640-646

Schrauwen P, Hoppeler H, Billeter R, Bakker A, Pendergast D (2001a) Fiber type dependent upregulation of human skeletal muscle UCP2 and UCP3 mRNA expression by high-fat diet. Int J Obes Relat Metab Disord 25:449-456

Schrauwen P, Saris WH, Hesselink MK (2001b) An alternative function for human uncoupling protein 3: protection of mitochondria against accumulation of non-esterified fatty acids inside the mitochondrial matrix. FASEB J 13:2497-502

Schultz E, McCormick KM (1994) Skeletal muscle satellite cells. Rev Physiol Biochem Pharmacol 123:213-257

Seals DR, Enoka RM (1989) Sympathetic activation is associated with increases in EMG during fatiguing exercise. J Appl Physiol 66:88-95

Seedorf U, Leberer E, Kirschbaum BJ, Pette D (1986) Neural control of gene expression in skeletal muscle Effect of chronic stimulation on lactate dehydrogenase isoenzymes and citrate synthase. Biochem J 239:115-120

Seko Y, Takahashi N, Tobe K, Kadowaki T, Yazaki Y (1997) Hypoxia and hypoxia/reoxygenation activate p65PAK, p38 mitogen-activated protein kinase (MAPK), and stress-activated protein kinase (SAPK) in cultured rat cardiac myocytes. Biochem Biophys Res Commun 239:840-844

Semenza GL (1999) Regulation of mammalian O2 homeostasis by hypoxia-inducible factor 1. Annu Rev Cell Dev Biol 15:551-578

Shadel GS, Clayton DA (1997) Mitochondrial DNA maintenance in vertebrates. Annu Rev Biochem 66:409-435

Sherwood DJ, Dufresne SD, Markuns JF, et al (1999) Differential regulation of MAP kinase, p70(S6K), and Akt by contraction and insulin in rat skeletal muscle. Am J Physiol 276:E870-E878

Simi B, Sempore B, Mayet MH, Favier RJ (1991) Additive effects of training and high-fat diet on energy metabolism during exercise. J Appl Physiol 71:197-203

Smol E, Zernicka E, Czarnowski D, Langfort J (2001) Lipoprotein lipase activity in skeletal muscles of the rat: effects of denervation and tenotomy. J Appl Physiol 90:954-960

Snow MH (1990) Satellite cell response in rat soleus muscle undergoing hypertrophy due to surgical ablation of synergists. Anat Rec 227:437-446

Sobel BE, Kaufman S (1970) Enhanced RNA polymerase activity in skeletal muscle undergoing hypertrophy. Arch BioChem Biophys. 137:469-476

Sodhi A, Montaner S, Miyazaki H, Gutkind JS (2001) MAPK and Akt act cooperatively but independently on hypoxia inducible factor-1alpha in rasV12 upregulation of VEGF. Biochem Biophys Res Commun 287:292-300

Soeta C, Suzuki M, Suzuki S, Naito K, Tachi C, Tojo H (2001) Possible role for the c-ski gene in the proliferation of myogenic cells in regenerating skeletal muscles of rats. Dev Growth Differ 43:155-164

Sorkin A, Waters CM (1993) Endocytosis of growth factor receptors. Bioessays 15:375-382

Sreter FA, Lopez JR, Alamo L, Mabuchi K, Gergely J (1987) Changes in intracellular ionized Ca concentration associated with muscle fiber type transformation. Am J Physiol 253:C296-C300

Stadtman ER (1990) Metal ion-catalyzed oxidation of proteins: biochemical mechanism and biological consequences. Free Radic Biol Med 9:315-325

Staron RS, Karapondo DL, Kraemer WJ, et al (1994) Skeletal muscle adaptations during early phase of heavy-resistance training in men and women. J Appl Physiol 76:1247-1255

Stephens TJ, Chen ZP, Canny BJ, Michell BJ, Kemp BE, Mcconell GK (2002) Progressive increase in human skeletal muscle AMPKalpha2 activity and ACC phosphorylation during exercise. Am J Physiol Endocrinol Metab 282:E688-E694

Stevens L, Sultan KR, Peuker H, Gohlsch B, Mounier Y, Pette D (1999) Time-dependent changes in myosin heavy chain mRNA and protein isoforms in unloaded soleus muscle of rat. Am J Physiol 277:C1044-C1049

Stroka DM, Burkhardt T, Desbaillets, et al (2001) HIF-1 is expressed in normoxic tissue and displays an organ specific regulation under systemic hypoxia. FASEB J 15:2445-2453

Strollo F (1999) Hormonal changes in humans during spaceflight. Adv Space Biol Med 7:99-129

Sutrave P, Kelly AM, Hughes SH (1990) ski can cause selective growth of skeletal muscle in transgenic mice. Genes Dev 4:1462-1472

Takahashi M, Hood DA (1996) Protein import into subsarcolemmal and intermyofibrillar skeletal muscle mitochondria. Differential import regulation in distinct subcellular regions. J Biol Chem 271:27285-27291

Talmadge RJ (2000) Myosin heavy chain isoform expression following reduced neuromuscular activity: potential regulatory mechanisms. Muscle Nerve 23:661-679

Tang H, Kerins DM, Hao Q, Inagami T, Vaughan DE (1998) The urokinase-type plasminogen activator receptor mediates tyrosine phosphorylation of focal adhesion proteins and activation of mitogen-activated protein kinase in cultured endothelial cells. J Biol Chem 273:18268-18272

Terrados N, Melichna J, Sylven C, Jansson E, Kaijser L (1988) Effects of training at simulated altitude on performance and muscle metabolic capacity in competitive road cyclists. Eur J Appl Physiol 57:203-209

Terrados N, Jansson E, Sylven C, Kaijser L (1990) Is hypoxia a stimulus for synthesis of oxidative enzymes and myoglobin? J Appl Physiol 68:2369-2372

Thomason DB, Booth FW (1990) Atrophy of the soleus muscle by hindlimb unweighting. J Appl Physiol 68:1-12

Thompson HS, Scordilis SP, Clarkson PM, Lohrer WA (2001) A single bout of eccentric exercise increases HSP27 and HSC/HSP70 in human skeletal muscle. Acta Physiol Scand 171:187-193

Trappe S, Williamson D, Godard M, Porter D, Rowden G, Costill D (2000) Effect of resistance training on single muscle fiber contractile function in older men. J Appl Physiol 89:143-152

Trounce I, Byrne E, Marzuki S (1989) Decline in skeletal muscle mitochondrial respiratory chain function: possible factor in ageing. Lancet 1:637-639

Tunstall RJ, Mehan KA, Hargreaves M, Spriet LL, Cameron-Smith D (2002a) Fasting activates the gene expression of UCP3 independent of genes necessary for lipid transport and oxidation in skeletal muscle. Biochem Biophys Res Commun 294:301-308

Tunstall RJ,Mehan KA, Wadley GD, Collier GR, Bonen A, Hargreaves M, Cameron-Smith D(2002b) Exercise training increases lipid metabolism gene expression in human skeletal muscle. Am J Physiol Endocrinol Metab 283:E66-E72

Tuominen JA, Ebeling P, Laquier FW, Heiman ML, Stephens T, Koivisto VA (1997) Serum leptin concentration and fuel homeostasis in healthy man. Eur J Clin Invest 27:206-211

Tyml K, Mathieu-Costello O (2001) Structural and functional changes in the microvasculature of disused skeletal muscle. Front Biol Sci 6:D45-D52

Van Loon LJ, Greenhaff PL, Constantin-Teodosiu D, Saris WH, Wagenmakers AJ (2001) The effects of increasing exercise intensity on muscle fuel utilisation in humans. J Physiol 536:295-304

VandenBorne K, Elliott MA, Walter GA, et al (1998) Longitudinal study of skeletal muscle adaptations during immobilization and rehabilitation. Muscle Nerve 21:1006-1012

Vandervoort AA (2002) Aging of the human neuromuscular system. Muscle Nerve 25:17-25

Vidal-Puig AJ, Considine RV, Jimenez-Linan M, et al (1997) Peroxisome proliferator-activated receptor gene expression in human tissues. Effects of obesity, weight loss, and regulation by insulin and glucocorticoids. J Clin Invest 99:2416-2422

Virbasius JV, Scarpulla RC (1994) Activation of the human mitochondrial transcription factor A gene by nuclear respiratory factors: a potential regulatory link between nuclear and mitochondrial gene expression in organelle biogenesis. Proc Natl Acad Sci U S A. 91:1309-1313

Vogt M, Puntschart A, Geiser J, Zuleger C, Billeter R, Hoppeler H (2001) Molecular adaptations in human skeletal muscle to endurance training under simulated hypoxic conditions. J Appl Physiol 91:173-182

Walters EH, Stickland NC, Loughna PT (2000) The expression of the myogenic regulatory factors in denervated and normal muscles of different phenotypes. J Muscle Res Cell Motil 21:647-653

Weibel ER (1984) The pathway for oxygen. Harvard University Press, Cambridge, MA, p 100

Weibel ER (2002) Physiology: the pitfalls of power laws. Nature 417:131-132

Weindruch R (1995) Interventions based on the possibility that oxidative stress contributes to sarcopenia. J Gerontol A Biol Sci Med Sci 50 [Spec No]:157-161

Weindruch R, Kayo T, Lee CK, Prolla TA (2002) Gene expression profiling of aging using DNA microarrays. Mech Ageing Dev 123:177-193

Weis J, Kaussen M, Calvo S, Buonanno A (2000) Denervation induces a rapid nuclear accumulation of MRF4 in mature myofibers. Dev Dyn 218:438-451

Wenger RH, Bauer C (2001) Oxygen sensing: "hydroxy" translates "oxy". News Physiol Sci 16:195-196

Widegren U, Ryder JW, Zierath JR (2001) Mitogen-activated protein kinase signal transduction in skeletal muscle: effects of exercise and muscle contraction. Acta Physiol Scand 172:227-238

Widrick JJ, Romatowski JG, Bain JL, Trappe SW, Trappe TA, Thompson JL, Costill DL, Riley DA, Fitts RH (1997) Effect of 17 days of bed rest on peak isometric force and unloaded shortening velocity of human soleus fibers. Am J Physiol 273:C1690-C1699

Widrick JJ, Knuth ST, Norenberg KM, Romatowski JG, Bain JL, Riley DA, Karhanek M, Trappe SW, Trappe TA, Costill DL, Fitts RH (1999) Effect of a 17-day spaceflight on contractile properties of human soleus muscle fibres. Res J Physiol 516:915-930

Wilkinson KD (2000) Ubiquitination and deubiquitination: targeting of proteins for degradation by the proteasome. Semin Cell Dev Biol 11:141-148

Williamson DL, Gallagher PM, Carroll CC, Raue U, Trappe SW (2001) Reduction in hybrid single muscle fiber proportions with resistance training in humans. J Appl Physiol 91:1955-1961

Winder WW, Holmes BF, Rubink DS, Jensen EB, Chen M, Holloszy JO (2000) Activation of AMP-activated protein kinase increases mitochondrial enzymes in skeletal muscle. J Appl Physiol 88:2219-2226
Wittwer M, Flück M, Hoppeler H, Muller S, Desplanches D, Billeter R (2002a) Prolonged unloading of rat soleus muscle causes distinct adaptations of the gene profile. FASEB J 16:884-886
Wojtaszewski JF, Nielsen P, Hansen BF, Richter EA, Kiens B (2000) Isoform-specific and exercise intensity-dependent activation of 5'-AMP- activated protein kinase in human skeletal muscle. J Physiol 528:221-226
Wong TS, Booth FW (1990) Protein metabolism in rat gastrocnemius muscle after stimulated chronic concentric exercise. J Appl Physiol 69:1709-1717
Wu Z, Puigserver P, Andersson U, et al (1999) Mechanisms controlling mitochondrial biogenesis and respiration through the thermogenic coactivator PGC-1. Cell 98:115-124
Wu H, Rothermel B, Kanatous S, et al (2001) Activation of MEF2 by muscle activity is mediated through a calcineurin-dependent pathway. EMBO J 20:6414-6423
Wu H,Kanatous SB, Thurmond FA, et al (2002) Regulation of mitochondrial biogenesis in skeletal muscle by CaMK. Science 296:349-352
Xia Y, Buja LM, Scarpulla RC, McMillin JB (1997) Electrical stimulation of neonatal cardiomyocytes results in the sequential activation of nuclear genes governing mitochondrial proliferation and differentiation. Proc Natl Acad Sci U S A. 94:11399-11404
Yan Z, Salmons S, Jarvis J, Booth FW (1995) Increased muscle carnitine palmitoyltransferase II mRNA after increased contractile activity. A J Physiol Endocrinol Metabol 31:E277-E281
Yan Z, Salmons S, Dang YI, Hamilton MT, Booth FW (1996) Increased contractile activity decreases RNA-protein interaction in the 3'-UTR of cytochrome c mRNA. Am J Physiol 271:C1157-C1166
Yarasheski KE, Campbell JA, Smith K, Rennie MJ, Holloszy JO, Bier DM (1992) Effect of growth hormone and resistance exercise on muscle growth in young men. Am J Physiol 262:E261-E267
Yarasheski KE, Zachweija JJ, Angelopoulos TJ, Bier DM (1993a) Short-term growth hormone treatment does not increase muscle protein synthesis in experienced weight lifters. J Appl Physiol 74:3073-3076
Yarasheski KE, Zachwieja JJ, Bier DM (1993b) Acute effects of resistance exercise on muscle protein synthesis rate in young and elderly men and women. Am J Physiol 265:E210-E214
Yu M, Blomstrand E, Chibalin AV, Krook A, Zierath JR (2001) Marathon running increases ERK1/2 and p38 MAP kinase signalling to downstream targets in human skeletal muscle. J Physiol 536:273-282
Zainal TA, Oberley TD, Allison DB, Szweda LI, Weindruch R (2000) Caloric restriction of rhesus monkeys lowers oxidative damage in skeletal muscle. FASEB J 14:1825-1836
Zelzer E, Levy Y, Kahana C, Shilo BZ, Rubinstein M, Cohen B (1998) Insulin induces transcription of target genes through the hypoxia-inducible factor HIF-1alpha/ARNT. EMBO J 17:5085-5094
Zhou M, Lin BZ, Coughlin S, Vallega G, Pilch PF (2000) UCP-3 expression in skeletal muscle: effects of exercise, hypoxia, and AMP-activated protein kinase. Am J Physiol Endocrinol Metab 279:E622-E629
Zhu H, Bunn HF (1999) Oxygen sensing and signaling: impact on the regulation of physiologically important genes. Respir Physiol 115:239-247
Zou Y, Hu Y, Metzler B, Xu Q (1998) Signal transduction in arteriosclerosis: mechanical stress-activated MAP kinases in vascular smooth muscle cells. Int J Mol Med 1:827-834
Zumstein A, Mathieu O, Howald H, Hoppeler H (1983) Morphometric analysis of the capillary supply in skeletal muscles of trained and untrained subjects-its limitations in muscle biopsies. Pfluegers Arch 397:277-283

Instructions for authors

1 Legal requirements

The author(s) guarantee(s) that the manuscript will not be published elsewhere in any language without the consent of the copyright holders, that the rights of third parties will not be violated, and that the publisher will not be held legally responsible should there be any claims for compensation.

Authors wishing to include figures or text passages that have already been published elsewhere are required to obtain permission from the copyright holder(s) and to include evidence that such permission has been granted when submitting their papers. Any material received without such evidence will be assumed to originate from the authors.

Manuscripts must be accompanied by the "Copyright Transfer Statement".

Please include at the end of the acknowledgements a declaration that the experiments comply with the current laws of the country in which they were performed.

2 Editorial procedure

Manuscripts should be submitted in English, together with one set of illustrations and a complete pdf file, to the editor in charge.

The author is responsible for the accuracy of the references.

3 Manuscript preparation

To help you prepare your manuscript, Springer offers a template that can be used with Winword 7 (Windows 95), Winword 6 and Word for Macintosh.

For details see point 4.

All manuscripts are subject to copy editing.

■ **Title page**
- The name(s) of the author(s)
- A concise and informative title
- The affiliation(s) and address(es) of the author(s)
- The e-mail address, telephone and fax numbers of the communicating author

■ **Abstract.** Each paper must be preceded by an abstract presenting the most important results and conclusions.

■ **Abbreviations** should be defined at first mention in the abstract and again in the main body of the text and used consistently thereafter

A list of **symbols** should follow the abstract if such a list is needed. Symbols must be written clearly. The international system of units (SI units) should be used. The numbering of chapters should be in decimal form.

Footnotes on the title page are not given reference symbols. Footnotes to the text are numbered consecutively; those to tables should be indicated by superscript lower-case letters (or asterisks for significance values and other statistical data).

Acknowledgements. These should be as brief as possible. Any grant that requires acknowledgement should be mentioned. The names of funding organizations should be written in full.

Funding. Authors are expected to disclose any commercial or other associations that might pose a conflict of interest in connection with submitted material. All funding sources supporting the work and institutional or corporate affiliations of the authors should be acknowledged.

■ **References**
The list of References should only include works that are cited in the text and that have been published or accepted for publication. Personal communications should only be mentioned in the text.

In the text, references should be cited by author and year (e.g. Hammer 1994; Hammer and Sjöqvist 1995; Hammer et al. 1993) and listed in alphabetical order in the reference list.

Examples:

Monographs:
Snider T, Grand L (1982) Air pollution by nitrogen oxides. Elsevier, Amsterdam

Anthologies and proceedings:
Noller C, Smith VR (1997) Ultraviolet selection pressure on earliest organisms. In: Kingston H, Fulling CP (eds) Natual environment background analysis. Oxford University Press, Oxford, pp 211–219

Journals:
Meltzoff AN, Moore MK (1977) Imitation of facial and manual gestures by human neonates. Science 198:75–78

If available the Digital Object Identifier (DOI) of the cited literature should be added at the end of the reference in question.

■ **Illustrations and Tables**
All figures (photographs, graphs or diagrams) and tables should be cited in the text, and each numbered consecutively throughout. Figure parts should be identified by lower-case roman letters. The placement of figures and tables should be indicated in the left margin. For submission of figures in electronic form see below

Line drawings. Please submit good-quality prints. The inscriptions should be clearly legible.

Half-tone illustrations (black and white and color). Please submit well-contrasted photographic prints with the top indicated on the back. Magnification should be indicated by scale bars.

Figure legends must be brief, self-sufficient explanations of the illustrations. The legends should be placed at the end of the text.

Tables should have a title and a legend explaining any abbreviation used in that table. Footnotes to tables should be indicated by superscript lower-case letters (or asterisks) for significance values and other statistical data.

4 Electronic submission of final version

Please send only the final version of the article, as accepted by the editors.

Preparing your manuscript

The template is available:

→ via ftp:
  Address: ftp.springer.de/
  User ID: ftp
  Password: your own e-mail address
  - Directory: /pub/Word/journals
  - File names: either sv-journ.zip or sv-journ.doc and sv-journ.dot

→ via browser
  - http://www.springer.de/author/index.html

The zip file should be sent uuencoded.

Layout guidelines

1. Use a normal, plain font (e.g., Times Roman) for text.
   Other style options:
   - for textual emphasis use italic types.
   - for special purposes, such as for mathematical vectors, use boldface type.
2. Use the automatic page numbering function to number the pages.
3. Do not use field functions.
4. For indents use tab stops or other commands, not the space bar.
5. Use the table functions of your word processing program, not spreadsheets, to make tables.
6. Use the equation editor of your word processing program or MathType for equations.
7. Place any figure legends or tables at the end of the manuscript.
8. Submit all figures as separate files and do not integrate them within the text.

Data formats

Save your file in two different formats:

1. RTF (Rich Text Format) or Word compatible Word 95/97
2. pdf (a single pdf file including text, tables and figures)

Illustrations

The preferred figure formats are EPS for vector graphics exported from a drawing program and TIFF for halftone illustrations. EPS files must always contain a preview in TIFF of the figure. The file name (one file for each figure) should include the figure number. Figure legends should be included in the text and not in the figure file.

Scan resolution: Scanned line drawings should be digitized with a minimum resolution of 800 dpi relative to the final figure size. For digital halftones, 300 dpi is usually sufficient.

Color illustrations: Store color illustrations as RGB (8 bits per channel) in TIFF format.

General information on data delivery

Please send us a zip file (text and illustrations in separate files) either:

→ Via ftp.springer.de
  (to our ftp.server; log-in "anonymous"; password: your e-mail address; further information in the readme file on the server)

→ By e-mail
  (only suitable for small volumes of data)

→ or on any of the following media:
  - On a diskette [you may use .tar, .zip, .gzip (.gz), .sit, and compress (.Z)]
  - On a ZIP cartridge
  - On a CD-ROM

Please always supply the following information with your data: journal title, operating system, word processing program, drawing program, image processing program, compression program.

The file name should be memorable (e.g., author name), have no more than 8 characters, and include no accents or special symbols. Use only the extensions that the program assigns automatically.

5 Proofreading

Authors should make their proof corrections on a printout of the pdf file supplied, checking that the text is complete and that all figures and tables are included. After online publication, further changes can only be made in the form of an Erratum, which will be hyperlinked to the article. The author is entitled to formal corrections only. Substantial changes in content, e.g. new results, corrected values, title and authorship are not allowed without the approval of the editor in charge. In such a case please contact the Editor in charge before returning the proofs to the publisher.

6 Offprints, Free copy

You are entitled to receive a pdf file of your article for your own personal use. Orders for offprints can be placed by returning the order form with the corrected proofs. One complimentary copy of the issue in which your article appears is supplied.